Symposium on Materials and Joining Technology 2020

IOP Conference Series: Materials Science and Engineering
Volume 882

Magdeburg, Germany
7 - 8 September 2020

Editors:

Thorsten Halle
Sven Juttner
Thomas Kannengieber
Manja Kruger
Gerhard Mook
Michael Rhode
Michael Scheffler

ISBN: 978-1-7138-2809-9
ISSN: 1757-8981

Printed from e-media with permission by:

Curran Associates, Inc.
57 Morehouse Lane
Red Hook, NY 12571

Some format issues inherent in the e-media version may also appear in this print version.

Institute of Physics
Dirac House, Temple Back
Bristol BS1 6BE UK

Phone: 44 1 17 929 7481
Fax: 44 1 17 920 0979

techtracking@iop.org

Additional copies of this publication are available from:

Curran Associates, Inc.
57 Morehouse Lane
Red Hook, NY 12571 USA
Phone: 845-758-0400
Fax: 845-758-2633
Email: curran@proceedings.com
Web: www.proceedings.com

TABLE OF CONTENTS

IOP Conf. Series: Materials Science and Engineering **882** (2020) 011001 doi:10.1088/1757-899X/882/1/011001

Dear Alumni, dear friends and colleagues and dear guests of the IWF,

We are pleased to invite you to this year's summer symposium, which traditionally takes place every three years. On September 7 and 8, the Institute of Materials and Joining Technology cordially invites you to the Symposium Materials and Joining Technology in Magdeburg.

We always strive to bring together scientists and engineers from science and industry. This is also reflected in this year's symposium, which describes many different topics on various materials, processes and analytical or numerical methods at different size scale.

For the first time since the series of symposia has started as a summer school in 1972, we have chosen the IOP conference series: Materials Science and Engineering to give all contributors the opportunity to publish their papers in an international, peer-reviewed journal. Therefore, this volume provides an excellent overview of the main topics of our seminar.

We are pleased about the numerous contributions, which unfortunately cannot all be accompanied by an oral presentation during the symposium. Therefore, we would like to explicitly refer to the scientific papers in this volume of the IOP Conference Series: Materials Science and Engineering, which contains all submitted and accepted papers.

M. Scheffler
Institute Director

Magdeburg, June 2020

IOP Conf. Series: Materials Science and Engineering **882** (2020) 011002 doi:10.1088/1757-899X/882/1/011002

List of Editors

Prof. Dr.-Ing. habil. Thorsten Halle
thorsten.halle@ovgu.de
+49 391 67-54580

Prof. Dr.-Ing. Sven Jüttner
sven.juettner@ovgu.de
+49 391 67-58741

Prof. Dr.-Ing. habil. Thomas Kannengießer
thomas.kannengiesser@ovgu.de
+49 391 67-58137

Prof. Dr.-Ing. habil. Manja Krüger
manja.krueger@ovgu.de
+49 391 67-54516

Prof. Dr.-Ing. habil. Gerhard Mook
gerhard.mook@ovgu.de
+49 391 67-54555

Jun.-Prof. Dr.-Ing. Michael Rhode
michael.rhode@ovgu.de
+49 391 67-58137

Prof. Dr. rer. nat. Michael Scheffler
m.scheffler@ovgu.de
+49 391 67-54596

Institution address:
Otto von Guericke University Magdeburg
Institute of Materials and Joining Technology
Universitaetsplatz 2
D-39106 Magdeburg

Peer review statement

All papers published in this volume of *IOP Conference Series: Materials Science and Engineering* have been peer reviewed through processes administered by the proceedings Editors. Reviews were conducted by expert referees to the professional and scientific standards expected of a proceedings journal published by IOP Publishing.

IOP Conf. Series: Materials Science and Engineering **882** (2020) 012001 doi:10.1088/1757-899X/882/1/012001

The effects of Zr and La_2O_3 additions on the microstructural and mechanical properties of a Mo-6Si-5B alloy

J Becker, M Giese and M Krüger

Institute of Materials and Joining Technology, Faculty of Mechanical Engineering, Otto-von-Guericke-University Magdeburg, Universitätsplatz 2, 39106 Magdeburg, Germany

julia.becker@ovgu.de

Abstract. In order to improve the application-related properties, as fracture toughness, creep response and oxidation behavior of Mo-Si-B alloys, certain alloying strategies can be applied. Promising alloying partners like Zr and La_2O_3 may help to enhance the compressive and flexural strength of Mo-Si-B materials. Accordingly, a combination of 1 at.% Zr and 0.5 wt.% La_2O_3 was chosen in order to improve the characteristic properties of a powder metallurgical (PM) Mo-6Si-5B alloy. The influence of Zr and La_2O_3 on the phase distribution and the mechanical properties will be discussed by means of three-point-flexure at room temperature as well as uniaxial compressive creep tests at elevated temperatures (> 1000 °C). Compared with other PM Mo-Si-B alloys the present Zr-La_2O_3-strengthened alloy shows highly improved fracture toughness at room temperature.

1 Introduction

Mo-Si-B alloys have been identified as high-potential materials in numerous comprehensive research studies and are thus promising candidates to replace nickel-based superalloys which are limited in their operating temperatures to a maximum of 1150 °C. Prior investigations show that certain application-related properties, as fracture toughness, creep response and oxidation behavior of those Mo-Si-B alloys [1–3], can be directly affected by their chemical composition according to the ternary phase diagram [4–6] and via different existing manufacturing processes [7–10].

Properties of Mo-Si-B alloys consisting of both brittle as well as ductile phases (silicides and Mo solid solution phase, respectively), are significantly affected by their volume fractions. The thermomechanical stable silicide phases Mo_3Si and Mo_5SiB_2 are mainly responsible for the excellent creep and good oxidation resistance of this group of alloys [2,3,7]. In contrast, the bcc Mo solid solution phase (Mo_{ss}) is responsible for the fracture resistance and brittle-to-ductile transition temperature [1,11–13].

Therefore, the Mo_{ss} phase is a decisive microstructural constituent since it was found to affect the properties of multi-phase Mo-Si-B alloys significantly by its volume fraction, homogeneity and chemical composition [14]. Accordingly, it was reported that Mo-Si-B alloys consisting of a continuously formed Mo_{ss} matrix phase with homogeneously distributed silicides (Mo_3Si and Mo_5SiB_2) show a balanced relation between the application-related properties [7]. With regard to further improvement of the crack and fracture tolerance, recent research focused on avoiding the embrittlement of the Mo_{ss} phase by interstitial as well as solid solution elements. The most critical embrittling element is Si which is dissolved up to concentrations about 3 at.% in technical alloys, reported in [7,13,15].

A promising alloying strategy was developed to improve the ductility and fracture toughness of the Mo_{ss} phase by alloying with Zr, which is mainly due to its getter effect on the detrimental oxygen and the suppressing of Si segregation as well as the formation of ZrO_2 particles and the deflection of cracks

IOP Conf. Series: Materials Science and Engineering **882** (2020) 012001 doi:10.1088/1757-899X/882/1/012001

at the phase boundaries [1,11,16,17]. Furthermore, an improvement of hardness and strength was reported due to the formation of nanoscale zirconia particles affecting the microstructure by dispersion strengthening, stabilization of a very fine-grained microstructure as well as increased grain and phase boundary cohesion due to reduced interstitial contamination [18]. Another positive effect was found by alloying with La_2O_3, which predominantly caused grain refinement and particles dispersion strengthening improving both compression and flexure strength of Mo-Si-B alloys [19].

Combining the promising effects of Zr and La_2O_3, this article reports on the impact of alloying additions of 1 at.% Zr and 0.5 wt.% La_2O_3 on the mechanical properties of a Mo-6Si-5B alloy. Microstructural investigations, room temperature three-point-bending tests using notched samples and compressive creep tests at 1093 °C will be applied in order to evaluate the effects on the application-related properties as fracture toughness as well as high temperature creep response. In addition to the reference alloy Mo-6Si-5B, the alloys Mo-9Si-8B and Mo-9Si-8B-1Zr were used to evaluate the creep performance, since these provide similar microstructures and promising creep properties.

2 Materials and Methods

For the present investigations, a Mo-6Si-5B material was alloyed with 1 at.% Zr and 0.5 wt.% La_2O_3, which in the following will be mentioned as the Mo-6Si-5B-1Zr-La_2O_3 alloy. The material was processed by a powder metallurgical (PM) route that included a mechanical alloying (MA) step to achieve fine-grained and chemically homogeneous powder particles. MA was carried out under protective (argon) atmosphere in a planetary ball mill (Retsch PM 400) with a speed of 200 rpm followed by a CIP-sinter-HIP process, detailed described elsewhere [7,20]. For reasons of comparison a similarly processed alloy Mo-6Si-5B was used [7,20].

For microstructure investigations the samples were wet-grinded from 500 down to 1200 grit and the specimen surface was finished by polishing with 3 µm and 1 µm diamond suspension. The microstructural analysis was applied by means of X-ray diffraction measurements using an X'Pert X-ray diffractometer (PANalytical) with Bragg-Brentano geometry and Co - $K_{\alpha 1,2}$ radiation ($\alpha_1 = 1.789$ and $\alpha_2 = 1.793$) and scanning electron microscopy (SEM-FEI ESEM XL30 FEG equipped with EDS, while images were typically obtained in the BSE mode. Quantitative phase analysis was performed using the Rietveld method with the Topas Academic V5 program package (A.A. Coelho, Topas Academic V5, Coelho Software) and the crystal structures of Mo, Mo_3Si, and Mo_5SiB_2 as starting parameters, detailed described elsewhere [21].

The room-temperature fracture strength was applied performing three-point-bending tests on notched bending specimen with dimensions of 2 mm x 2 mm x 18 mm (DIN EN ISO 23146). In a first step, the ground and polished bending specimen were provided with a pre-notch using electrical discharge machining (EDM), ending up with a notch geometry of 0.8 mm in depth and less than 0.2 mm in width. A "sharp crack" with typical depth of 200 µm and a tip radius of around 10 µm was produced using a self-built cutting device equipped with a razor blade working with 1 µm diamond paste. In order to control the notch depth and tip-radius, the samples were examined at defined intervals (15 – 20 min) using an optical microscope.

Fracture toughness measurements were then performed applying a minimum number of five samples having similar notch base radii of around 10 µm. Three-point flexure was realized using a microtest cell (Materials Testing Nano Tomography Version 3.2, Deben) at ambient temperature with a load rate of 0.5 mm/min after preloading with 2 N. The determination of fracture toughness was carried out as described elsewhere [21]. The compressive creep behavior was determined on EDM-cut samples with dimensions of 2 mm x 2 mm x 3.5 mm, tested under constant true stress using a Zwick electromechanical testing device equipped with a Maytec furnace at a temperature of 1093 °C. This temperature was chosen to ensure the comparability to other Mo-Si-B alloys tested at 2000 °F (= 1093 °C).

3 Results and Discussion

3.1 Microstructural investigations

Representative micrographs of the PM alloys are shown in Figure 1. In case of the Zr-La_2O_3-added alloy, a finer-grained microstructure can be observed with an average grain size of 0.7 μm, which represents a reduction of 56 % compared with alloy Mo-6Si-5B [7] having an average grain size of 1.6 μm.

According to X-ray diffraction (Figure 1c and [7]), both alloys exhibit the typical three-phase microstructure of Mo_{ss}-Mo_3Si-Mo_5SiB_2. However, the additions of 1 at.% Zr and 0.5 wt.% La_2O_3 are of a very low quantity in order to detect them via XRD measurements.

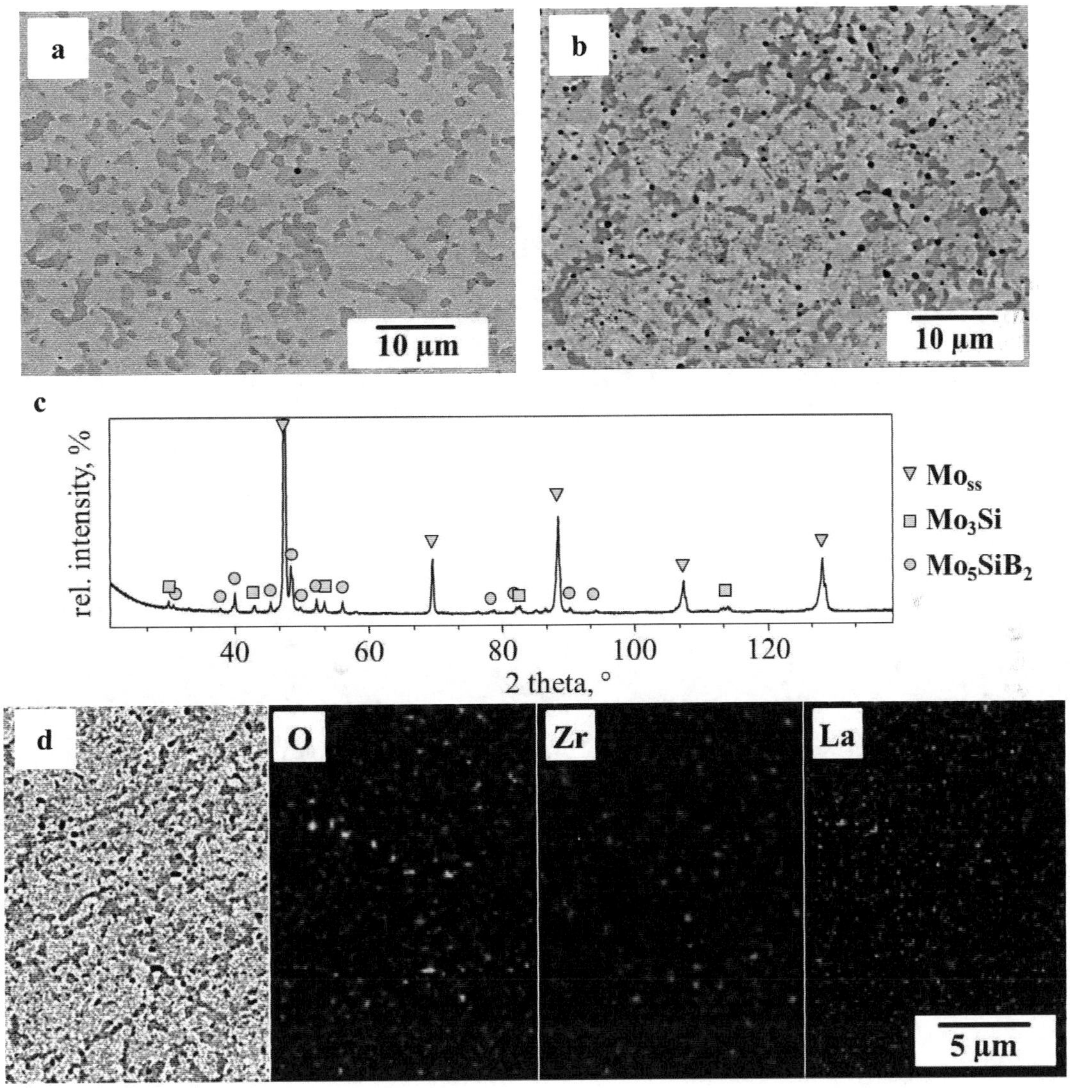

Figure 1. BSE micrograph of alloy (a) Mo-6Si-5B, (b) Mo-6Si-5B-1Zr-La_2O_3 (c) X-ray diffraction patterns of Mo-6Si-5B-1Zr-La_2O_3 showing the existence of three phases: Mo_{ss}, Mo_3Si and Mo_5SiB_2 and (d) EDS element mapping for O, Zr and La_2O_3 in Mo-6Si-5B-1Zr-La_2O_3.

Apart from that, EDS element mapping for the elements O, Zr and La, depicted in Figure 1d, illustrate that the black particles within the microstructure of Mo-6Si-5B-1Zr-La_2O_3 are mainly ZrO_2. However, also a small amount of SiO_2 particles might be present since not every individual oxygen-containing particle could be identified either as Zr or La oxide. So, this illustrates the oxygen gettering effect by Zr, since in Mo-6Si-5B all the black particles are SiO_2. Due to the mechanical alloying process, most of the La_2O_3 particles are very small and therefore, difficult to be detected by EDS. From the EDS mapping in Figure 1d it can be assumed that La_2O_3 particles are finely dispersed within the microstructure. Both materials, the reference alloy and the Zr-La_2O_3-modified alloy exhibit continuously formed Mo_{ss} matrices with finely distributed intermetallic phases Mo_3Si and Mo_5SiB_2. With 74 vol.% Mo_{ss} and 26 vol.% intermetallic phases (IP) in Mo-6Si-5B-1Zr-La_2O_3, the amounts of the corresponding phases are similar to the Mo-6Si-5B alloy (70 vol.% Mo_{ss} : 30 vol.% IP).

3.2 Impact of Zr- and La_2O_3- alloying on the room temperature fracture toughness

Figure 2a shows a comparison of the fracture toughness of different Mo-Si-B alloys arranged by their increasing amount of continuous or discontinuous Mo_{ss} phase. The room temperature fracture toughness of the intermetallic phases with 1.5...3 $MPam^{1/2}$ [22] is significantly lower than for pure Mo (24 $MPam^{1/2}$ [15]). Therefore, fracture toughness values of Mo-Si-B alloys as reported in literature are typically in between 6 and 16 $MPam^{1/2}$.

In general, depending on the chemical composition (Si and B concentration) and the manufacturing process a continuous Mo_{ss} or silicide matrix (discontinuous Mo_{ss}) forms. As illustrated in Figure 2a, the fracture toughness of Mo-Si-B alloys can be divided into two groups: The alloys having a silicide matrix (black symbols - discontinuous Mo_{ss}) exhibit typical fracture toughness values between 6 and 7.4 $MPam^{1/2}$ and alloys having a continuous Mo_{ss} matrix (open symbols) provide values between 11 and 18.8 $MPam^{1/2}$. The Mo-6Si-5B-1Zr-La_2O_3 alloy has a nearly continuous Mo_{ss} matrix phase and the fracture toughness of 18.8 $MPam^{1/2}$ exceeds the values for this type of alloys given in the literature. This is assumed due to the following reasons: (I) the high amount of Mo_{ss} phase, (II) and its continuous distribution, (III) the very fine microstructure of Mo_{ss} phase after alloying with Zr and La_2O_3, which according to Schneibel et al. [23] also affects the fracture toughness and (IV) reduced embrittlement of grain boundaries by gettering the detrimental oxygen accompanied with the formation of ZrO_2 [17].

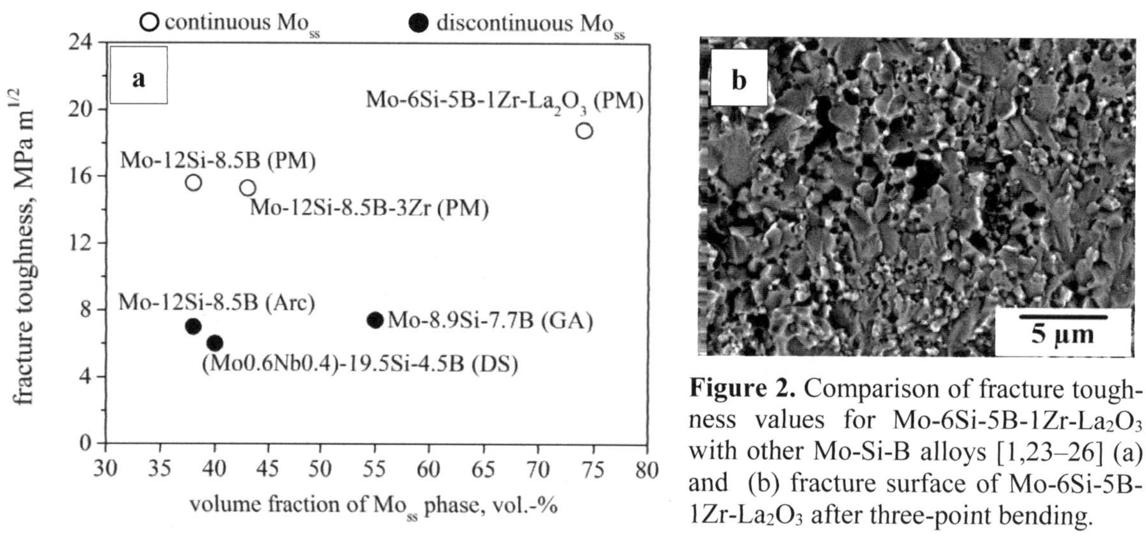

Figure 2. Comparison of fracture toughness values for Mo-6Si-5B-1Zr-La_2O_3 with other Mo-Si-B alloys [1,23–26] (a) and (b) fracture surface of Mo-6Si-5B-1Zr-La_2O_3 after three-point bending.

As depicted in Figure 2b alloy Mo-6Si-5B-1Zr-La_2O_3 shows a mixed fracture mode of predominantly transgranular fracture in Mo_{ss} but also intergranular fracture, the latter within the silicide phases. Metallographic observations of cracked samples show interactions between the Mo_{ss} phase and the intermetallic phases, i.e. crack deflection and crack interception which were also identified in other fine-grained Mo-Si-B materials with homogeneous microstructures, e.g. in [1]. These effects are due to a high probability of the interaction of the crack path with the ductile Mo_{ss} phases.

3.3 Effects of Zr and La_2O_3 on the creep properties of Mo-6Si-5B

The high temperature creep performance of the Mo-6Si-5B-1Zr-La_2O_3 alloy at 1093 °C was comparatively investigated with other PM processed alloys, consisting of a Mo_{ss}-Mo_3Si-Mo_5SiB_2 three-phase microstructure and continuous Mo_{ss} matrix phase. The corresponding Norton-plot is illustrated in Figure 3 at a temperature of 1093 °C, as a typical service temperature in high pressure turbines. For reasons of comparison previously analyzed creep data are included.

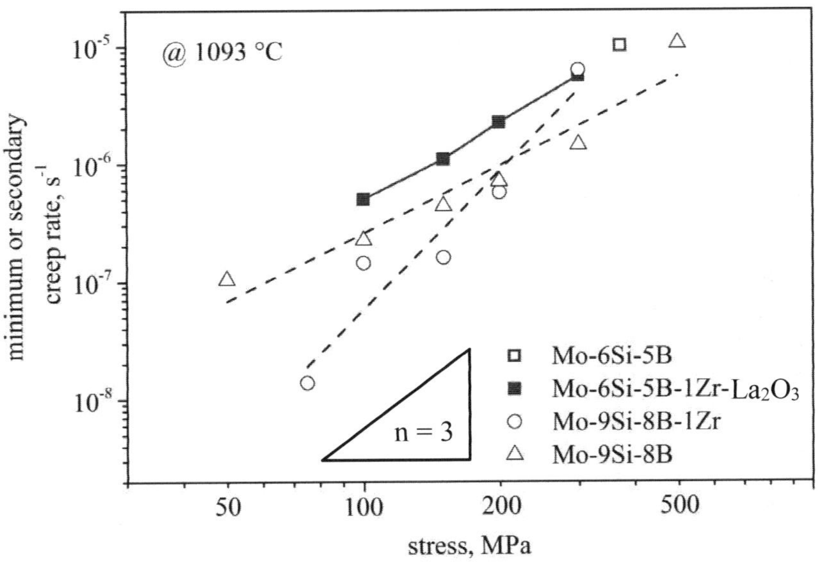

Figure 3. Double-logarithmic plot of the creep rate vs. constant stress at 1093°C for Mo-6Si-5B-1Zr-La_2O_3 compared with other Mo-Si-B alloys.

For the reference alloy Mo-6Si-5B there is only one data point available, which at least allows the assumption, that the creep performance in case of Mo-6Si-5B-1Zr-La_2O_3 may be slightly improved by the effects of alloying additions Zr and La_2O_3. In case of alloy Mo-6Si-6B-1Zr-La_2O_3 this might be explained by the mutual impact of strengthening particles which are thermally stable and reduce the creep deformation by providing obstacles to dislocation movement at high temperatures. On the other hand, the Mo-6Si-5B benefits from the larger grain size providing less creep deformation in case of grain boundary sliding.

Compared to the reference alloy Mo-9Si-8B (HIP) the Mo-6Si-5B-1Zr-La_2O_3 alloy is slightly weaker under similar creep loading. Assuming power law creep, the stress exponent is expressed as the slope $n = \Delta\log \dot{\varepsilon} / \Delta\log \sigma$ of straight lines in Figure 3. The stress exponent for Mo-6Si-5B-1Zr-La_2O_3 was found to be 2.2 and 1.9 for Mo-9Si-8B, which indicates grain boundary sliding and dislocation climb-controlled creep, which can be rationalized by the fine-grained microstructure of this type of PM alloys. Accordingly, the creep resistance may be improved by a grain growth heat treatment [27]. In comparison, the alloy Mo-9Si-8B-1Zr, which yields a stress exponent of 3.9, shows a faster weakening especially above 150 MPa. The reasons of more significant weakening of Mo-9Si-8B and Mo-9Si-8B-1Zr at higher stresses are not clarified, yet.

At temperatures lower than 1100 °C crept Mo-Si-B samples typically show dislocation activities only in the Mo_{ss} phase, which has the lowest creep resistance in this multi-phase system [28]. Since the amount of Mo_{ss} is comparably higher (74 vol.%) than in Mo-9Si-8B (~ 50 vol.%) and Mo-9Si-8B-1Zr the creep response is mainly driven by the plastic deformation of the Mo_{ss} matrix. Considering those conditions, the creep response of the Zr-La_2O_3-modified alloy is comprehensible.

IOP Conf. Series: Materials Science and Engineering **882** (2020) 012001 doi:10.1088/1757-899X/882/1/012001

4 Summary and Conclusions

The present investigations showed how small additions of Zr and La_2O_3 can affect the room and high temperature properties of PM Mo-Si-B alloys. Finally, the following conclusions can be drawn:

1. The addition of 1 at.% Zr and 0.5 wt..% La_2O_3 to a Mo-6Si-5B alloy leads to a grain refinement of about 56 %. Instead of SiO_2 particles, ZrO_2 formed, due to the getter effect on the detrimental oxygen. The La_2O_3 particles are finely dispersed within the microstructure.
2. The fracture toughness could be highly improved up to 18.8 $MPam^{1/2}$ due to the high amount (74 vol.%) of Mo_{ss} matrix phase, but also caused by the grain refinement as well as grain boundary strengthening due to ZrO_2 and La_2O_3 particles.
3. The creep response of Mo-6Si-5B-1Zr-La_2O_3 can be well explained in comparison with other PM Mo-Si-B alloys. The Zr- and La_2O_3-modified alloy shows comparable creep exponents, indicating grain boundary sliding and dislocation climb-controlled creep. Since the amount of Mo_{ss} is higher than in other PM Mo-Si-B alloys used for comparison, the creep response is mainly driven by the plastic deformation of the Mo_{ss} matrix.

Acknowledgements

We acknowledge for funding by the DFG graduate school 1554 "Micro-macro interactions in structured media and particle systems" (funding number: 83477795).

References

[1] Li R, Li B, Wang T, Ren S, Chen X, Wang J and Zhang G 2018 Improved fracture toughness of a Mo-12Si-8 . 5B-3Zr alloy by grain coarsening and its multiple toughening mechanisms *J. Alloys Compd.* **743** 716–27

[2] Azim M A, Schliephake D, Hochmuth C, Gorr B, Christ H J, Glatzel U and Heilmaier M 2015 Creep Resistance and Oxidation Behavior of Novel Mo-Si-B-Ti Alloys *JOM* **67** 2621–8

[3] Rioult F A, Imhoff S D, Sakidja R and Perepezko J H 2009 Transient oxidation of Mo-Si-B alloys: Effect of the microstructure size scale *Acta Mater.* **57** 4600–13

[4] Nowotny E, Dimakopoulou and Kudielka H 1957 Untersuchungen in den Dreistoffsystemen : Molybdän-Silizium-Bor, Wolfram-Silizium-Bor und in dem System : VSi2-TaSi2 *Monatshefte für Chemie und verwandte Teile anderer Wissenschaften* **88** 180–92

[5] Hasemann G, Ida S, Zhu L, Iizawa T, Yoshimi K and Krüger M 2020 Experimental assessment of the microstructure evolution and liquidus projection in the Mo-rich Mo–Si–B system *Mater. Des.* **185** 108233

[6] Soboyejo W O and Srivatsan T S 2006 *Advanced Structural Materials: Properties, Design Optimization, and Applications* (Boca Raton: CRC Press)

[7] Krüger M, Jain P, Kumar K S and Heilmaier M 2014 Correlation between microstructure and properties of fine grained Mo-Mo_3Si-Mo_5SiB_2 alloys *Intermetallics* **48** 10–8

[8] Krüger M, Hasemann G, Bogomol I and Loboda P I 2012 Multiphase Mo-Si-B alloys processed by diractional soldification.pdf 303–8

[9] Ihara K, Ito K, Tanaka K and Yamaguchi M 2002 Mechanical properties of Mo 5 SiB 2 single crystals *Mater. Sci. Eng. A* **329–331** 222–7

[10] Schmelzer J, Rittinghaus S-K, Weisheit A, Stobik M, Paulus J, Gruber K, Wessel E, Heinze C and Krüger M 2019 Printability of gas atomized Mo-Si-B powders by laser metal deposition *Int. J. Refract. Met. Hard Mater.* **78** 123–6

[11] Saage H, Krüger M, Sturm D, Heilmaier M, Schneibel J H, George E, Heatherly L, Somsen C, Eggeler G and Yang Y 2009 Ductilization of Mo-Si solid solutions manufactured by powder metallurgy *Acta Mater.* **57** 3895–901

[12] Moriyama T, Yoshimi K, Zhao M, Masnou T, Yokoyama T, Nakamura J, Katsui H and Goto T 2017 Room-temperature fracture toughness of MoSiBTiC alloys *Intermetallics* **84** 92–102

[13] Jain P and Kumar K S 2010 Dissolved Si in Mo and its effects on the properties of Mo-Si-B alloys *Scr. Mater.* **62** 1–4

IOP Conf. Series: Materials Science and Engineering **882** (2020) 012001 doi:10.1088/1757-899X/882/1/012001

[14] Becker J and Krüger M 2015 Impact of Phase Distribution on the Fracture Toughness of High Temperature Resistant Mo-Si-B Alloys Der Einfluss der Phasenverteilung auf die Bruchzähigkeit von hochtemperatur- beständigen Mo-Si-B-Legierungen *Pract. Metallogr.* **52** 295–313
[15] Sturm D, Heilmaier M, Schneibel J H, Jéhanno P, Skrotzki B and Saage H 2007 The influence of silicon on the strength and fracture toughness of molybdenum *Mater. Sci. Eng. A* **463** 107–14
[16] Geller C B, Smith R W, Hack J E, Saxe P and Wimmer E 2005 A computational search for ductilizing additives to Mo *Scr. Mater.* **52** 205–10
[17] Krüger M, Kauss O, Naumenko K, Burmeister C, Wessel E and Schmelzer J 2019 Intermetallics The potential of mechanical alloying to improve the strength and ductility of Mo – 9Si – 8B – 1Zr alloys – experiments and simulation *Intermetallics* **113** 106558
[18] Krüger M, Schliephake D, Jain P and Kumar K S 2013 Effects of Zr Additions on the Microstructure and the Mechanical Behavior of PM Mo-Si-B Alloys *JOM* **65** 301–6
[19] Zhang G, Zha Y, Li B, He W and Sun J 2013 Effects of lanthanum oxide content on mechanical properties of mechanical alloying Mo – 12Si – 8 . 5B (at .%) alloys *Int . J. Refract. Met. Hard Mater.* **41** 585–9
[20] Krüger M, Franz S, Saage H, Heilmaier M, Schneibel J H, Jéhanno P, Böning M and Kestler H 2008 Mechanically alloyed Mo-Si-B alloys with a continuous $α$-Mo matrix and improved mechanical properties *Intermetallics* **16** 933–41
[21] Becker J, Betke U, Hoffmeister M and Krüger M 2018 Density Reduction of Mo-Si-B Alloys by Vanadium Alloying *JOM* **70** 2574–2581
[22] Mitra R 2006 Mechanical behaviour and oxidation resistance of structural silicides *Int. Mater. Rev.* **51** 13–64
[23] Schneibel J H, Kramer M J and Easton D S 2002 A Mo-Si-B intermetallic alloy with a continuous α-Mo matrix *Scr. Mater.* **46** 217–21
[24] Jéhanno P, Heilmaier M and Kestler H 2004 Characterization of an industrially processed Mo-based silicide alloy *Intermetallics* **12** 1005–9
[25] Choe H, Chen D, Schneibel J H and Ritchie R O 2001 Ambient to high temperature fracture toughness and fatigue-crack propagation behavior in a Mo - 12Si - 8.5B (at.%) intermetallic *Intermetallics* **9** 319–29
[26] Ito K, Kumagai M, Hayashi T and Yamaguchi M 2003 Room temperature fracture toughness and high temperature strength of T2/Moss and (Mo, Nb)ss/T1/T2 eutectic alloys in the Mo-Si-B system *Scr. Mater.* **49** 285–90
[27] Jéhanno P, Heilmaier M, Saage H, Böning M, Kestler H, Freudenberger J and Drawin S 2007 Assessment of the high temperature deformation behavior of molybdenum silicide alloys *Mater. Sci. Eng. A* **463** 216–23
[28] Jain P and Kumar K S 2010 Tensile creep of Mo – Si – B alloys *Acta Mater.* **58** 2124–42

IOP Conf. Series: Materials Science and Engineering **882** (2020) 012002 doi:10.1088/1757-899X/882/1/012002

Oxidation response of additively manufactured eutectic Mo-Si-B alloys

J Becker[1], D Fichtner[2], S Schmigalla[3], S Schultze[3], C Heinze[4], Y Küsters[2], G Hasemann[1], J Schmelzer[1] and M Krüger[1]

[1] Institute of Materials and Joining Technology, Faculty of Mechanical Engineering, Otto-von-Guericke-University Magdeburg, Universitätsplatz 2, 39106 Magdeburg, Germany

[2] Siemens AG, Corporate Technology Research in Energy and Electronics, Coatings & Additive Manufacturing, Siemensdamm 50, 13629 Berlin, Germany

[3] Institut für Korrosions- und Schadensanalyse, 39124 Magdeburg, Germany

[4] Siemens AG, Energy Generation Large Rotating Equipment Additive Manufacturing Materials and Technology, Huttenstraße 12, 10553 Berlin

julia.becker@ovgu.de

Abstract. Previous studies showed the technical feasibility of additive manufacturing (AM) of Mo-Si-B alloys using the laser powder bed fusion (L-PBF) process. The competitiveness of the properties of AM alloys was demonstrated by a comparison to literature data from conventionally processed Mo-Si-B materials. In addition to the excellent mechanical properties at ambient and elevated temperatures, the oxidation resistance in a wide temperature range is another decisive aspect for the potential use of AM Mo-Si-B alloys. The present work shows investigations on the cyclic oxidation performance of a eutectic AM Mo-16.5Si-7.5B alloy. Depending on the temperature (800 °C, 1100 °C, 1300 °C), the oxidation mechanisms are different, which is due to different reactions at the surface of the alloys accompanied with mass changes of samples. These mass changes can be explained on the basis of microstructural investigations. However, compared to a directionally solidified (DS) alloy, the AM alloy shows improved oxidation resistance.

1 Introduction

Recent research has demonstrated that Mo-Si-B alloys from the three-phase field Mo_{ss}-Mo_3Si-Mo_5SiB_2 are promising candidates for high temperature applications, e.g. in gas or aircraft turbines or solar receivers [1–4]. This class of materials combines different key capabilities, like excellent ultra-high temperature creep resistance, sufficient fracture toughness and good oxidation resistance. It is also known that all properties, specifically the oxidation resistance, depend strongly on the phase fraction and size of the Mo solid solution (Mo_{ss}) phase [5].

In general, there are some established production routes that result in dense and crack-free Mo-Si-B bulks, which are powder metallurgical methods based on solid-solid reactions [6,7] and techniques based on melting and subsequent cooling. Examples for the latter are drop-casting and annealing [8] or directional solidification [9]. These processes lead to either typical ultra-fine-grained and homogeneous microstructures in case of using powder metallurgy, typical solidification features, like dendrites of the primarily solidified phase, in case of cooling from a molten phase, or elongated constituents in case of directional solidification. Furthermore, depending on the processing route and the alloy composition, the materials behavior may be tailored in terms of balancing the ambient and high temperature properties and providing isotropic or anisotropic properties.

In the present approach we use the current knowledge on the solidification paths of Mo-Si-B alloys and a new additive processing method. Additive manufacturing (AM) allows the production of various geometries as well as the integration of complex cooling structures. This combination of a promising type of material and improved manufacturing technique will result in more efficient parts, e.g. for modern turbine blades. However, additive manufacturing of Mo-Si-B alloys is very challenging due to the ultra-high melting temperatures of this type of alloys ($T_M > 2000$ °C), the requirement of spherical powders for providing a good flowability in the AM process, and the affinity for oxygen. Oxygen is known to embrittle Mo alloys significantly, even if present in ppm concentrations [10].

Additionally, there is a complex temperature-dependent oxidation mechanism working in multi-phase Mo-Si-B alloys. This is schematically drawn in Figure 1. While heating a Mo-Si-B sample in air there will be a slight mass gain at temperatures > 500 °C, which is due to the formation of a B-Si-Mo-O scale at the surface. A MoO_3 phase starts to form at these comparatively low temperatures. Around 700 °C the MoO_3 phase begins to evaporate from the surface, which then results in a porous B-Si-O scale. In this temperature window a catastrophic mass loss may lead to entire degradation of the Mo alloy, which is called the "pesting" phenomenon. At 800 °C MoO_3 transiently evaporates and permeates out, which results in a significant mass loss as shown in Figure 1. At higher temperatures above 1000 °C the mass change is then reduced due to the formation of a borosilicate glass layer, which protects the surface from further oxidation. With increasing the temperature up to 1300 °C the evaporation of B_2O_3 will result in a B-lean silica scale [5,11–13].

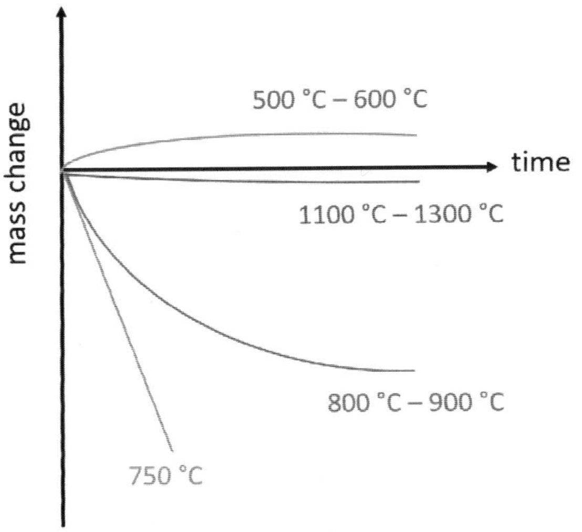

Figure 1. Schematic representation of the temperature-dependent oxidation behavior of Mo-Si-B alloys.

In order to use Mo-Si-B alloys as structural materials at ambient and high temperatures, a good oxidation resistance at elevated temperatures is essential. Cyclic oxidation tests provide an excellent imitation of real conditions including heating and cooling comparable to start and shut-down processes in high temperature turbines. In the present work it will be shown how a AM Mo-Si-B alloy behaves during cyclic oxidation treatment at three different temperatures: (1) 800 °C, which is within the "pesting" regime, (2) 1050 °C - 1100 °C, which is a typical application temperature of state-of-the-art Ni-based superalloys used in gas turbines and (3) 1300 °C as a potential application temperature of Mo-Si-B alloys. For this study a eutectic AM alloy Mo-16.5Si-7.5B is used, which is selected on the basis of a recently published liquidus projection identifying the eutectic point [14]. The oxidation response of this material is compared to a directionally solidified (DS) alloy with a very similar chemical composition of Mo-17.5Si-8B.

2 Materials and Methods

The eutectic alloy composition of Mo-16.5Si-7.5B (in at. %) was chosen due to the lowest melting point of the Mo-rich Mo-Si-B materials, which is slightly above 2000 °C. This may be beneficial for the gas atomization process and laser additive processing as compared to Mo-Si-B alloys with even higher Mo concentrations, since the melting point of Mo is ~2600 °C. The details of gas atomization yielding in spherical pre-alloyed powders are described elsewhere [15]. For additive manufacturing the process of laser powder bed fusion (L-PBF) is used, which allows to build complex geometries by using high pre-heating temperatures. The latter is important since the brittle-to-ductile transition temperature of this type of materials is around 1000 °C and heating is essential to avoid cracks during cooling [16]. For comparison, a directionally solidified (DS) alloy Mo-17.5Si-8B processed via a crucible-free float-zone technique was used. DS samples with a typical diameter of 6 mm were grown in a high purity He atmosphere using a solidification rate of 1 mm/min [17].

Microstructural investigations of the L-PBF and DS materials in their as-processed state and after oxidation were carried out on samples cut via electrical discharge machining (EDM), subsequently embedded in a hot mounting polymer (Struers PolyFast). After wet-grinding from 500 down to 1200 grit, the specimens were polished with a 3 μm and 1 μm diamond suspension and finished by a colloidal silica suspension (OPS) [16,17]. The microstructural observations were accomplished by scanning electron microscopy (SEM - FEI ESEM XL30 FEG and Zeiss EVO 15 equipped with EDS). For the AM Mo-16.5Si-7.5B alloy, electron backscatter diffraction (EBSD) analysis was performed to identify the phases using a Zeiss Merlin SEM equipped with a Nordlys EBSD camera and Aztec software package (Oxford Instruments) provided by the Forschungszentrum Jülich, Germany.

Samples for cyclic oxidation tests were cut from the bulks in dimensions of 3 mm x 3 mm x 5 mm for L-PBF and 3 mm in diameter and 5 mm in height for the DS material. For cyclic oxidation certain temperature regimes and exposure times were chosen: (1) 800 °C/ 500 h as well as 800 °C/ 100 h on pre-oxidized (pre-oxidation at 1300 °C/ 1 h, according to [13]) samples, (b) 1050 °C/ 500 h and 1100 °C/ 100 h and (c) 1300 °C/ 100 h.

To study the oxidation kinetics, the exposure of samples in static air at certain temperatures was typically accomplished at defined cycles. For the AM L-PBF Mo-16.5Si-7.5B alloy, the intermediate weighing of samples was performed every 10 minutes during the first hour, every 60 minutes up to 8 h and finally every 24 hours until the total exposure time (100 h or 500 h, respectively) was reached. The experimental work was done at the IfKorr (Institut für Korrosions- und Schadensanalyse, Magdeburg, Germany) using a LF15/14 Nabertherm furnace. The exposure times per cycle for alloy DS Mo-17.5Si-8B were 10 minutes during the initial oxidation stage (< 1 h), subsequently increased to 1 h cycles up to 20 h, followed by 5 h cycles up to 50 h and 10 h cycles until the final 100 h of total oxidation exposure was reached. The experiments were performed under static air in a box furnace at 1100 °C (Thermconcept KLS 15/14).

Before placing the AM L-PBF and DS samples within the pre-heated furnace, their initial weight and dimensions were exactly determined. At the end of each cycle the samples were removed from the furnace, cooled down to room temperature under atmospheric conditions and weighed carefully to determine their weight loss after certain exposure times. Afterwards the samples were re-exposed to the particular temperature regime described above. Finally, the macroscopic sample morphology and shape retention after finishing the total exposure time was recorded using stereomicroscopes (Olympus SZX7 and Zeiss Stemi 2000 C).

The resulting weight change curves were plotted vs. the exposure time, representing the predominant oxidation mechanisms. Additionally, in order to gain a better understanding of the temperature-dependent oxidation mechanisms, the microstructural features of the oxidized samples were studied via SEM. Therefore, the samples were prepared as described above.

3 Results and Discussion

3.1 Microstructures after processing

The microstructure of alloy AM L-PBF Mo-16.5Si-7.5B is depicted in Figure 2a, representing Mo_{ss} dendrites embedded in a silicide matrix. The SEM-BSE image reveals that during AM processing, the liquid melt solidifies extremely fast which leads to the very fine microstructure in the lower µm scale. Additional EBSD analysis enabled the identification of the corresponding phases. The resulting EBSD phase map, illustrated in Figure 2b, shows the homogeneous distribution of the three phases Mo_{ss} (blue), Mo_3Si (yellow) and Mo_5SiB_2 (red). With 48 vol. % the Mo_3Si phase is dominantly represented whereas the phases Mo_{ss} and Mo_5SiB_2 occur with 30 vol. % and 22 vol. %, respectively.

In contrast, the DS Mo-17.5Si-8B alloy shows a cellular growth of ternary eutectic cells with a well-aligned microstructure orientation parallel to the growth direction. Figure 2c and d illustrate the DSed microstructure in which the typical eutectic cells can be observed. Ternary eutectic colonies with typical cell sizes ≤ 100 µm are clearly visible. The appearance of the ternary eutectic is similar to that prepared via AM, consisting of ~ 50 vol. % Mo_3Si eutectic matrix phase and small amounts of Mo_{ss} (~ 20 vol. %) and Mo_5SiB_2 (~ 30 vol. %).

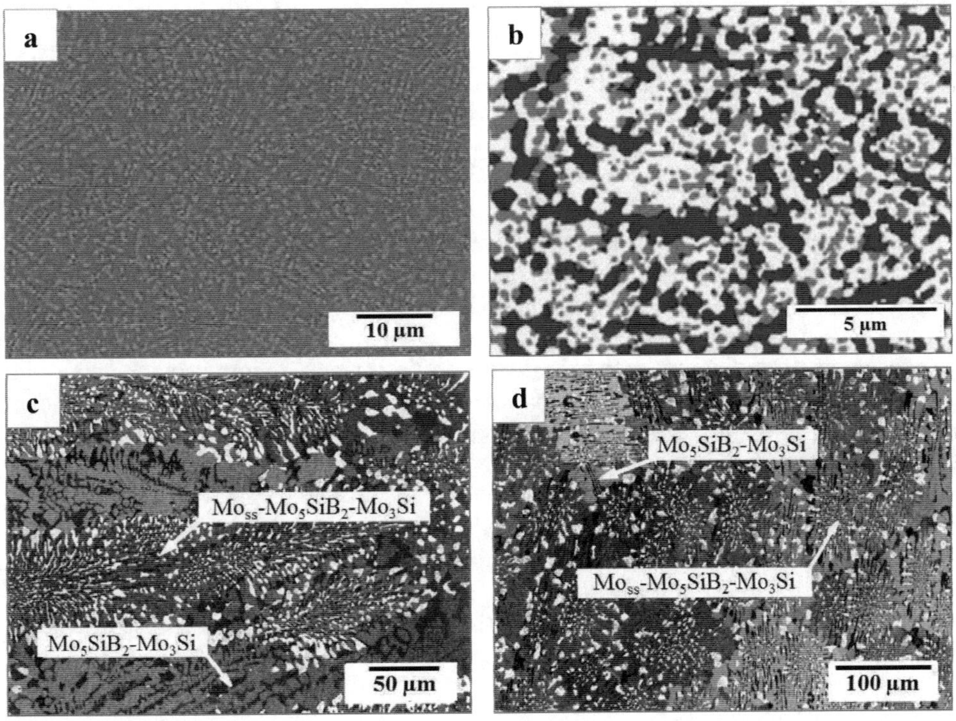

Figure 2. SEM-BSE micrographs of the near eutectic alloys (a) AM Mo-16.5Si-7.5B with (b) EBSD phase mapping of Mo_{ss} (blue), Mo_3Si (yellow), Mo_5SiB_2 (red) and DS Mo-17.5Si-8B in (c) longitudinal and (d) transversal direction.

3.2 Cyclic oxidation response at 800 °C

The following sections concentrate on the description of the oxidation properties of the eutectic alloy AM Mo-16.5Si-7.5B compared with the DS Mo-17.5Si-8B alloy having a very similar chemical composition. The mass loss curves and corresponding microstructures of the oxide scales are shown and discussed in order to understand the predominant oxidation mechanisms.

The temperature regime around 800 °C is the most challenging for Mo-Si-B alloys due to the occurrence of the "pesting" phenomenon characterized by catastrophic oxidation due to MoO_3

evaporation. The mass loss curve of alloy AM Mo-16.5Si-7.5B, being typical for Mo-rich alloys according to Parthasarathy et al. [13], is represented by the blue curve in Figure 3a.

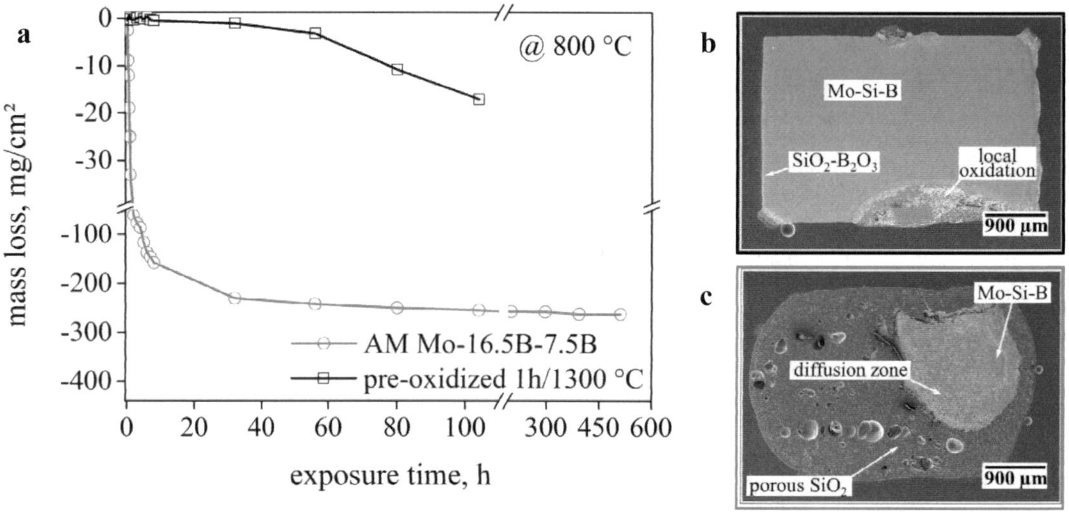

Figure 3. Mass change curves (a) of as-built and pre-oxidized AM Mo-16.5Si-7.5B oxidized at 800 °C and macroscopic cross sections of (b) the pre-oxidized sample after 100 h and (c) the sample without pre-treatment after 500 h.

This typical pesting behavior is characterized by significant weight change right from the beginning of exposure. After approximately 30 h almost no further mass loss than ~ -263 mg/cm² can be observed, which means that catastrophic oxidation failure occurred by nearly entire oxidation of the sample. After the total exposure time of around 500 h, the sample is nearly fully oxidized, as depicted in Figure 3c.

In order to improve the oxidation resistance at 800 °C the pre-treatment concept after Parthasarathy et al. [13] was used. Significantly slower oxidation kinetics was reported, when pre-oxidizing the samples (1300 °C for 1 h) allowing the formation of thermodynamically stable silica scales, which prevents the sample from MoO_3 formation and evaporation. Accordingly, the alloy AM Mo-16.5Si-7.5B was initially oxidized at 1300 °C for 1 h before the sample had been tested under cyclic conditions at 800 °C for a total exposure time of 100 h.

During the relatively short pre-oxidation time a protective SiO_2 should form relatively fast on the material's surface. After its formation MoO_3 volatilizes nearly instantaneously at this temperature, the materials surface is quickly saturated with Si and B forming the silica scale. Since also B_2O_3 slowly evaporates at 1300 °C, the B_2O_3-lean glass layer is highly viscous. As can be seen in Figure 3b, a protective SiO_2-B_2O_3 scale could be formed, resulting in a significantly improved oxidation resistance at 800 °C compared to the as-built AM Mo-16.5Si-7.5B alloy, depicted in Figure 3c. However, after approximately 60 h, significant local oxidation occurs.

An EDS element mapping of an area with beginning oxidation, depicted in Figure 3, shows the formation of a porous B-lean SiO_2 scale enabling further oxygen diffusion. It can be assumed, that due to the high fractions of Mo_3Si + Mo_{ss} phase in this alloy, but only 22 % of Mo_5SiB_2 phase, no significant reservoir for the formation B_2O_3 is present. Hence, the viscosity of the glass layer is reduced, impeding the formation of a continuous protective scale at intermediate temperatures. Accordingly, pre-oxidizing the sample just offers a temporary protection against oxidation at 800 °C as long as the previously formed borosilica scale is completely intact.

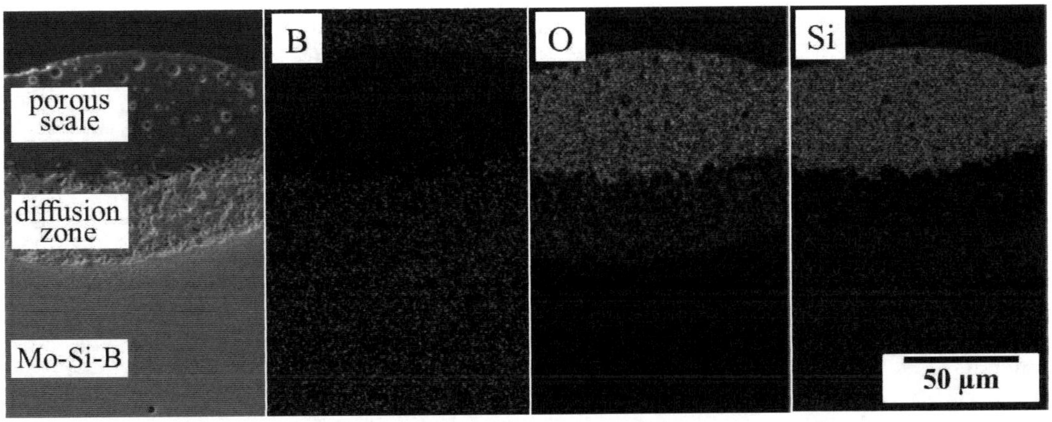

Figure 4. SEM-SE micrograph of the pre-oxidized AM Mo-16.5Si-7.5B alloy oxidized at 800 °C for 100 h combined with EDS element mapping for B, O and Si.

3.3 Oxidation resistance at 1050 °C and 1100 °C

The temperature regime around 1100 °C is very important since it displays a typical application temperature of current state-of-the-art Ni-based superalloys used in gas turbines. In contrast to the oxidation mechanism at 800 °C accompanied with a high mass loss, the oxidation behavior changes when the samples are exposed and oxidized at higher temperatures.

As depicted in Figure 5a, after an initial mass loss during the first hour at 1050 °C and 1100 °C, there is no further change of the mass of the samples in case of AM Mo-16.5Si-7.5B. Even if the exposure time is extended to 500 h, no further mass loss can be observed at 1050 °C. However, for alloy DS Mo-17.5B-8B, Figure 5a shows a higher initial mass loss compared to AM Mo-16.5Si-7.5B until a plateau-like behavior is reached. This slightly higher oxidation rate might be due to the larger sized Mo_{ss} phases in the DS alloy in comparison to the AM alloy (Figure 2), as it is known from Rioult et al. [5] that the microstructure size scale affects the mass losses in the transient stage of oxidation.

In good correlation with the initial sample geometry, the present alloys maintained their geometry reasonably well (Figure 5b, c, d), due to the formation of the protective SiO_2-B_2O_3 scale. However, it is assumed, that these protective scales exhibit different "qualities" since in contrast to the AM alloys, further slight mass loss over longer exposure time is observed for alloy DS Mo-17.5B-8B.

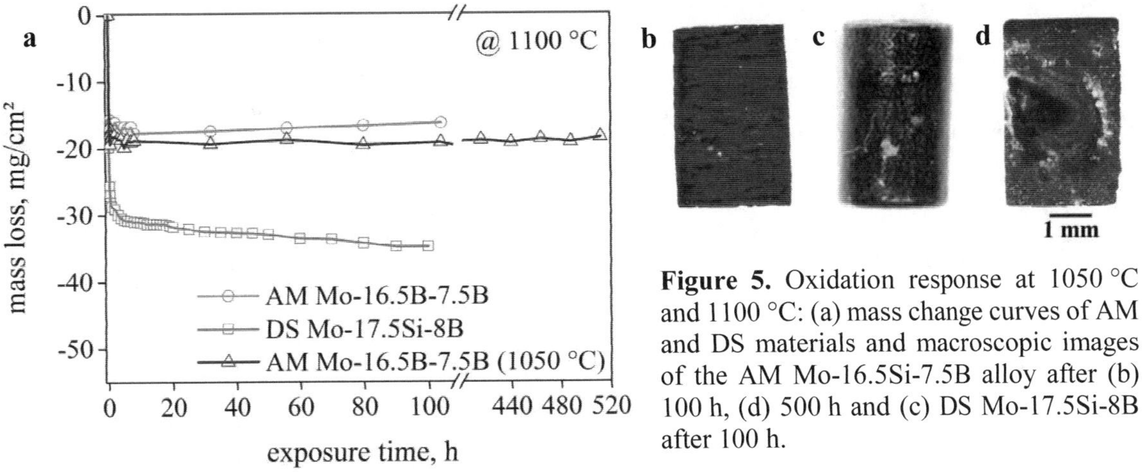

Figure 5. Oxidation response at 1050 °C and 1100 °C: (a) mass change curves of AM and DS materials and macroscopic images of the AM Mo-16.5Si-7.5B alloy after (b) 100 h, (d) 500 h and (c) DS Mo-17.5Si-8B after 100 h.

This can be attributed to the different microstructures after AM and DS processing which have a significant influence on the formation of the protective layer. As illustrated in Figure 2, the ternary eutectic Mo_{ss}-Mo_3Si-Mo_5SiB_2 microstructure in the AM samples is much finer than the DS material. Consequently, the diffusion paths for the reaction of O_2 with the constituents Mo_{ss}, Mo_3Si or Mo_5SiB_2 are much smaller due to the fine microstructure, so a protective borosilicate layer can be formed very rapidly, as can be seen in Figure 6a and c. Hence, with scale thicknesses of ~10 µm after 100 h at 1100 °C and ~20 µm after 500 h at 1050 °C, thin but continuous layers formed.

Furthermore, the microstructure scale effect explains the higher specific mass loss of alloy DS Mo-17.5Si-8B relative to its AM processed counterparts. Even if the eutectic structures may have similar phase scales, the directionally grown microstructure is much coarser in the direction of crystal growth as compared to the AM alloys. Hence, the boundaries between the different eutectic cells are much larger having coarser phases of Mo_{ss}, Mo_5SiB_2, and Mo_3Si and thus, lead to a slightly increased oxidation rate in the initial stage of the reaction. Furthermore, this leads to varying scale thicknesses between ~20 µm and 50 µm in dependence on the individual phase located near the surface. As visualized in Figure 6b, the surface scale of the DS alloy exhibits a higher porosity compared to the ones of AM Mo-16.5Si-7.5B. Due to the evaporation of B_2O_3, MoO_3 may partially volatilize through small pores in the protective glass layer. Therefore, a small mass loss over longer exposure time is observed within the steady-state period for alloy DS Mo-17.5B-8B at 1100 °C.

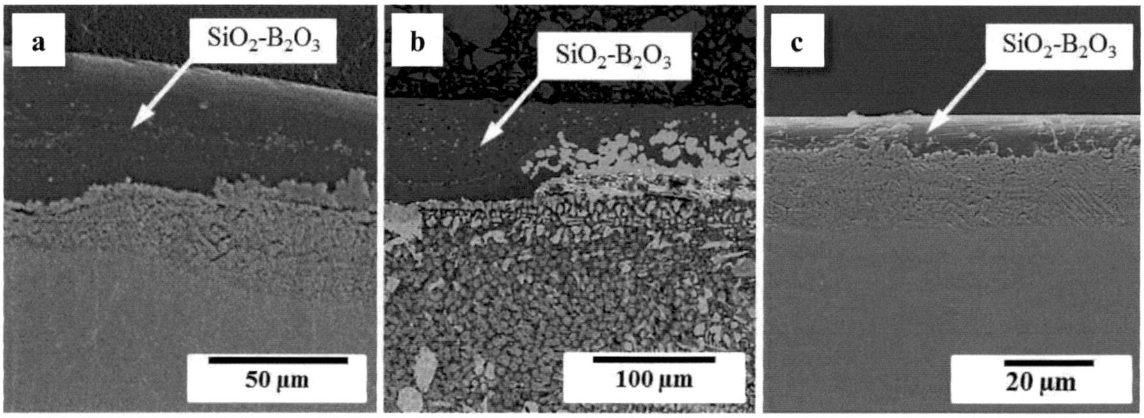

Figure 6. SEM micrographs of the oxidized alloys after exposure; (a) AM Mo-16.5Si-7.5B and (b) DS Mo-17.5Si-8B after 100 h at 1100 °C, respectively, and (c) AM Mo-16.5Si-7.5B after 500 h at 1050 °C.

3.4 Oxidation performance at 1300 °C

The AM Mo-16.5Si-7.5B showed promising oxidation behavior at 1100 °C, so in the next step the material is exposed at 1300 °C since this temperature demonstrates a potential application temperature of Mo-Si-B alloys. The mass loss curve shown in Figure 7a represents an initial mass loss of -10 mg/cm^2 which is even smaller than the initial mass loss at 1100 °C (-20 mg/cm^2). The fluctuations in the initial stage of the curve might be explained by increased diffusion processes at 1300 °C, which enable a rapid surface coverage. The minimum thickness of the protective SiO_2-B_2O_3 scale was about 5 µm. With increasing exposure time, the mass loss reached a steady level. At the end of exposure, after 100 h, the sample geometry is still in good correlation with the initial sample geometry, as depicted in Figure 7b. In Figure 7c the typical surface layers are observed: a boron-lean interdiffusion zone (comparable to the oxidation mechanism at 1100 °C, Fig. 4), which is covered by a continuous borosilica scale.

IOP Conf. Series: Materials Science and Engineering **882** (2020) 012002 doi:10.1088/1757-899X/882/1/012002

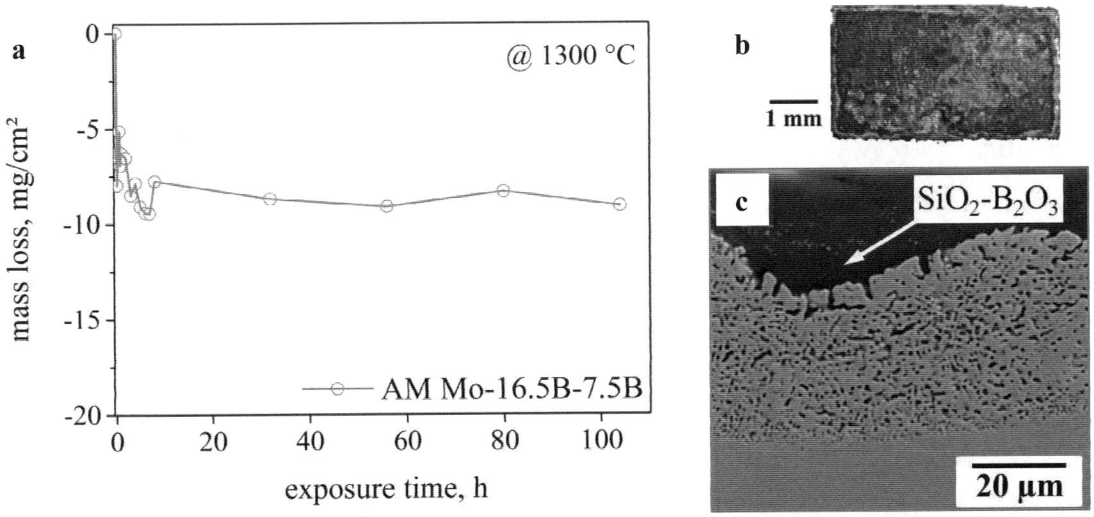

Figure 7. Mass change curve of AM Mo-16.5Si-7.5B at 1300 °C (a) and macroscopic image of the AM Mo-16.5Si-7.5B alloy after 100 h of cyclic oxidation (b) as well as a cross section (c) showing the protective glass layer.

4 Summary and Conclusions

The cyclic oxidation tests on the eutectic alloys AM Mo-16.5Si-7.5B and DS Mo-17.5Si-8B demonstrated the influence of different temperature regimes and microstructures on the oxidation performance of Mo-Si-B alloys. To this end, the following conclusions can be drawn:

1. Despite of the similar composition of the AM Mo-16.5Si-7.5B and DS Mo-17.5Si-8B, the respective processing routes created different microstructures. The alloy AM Mo-16.5Si-7.5B showed a finer and more homogeneous microstructure compared with DS Mo-17.5Si-8B, which exhibits an anisotropic microstructure with larger Mo_{ss} regions.
2. At 800 °C, a "pesting" behavior was observed for the AM Mo-16.5Si-7.5B alloy as it is typical for Mo alloys. A pre-oxidation procedure at 1300 °C for 1 h before starting the exposure at 800 °C helped to enhance the oxidation resistance of the sample temporarily. After ~60 h exposure time, again oxidation failure accompanied with a higher mass losses was observed.
3. Due to the much finer microstructure and homogeneity, the AM Mo-16.5Si-7.5B alloy showed an improved oxidation performance compared with the DS Mo-17.5Si-8B alloy. This was attributed to the much finer microstructure which offers much smaller diffusion paths for the reaction of O_2 with either Mo_{ss}, Mo_3Si or Mo_5SiB_2. So, the protective SiO_2-B_2O_3 glass formed very rapidly, which resulted in a high stability of sample geometry and small mass losses (-20 mg/cm²) over exposure time. Even if the exposure time is extended to 500 h at 1050 °C, no further mass losses could be observed.
4. The kinetics for the formation of the protective borosilicate layer is temperature-dependent, i.e. the layer forms faster at 1300 °C, significantly reducing the evaporation at the initial experimental procedure. This finally lead to a very promising oxidation performance in case of AM Mo-16.5Si-7.5B providing a high stability after slight initial mass losses of -10 mg/cm².

Acknowledgements
The research on AM materials was supported by the German Ministry of Education and Research BMBF in the framework of the cooperative program LextrA (funding number: 03XP0094). The authors acknowledge company Nanoval, Berlin, for processing the gas atomized powders. We thank Dr. Wessel, Forschungszentrum Jülich, for EBSD investigations of the AM samples. The oxidation experiments with reference alloy DS Mo-17.5Si-8B were performed in the context of a DFG project (German Research

IOP Conf. Series: Materials Science and Engineering **882** (2020) 012002 doi:10.1088/1757-899X/882/1/012002

Foundation, funding number: 317275772). J.B. thanks for funding by the DFG graduate school 1554 (funding number 83477795).

References

[1] Brindley K A, Priddy M W and Neu R W 2019 Integrative Materials Design of Three-Phase Mo-Si-B Alloys *Integr. Mater. Manuf. Innov.* **8** 1–16
[2] Morris D G, Lopez-Delgado A, Padilla I and Munoz-Morris M A 2015 Selection of high t emperature materials for concentrated solar power systems: Property maps and experiments *Sol. Energy* **112** 246–58
[3] Shifler D 2017 Department of the Navy SBIR / STTR Transition Program: Mechanical Property Characterization and Modeling for Structural Mo-Si-B Alloys for high Temperature Applications *ONR Approv. # 43 2203-16* **Topic # N1**
[4] Kauss O, Tsybenko H, Naumenko K, Hütter S and Krüger M 2019 Structural analysis of gas turbine blades made of Mo-Si-B under transient thermo-mechanical loads *Comput. Mater. Sci.* **165** 129–36
[5] Rioult F A, Imhoff S D, Sakidja R and Perepezko J H 2009 Transient oxidation of Mo-Si-B alloys: Effect of the microstructure size scale *Acta Mater.* **57** 4600–13
[6] Middlemas M R and Cochran J K 2008 Dense, fine-grain Mo-Si-B alloys from nitride-based reactions *JOM* **July** 19–24
[7] Krüger M, Jain P, Kumar K S and Heilmaier M 2014 Correlation between microstructure and properties of fine grained Mo-Mo_3Si-Mo_5SiB_2 alloys *Intermetallics* **48** 10–8
[8] Schneibel J H, Liu C T, Easton D S and Carmichael C A 1999 Microstructure and mechanical properties of Mo-Mo_3Si-Mo_5SiB_2 silicides *Mater. Sci. Eng. A* **261** 78–83
[9] Hasemann G, Bogomol I, Schliephake D, Loboda P I and Krüger M 2014 Microstructure and creep properties of a near-eutectic directionally solidified multiphase Mo-Si-B alloy *Intermetallics* **48** 28–33
[10] Northcott L and Meyer-Zürich H U 1956 *Molybdenum: Metallurgy of the rarer Metals* vol V, ed H M Finniston (London: Butterworths Scientific Publications)
[11] Burk S, Gorr B, Krüger M, Heilmaier M and Christ H J 2011 Oxidation behavior of Mo-Si-B-(X) alloys: Macro- and microalloying (X= Cr, Zr, La_2O_3) *JOM* **63** 32–6
[12] Burk S and Christ H-J 2011 High-Temperature Oxidation Performance of Mo-Si-B Alloys: Current Results, Developments and Opportunities *Adv. Mater. Res.* **278** 587–92
[13] Parthasarathy T A, Mendiratta M G and Dimiduk D M 2002 Oxidation mechanisms in Mo-reinforced Mo_5SiB_2 (T2)–Mo_3Si alloys *Acta Mater.* **50** 1857–68
[14] Hasemann G, Ida S, Zhu L, Iizawa T, Yoshimi K and Krüger M 2020 Experimental assessment of the microstructure evolution and liquidus projection in the Mo-rich Mo-Si-B system *Mater. Des.* **185** 108233
[15] Schmelzer J, Rittinghaus S-K, Weisheit A, Stobik M, Paulus J, Gruber K, Wessel E, Heinze C and Krüger M 2019 Printability of gas atomized Mo-Si-B powders by laser metal deposition *Int. J. Refract. Met. Hard Mater.* **78** 123–6
[16] Fichtner D, Schmelzer J, Wessel E, Yang W, Heinze C and Krüger M L-PBF of a near-eutectic Mo-Si-B alloy: Processing and resulting properties *Intermetallics* to be submitted
[17] Hasemann G, Kaplunenko D, Bogomol I and Krüger M 2016 Near-Eutectic Ternary Mo-Si-B Alloys: Microstructures and Creep Properties *JOM* **68** 2847–53

IOP Conf. Series: Materials Science and Engineering **882** (2020) 012003 doi:10.1088/1757-899X/882/1/012003

Novel Al/Al_2O_3 composite foams by direct oxidation conversion

X Chen, R Sun, U Betke and M Scheffler

Institute for Materials and Joining Technology, Faculty of Mechanical Engineering, Otto-von-Guericke-University Magdeburg, Universitätsplatz 2, 39106 Magdeburg, Germany

xiaodong.chen@ovgu.de

Abstract. In this paper we combined an investment casting technique and a direct oxidation conversion (DOC) method to fabricate novel Al/Al_2O_3 composite foams. We treated open cellular Al foams with dense struts in oxidizing atmosphere and obtained novel Al/Al_2O_3 composite foams. These novel-type foams had dense struts, and owned a ternary-layered structure within struts, which consisted of the inner layer of Al, the middle layer of α-Al_2O_3 oxidation scale and the outer layer of α-Al_2O_3 ceramic coating. The resulting foams were characterized with respect to micro and macro structure, phase composition and compression behaviour. Comparing with conventional reticulated porous ceramics (RPCs), the novel composite foams demonstrated a new deformation mode in compression, behaving in accumulative damages instead of catastrophic collapse, with more deformation capacity and better ductility. The DOC approach might inspire a new way to manufacture composite foams, and such foams might be alternatives to other cellular solids for specific applications.

1 Introduction

Ceramic foams have been utilized in a wide range of fields [1]. In the family of ceramic foams, the class of RPCs plays as a unique member. The geometrically reticulated structure gives them great permeability, high interconnectivity, high specific surface area, low thermal conductivity and low density. The nature of ceramic materials offers them high wear resistance and high resistance to chemical corrosion [2, 3]. They have found applications in metal melt filters [4], porous medium burners [5], catalytic substrates [6, 7] and volumetric solar absorbers [8, 9], etc.

The predominant processing route to RPCs is the technique developed by Schwartzwalder and Sommer in the early 1960s [10]. In this method, a polymeric sponge is usually used as a template to be coated with a ceramic slurry. After the burn-out of the sponges and high-temperature sintering, reticulated ceramic bodies remain, owning a similar structure as the sponge does. However, the pyrolysis of polymeric templates naturally leaves triangular voids within struts, tending to cause stress concentration, especially at the corners [3]. In addition, due to thermal expansion of sponges during the pyrolysis, longitudinal cracks may occur within the struts [11]. Both flaws aforementioned substantially impact the mechanical performances of RPCs, resulting in a limitation of their applications. By the nature of brittleness in ceramics, RPCs usually deformed in the mode of catastrophic collapse under extra loading, just allowing a very limited deformation, which indicates their low mechanical stability and reliability [12, 13].

Up to now, a great variety of approaches has been proposed to improve mechanical properties of RPCs. They can be totally classified into three categories according to their foci. The first strategy was focusing on pretreating the polymeric templates in order to modify the wetting behavior [14, 15]. The second one focused on improving ceramic coating slurries, either by optimizing its rheological properties [16, 17] or by adding reinforcements, like fibers or whiskers [18-20]. The third route was to post-treat the as-received foams, such as recoating or spraying the green bodies or pre-sintered foams to repair surface cracks [21-28], infiltrating the pre-sintered foams to fill the hollow struts [29-35], etc. Some of these approaches [30, 32-34] are suitable to enhance the compressive strength of RPCs substantially with only slight reduction in porosity. However, all these methods mainly focused on improving the mechanical strength of RPCs, the deformation mode of catastrophic collapse in RPCs did not change, and still needs to be addressed.

Here we propose a novel strategy that attempts to ameliorate the deformation behavior of RPCs and enhance the ductility for them. Based on replica technique, we used investment-cast metallic aluminum foams as original templates instead of polymeric ones. Unlike polymeric templates, the aluminum ones did not decompose to induce hollow struts, instead they would be partly oxidized during high-temperature sintering and stay in resulting struts. The struts obtained a ternary-layered structure, in which the unoxidized Al foams acted as the core, the surrounding oxidized products as the second layer, the ceramic coating as the outside layer. For the Al foams, the processing of high-temperature sintering can be called direct oxidation conversion (DOC), which was initially developed by Lanxide Corporation to fabricate bulky Al/Al_2O_3 composite materials [36]. As with that, our prepared foams were designated as Al/Al_2O_3 composite foams. They were characterized with respect to micro- and macro- structure, phase composition as well as mechanical behavior.

2 Experimental details

2.1 Materials preparation

The first step was to prepare ceramic slurries. An ethanolammonium citrate based deflocculant (Dolapix CE 64, Zschimmer and Schwarz Chemie GmbH, Lahnstein, Germany) was dissolved in deionized water, and then Al_2O_3 powder (CT 3000 SG, d50 = 0.5 μm, Almatis GmbH, Ludwigshafen, Germany) was added to the solution. The solution was mixed with a planetary centrifugal mixer (THINKY Mixer ARE-250, THINKY Corp., Tokyo, Japan) at 1500 rpm for 15 min. Afterwards, a polyvinylalcohol-based binder (Optapix PA 4G, Zschimmer and Schwarz Chemie, GmbH, Lahnstein, Germany) and a nonionic alkyl polyalkylene glycolether-based anti-foaming agent (Contraspum K1012, Zschimmer and Schwarz Chemie GmbH, Lahnstein, Germany) were added and the mixing proceeded with the mixer at 1500 rpm for another 15 min. The solid load of the ceramic slurry was adjusted to 70 wt.%, in which the weight ratios of dispersant, binder, and anti-foaming agents to the mass of Al_2O_3 power were 1.0 wt.%, 1.5 wt.% and 0.1 wt.%, respectively.

Subsequently, investment-cast aluminum foams (15*20*20 mm^3, 10 PPI (pores per inch), 99.7 wt.% Al, MAYSER, Lindenberg, Germany) were totally immersed in the as-prepared slurry. Afterwards, the soaked foams were pulled out slowly with a pair of tweezers for the consecutive drying process. Within the first two minutes, the contact face to the drying plate were switched every 15 seconds to offset the influence of gravity on the flowing of ceramic coating and to guarantee uniform coating. The following drying step lasted for 24 hours at ambient conditions. The process described above referred to the first coating cycle. To obtain varied coating thicknesses, this process was repeated for two, three and four cycles, respectively. To investigate the influence of the sintering temperature, dried green bodies were sintered by varying the temperature from 1200 °C to 1600 °C with 100-K-steps, and the dwelling time was fixed for 6 hours. To explore the influence of the dwelling time, another two sample series, fired at 1600 °C, were treated for 3 hours and 12 hours, respectively. All the heat treatments were carried out in an air furnace (KU 40/04/A, THERMCONCEPT Dr. Fischer GmbH, Bremen, Germany), with heating and cooling rates of 1 K/min and 3 K/min, respectively. The labels of specimens were named after a combination of the processing parameters, e.g. "3-1600-6h": "3" stands for the number of coating cycles, "1600" refers to the sintering/oxidation temperature and "6h" refers to the dwelling time, respectively.

2.2 Characterization
The geometrical density of Al/Al_2O_3 composite foams were the ratio of total mass to geometrical volume. The strut density was determined by the Archimedes principle. The total porosity was deduced from the geometrical density and the strut density. The composite foams were characterized by micro-computed tomography (μ-CT; GE Sensing and Inspection Technologies GmbH, Wunstorf, Germany). Their 3D structures were reconstructed with the software Phoenix Datos|X 2.0 (Phoenix/GE Sensing & Inspection) and visualized with myVGL V2.1 (Volume Graphics GmbH, Heidelberg, Germany). The structural calculations, in terms of mean strut thickness and its distribution, were conducted with the CTAnalyser (CTAn V1.16, Skyscan/Bruker microCT, Kontich, Belgium). The as-prepared foams were mechanically milled in a steel can with ZrO_2 balls as grinding media to prepare powder for XRD powder diffraction analysis (PANalytical GmbH, Kassel, Germany) with Cu-Kα radiation. The microstructure of selected foams was imaged with scanning electron microscopy (SEM; XL30 ESEM-FEG, FEI/Philips, Hillsboro, OR, USA), and uniaxial compression tests were performed with a uniaxial testing machine (TIRAtest 2825, Tira-GmbH, Schalkau, Germany), with a loading rate of 1 mm/min for ten specimens for each test. For the compressive strain-stress curves, the strain and stress corresponding to the maximum stress point were extracted as peak strain and compression strength, respectively. The Weibull modulus in compression strength for selected specimens was determined by linear regression.

3 Results and discussion

3.1 Macrostructure characterization

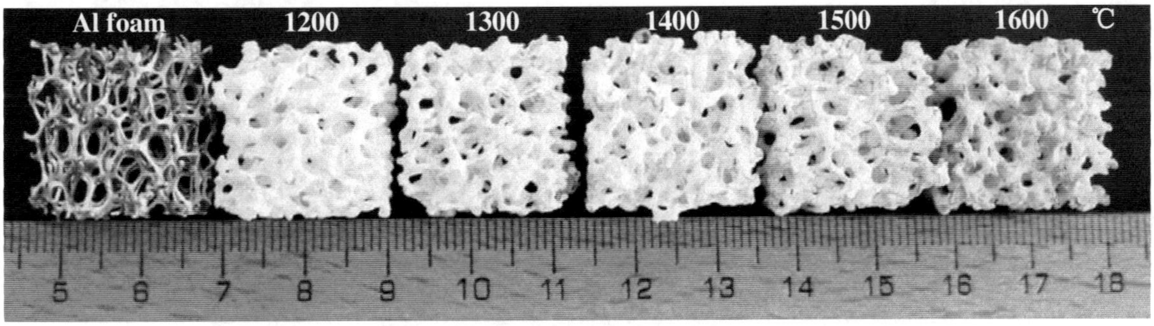

Figure 1. Optical images of the Al/Al_2O_3 composite foams with three coating cycles sintered at 1200-1600 °C for 6 hours.

Figure 1 shows the appearance of Al/Al_2O_3 composite foams sintered at varied temperatures for 6 hours. It is obvious that the composite foams survived severe high temperature sintering and retained their open-cellular structure. The outer ceramic coating tightly adhered to the Al substrates, and no spalling of them was found. Even though the clogging of cells took place occasionally within the resulting foams, they still showed high permeability, implying by the see-through of dark background. It also was observed that the as-received foams did not shrink significantly. This can be explained by the opposite evolution between Al substrates and Al_2O_3 coating. The oxidation of aluminum induced a volume expansion, because the volume of oxide was about 2.3 times as that of the aluminum consumed in the oxidation reaction [37]. By contrast, based on the study by Athena et al. [38], Al_2O_3 coating was estimated to have a volume shrinkage of around 30%.

The μ-CT 2D reconstructions in Figure 2 represent the cross sections of the resulting foams; the irregular green patterns represent the foam struts. Unlike conventional RPCs, novel Al/Al_2O_3 composite foams owned solid struts, having no triangular voids. This means that to replace the polymeric templates with metallic investment-cast Al foams was an effective approach to generate dense struts, which may also be achieved by infiltration techniques [30-34]. Under the sintering, metallic Al foams melted in liquid state with active mobility, but solid struts indicate that most of the molten aluminum remained within the Al_2O_3 ceramic shell, instead of splitting the ceramic coating to flow out. The flowing out of aluminum only occurred locally, generating voids in the struts. On the μ-CT slices, the voids in the struts are implied by the black close pores within the green bodies. Besides strut voids, surface cracks were also produced during direct oxidation conversion (DOC) processing, implied by the black lines within the green bodies. As shown in the images, the amount of surface cracks tend to increase with an increase in temperature, specifically in the specimens oxidized at 1500 °C and 1600 °C, in which cracks occurred massively. The occurrence of surface cracks can be attributed to thermal mismatch between the molten aluminum (coefficient of thermal expansion, $CTE_{Al} = 98 \times 10^{-6}\ K^{-1}$) [39] and Al_2O_3 ($CTE_{Al2O3} = 7.5 \times 10^{-6}\ K^{-1}$) [40] coating. The volume expansion of molten aluminum imposed stress to the shrinking ceramic shell, creating cracks on the outer ceramic coating.

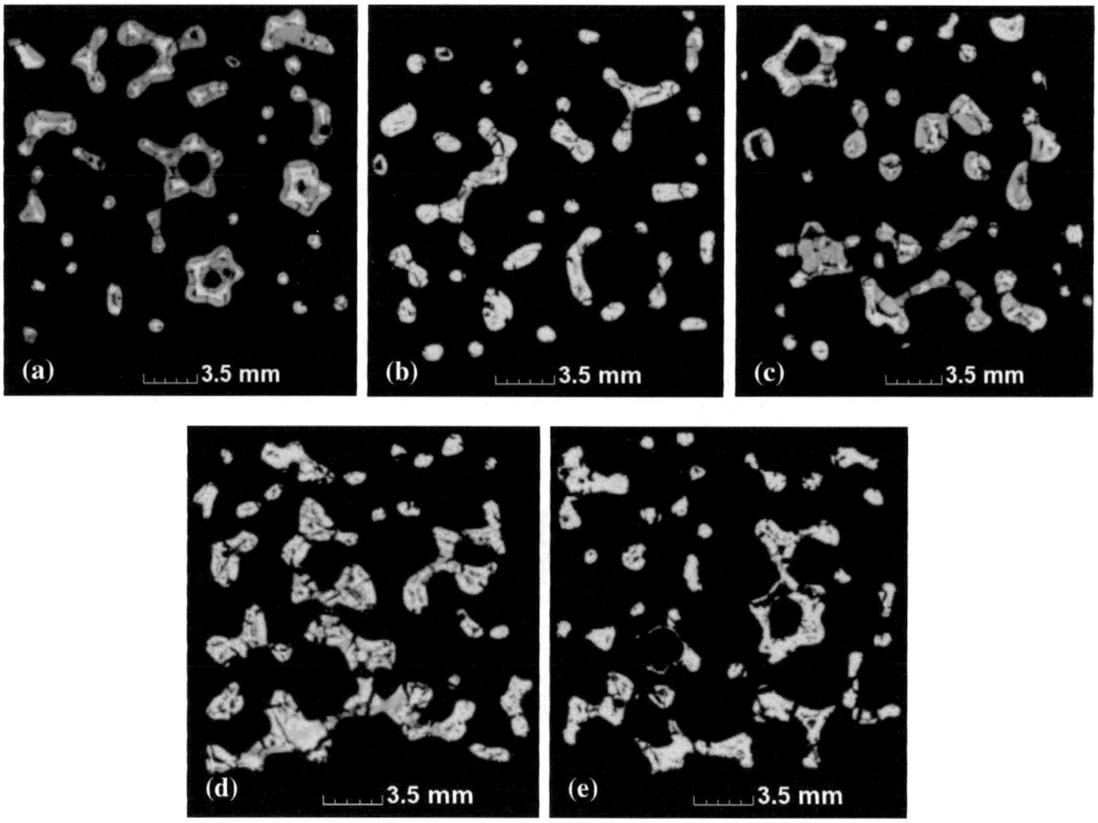

Figure 2. μ-CT 2D reconstructions of Al/Al_2O_3 composite foams with three coating cycles sintered at 1200 °C to 1600 °C for 6 hours: (a) 1200 °C, (b) 1300 °C, (c) 1400 °C, (d) 1500 °C, (e) 1600 °C.

The properties data of Al_2O_3 composite foams are collected in Table 1. They were also analysed specifically in following separate figures. As shown in Figure 3(a), the geometrical density of resuting foams stands around 0.5 g/cm^3, which is largely higher than that of the original density of Al foams. This data is comparable to that of conventional RPCs [30]. With varing sintering tempatures, the density of resulting foams was not influenced significantly, just with slight fluctuations. This also applied for the porostiy of the foams, which all valued around 84% for all foams.

Table 1. Properties of Al/Al_2O_3 composite foams.

Specimens	Geometrical Density [g/cm^3]	Total porosity [%]	Strut density [g/cm^3]	Mean strut thickness [mm]	Compression strength [MPa]	Weibull modulus m
Al foam	0.10±0.003	96.4	2.71±0.01	0.40	-	-
3-1200-6h	0.48±0.04	85.5	3.31±0.04	1.28	0.38±0.05	9.5
3-1300-6h	0.52±0.03	84.3	3.31±0.02	1.16	0.52±0.12	7.1
3-1400-6h	0.46±0.02	86.4	3.38±0.03	1.18	0.26±0.05	5.6
3-1500-6h	0.48±0.04	85.5	3.32±0.07	1.08	0.42±0.11	4.0
3-1600-6h	0.50±0.05	84.6	3.25±0.06	0.94	0.60±0.08	6.5
3-1600-3h	0.51±0.03	83.8	3.15±0.03	0.95	0.46±0.08	6.8
3-1600-12h	0.52±0.04	83.8	3.20±0.09	0.95	0.59±0.14	5.2
2-1600-6h	0.49±0.02	85.5	3.37±0.03	0.76	0.51±0.07	9.0
4-1600-6h	0.98±0.06	71.8	3.47±0.01	1.40	1.85±0.27	8.5

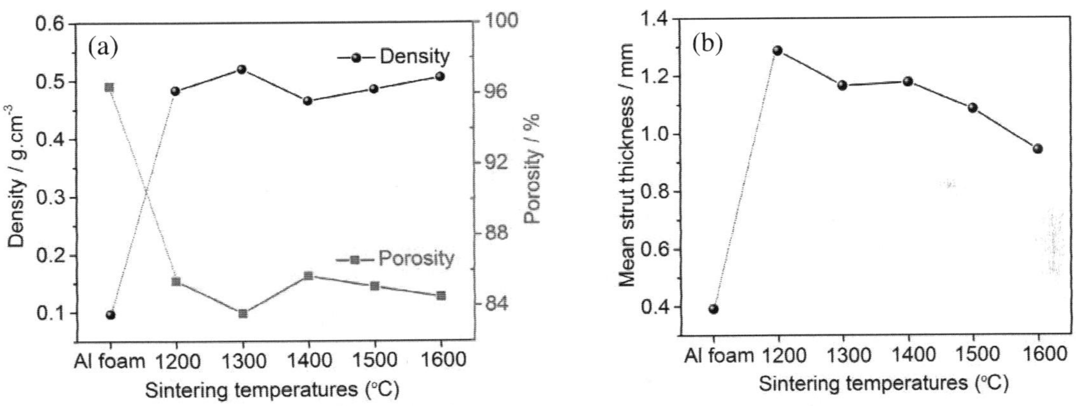

Figure 3. The influence of the sintering temperature on (a) density and porosity and (b) mean strut thickness of Al/Al_2O_3 composite foams oxidized for 6 hours.

Comparing with conventional RPCs [3, 32, 34], where the porosities were achieved above 90%, the porosity of Al_2O_3 composite foams was slightly lower, and this was due to the thickening of struts by repeated coating processes. In Figure 3(b), the mean strut thickness of resulting foams witnessed a decrease with an increasing sintering temperature. This is explained by varied densification of ceramic coating with an increasing sintering temperature, which was also observed in Al_2O_3 ceramic processing by slip casting [38]. The cearamic coating tended to be denser as sintering temperature rised, so the strut thickness of the foams showed a tendency to decrease.

To investigate the influence of the oxidation and sintering time on the macrostructure of the Al/Al_2O_3 foams, the sintering time was varied, as shown in Figure 4(a). It was observed, that the impact of sintering time on porosity and sturt thickness can be negelected, resulting in overall values of 84% and 0.95 mm, respectively.

In order to create Al/Al_2O_3 composite foams with different strut thicknesses, repeated coating was employed, as displayed in Figure 4(b). It clearly indicates that repeating the coating process can significantly thicken the struts of the foams, but with respect to the mean strut thickness, the thickening effect demontrated no propotional relation to the number of coating cycles, which was also found in recoated zirconia toughened alumina (ZTA) foams [30]. That may be attributed to the increasing surface area and the evolution of the surface morphology as the coating process was continually repeated.

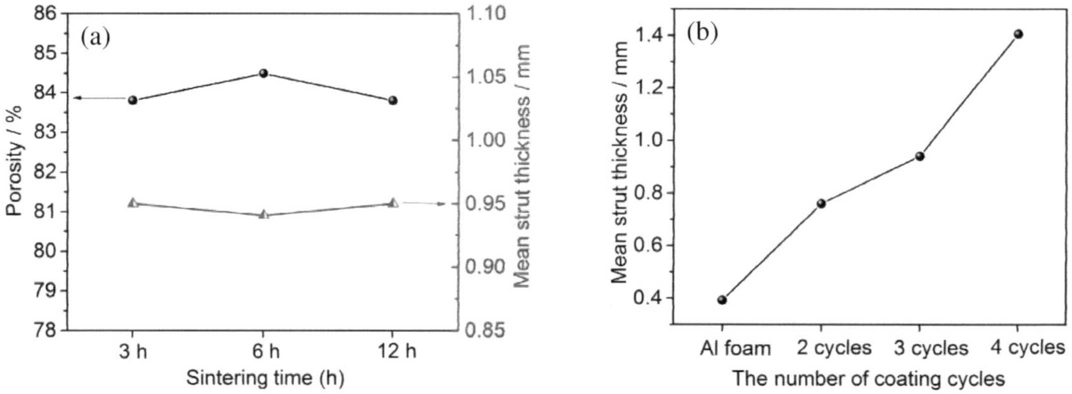

Figure 4. The influence of (a) the sintering time on the porosity and mean strut thickness of foams processed with three coating cycles at 1600 °C; and the influence of (b) the number of coating cycles on the mean strut thickness of foams treated at 1600 °C for 6 hours.

3.2 Microstructure characterization

In Figure 5, the microstructure of broken struts of Al/Al_2O_3 composite foams is displayed resulting from SEM imaging. It clearly indicates that the struts of novel foams owned solid formation without triangular voids, which was consistent with μ-CT results. More importantly, the struts attained a novel ternary-layered structure, as shown in Figure 5(c). By means of EDS (Figure 5 (e), (g), (f)) and XRD (Figure 6), the composition of each layer was determined.

The inner layer originated from the plain Al foam, which withstood the high-temperature sintering and oxidation process and retained mostly non-oxidized. The middle layer was comprised of α-Al_2O_3, closely encompassing the aluminum substrate. This layer resulted from incomplete oxidation of the metallic aluminum substrate, amounting from dozens of micrometers to even one hundred micrometers in thickness. The Al substrate was only partly oxidized even at up to1600 °C for 6 hours. The reason is that, as the temperature and the dwell time increased, the outer coated Al_2O_3 ceramic became denser and the already formed oxidation scale became thicker, they blocked the paths through which the oxygen diffused into the inner aluminum [38, 41]. It must be note that the grain size of α-Al_2O_3 in this layer reached dozens of micrometers, indicating abnormal grain growth. This may result from their growing conditions at elevated temperatures where they were soaked in liquid aluminum: Zhang et al. [42] reported that the introduced Al nanoparticles to Al_2O_3 ceramic were able to induce elongated growth of the Al_2O_3 grain. The outer layer was an Al_2O_3 ceramic coating. As shown in Figure 5(a), severe surface cracks appeared on the struts, the reason for that has been discussed in former sections. The existence of surface cracks allowed the flowing out of molten Al, generating cavities locally within the struts, as displayed in Figure 5 (d).

In summary, even though flaws exist, such novel Al/Al_2O_3 composite foams demonstrate unique ternary-layered struts, bridging and combing metal and ceramic. To the best of our knowledge, no such strut formation in ceramic foams has been reported in literature yet [1-3]. It may turn out that the direct oxidation conversion (DOC) might inspire a new way to fabricate ceramic composite foams.

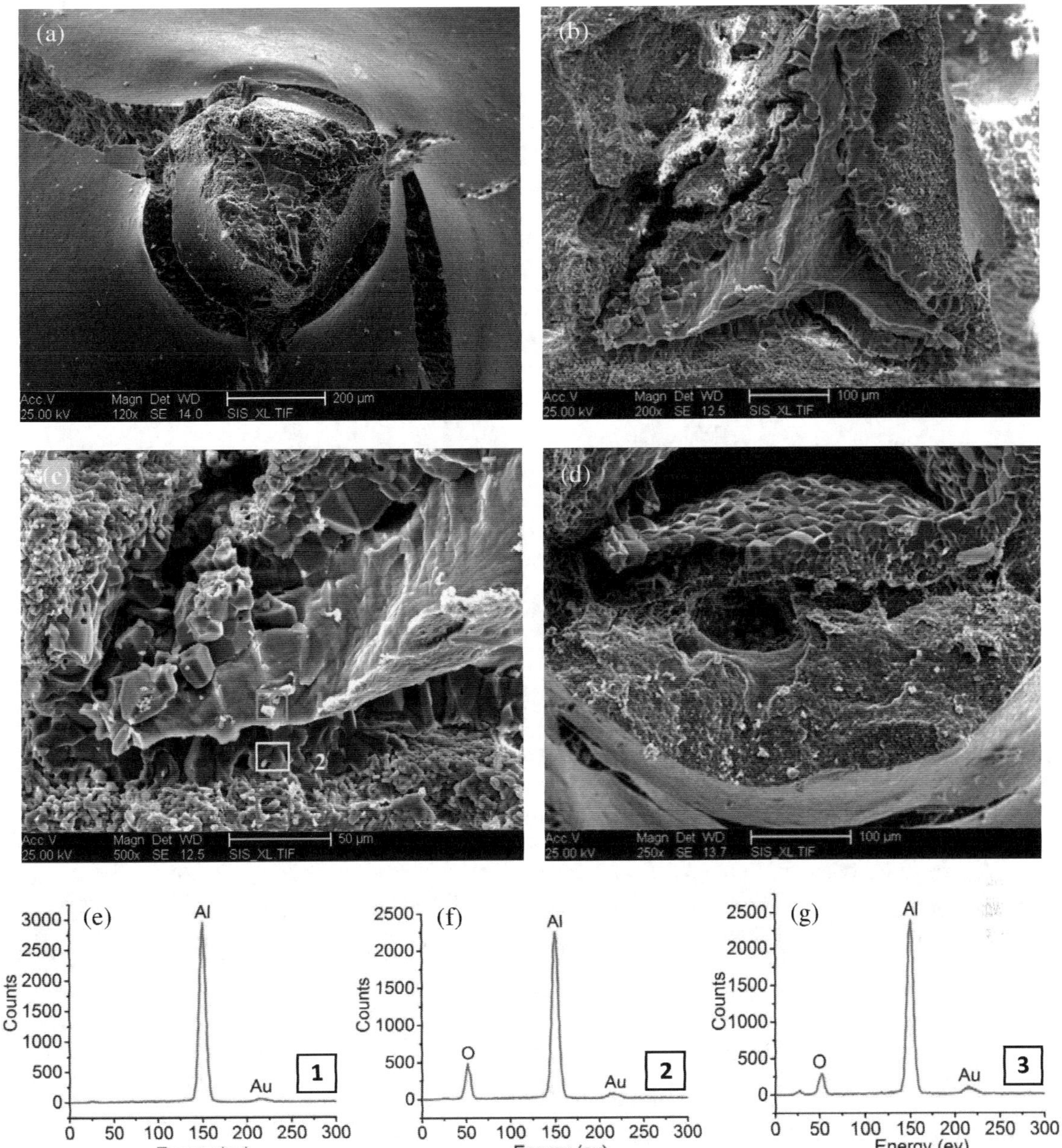

Figure 5. Typical SEM images for Al/Al_2O_3 foams: (a) 1500 °C for 6 hours, (b) and (c) 1600 °C for 6 hours, (d) 1600 °C for 3 hours; EDS measurements for image (c) in different zones, (e) zone 1, (f) zone 2, (g) zone 3.

The phase composition of the Al/Al_2O_3 foams is shown in Figure 6. The resulting foams consisted of α-Al_2O_3 and metallic aluminum, and only α-Al_2O_3 existed, no other types were detected. Note that the intensities of diffraction peaks for aluminum did not precisely represent the real fraction of aluminum in resulting foams. Because the samples were powdered for XRD by mechanical milling; considering the high degree of ductility of metallic Al, not all the aluminum may have been transferred to the XRD's sample holder.

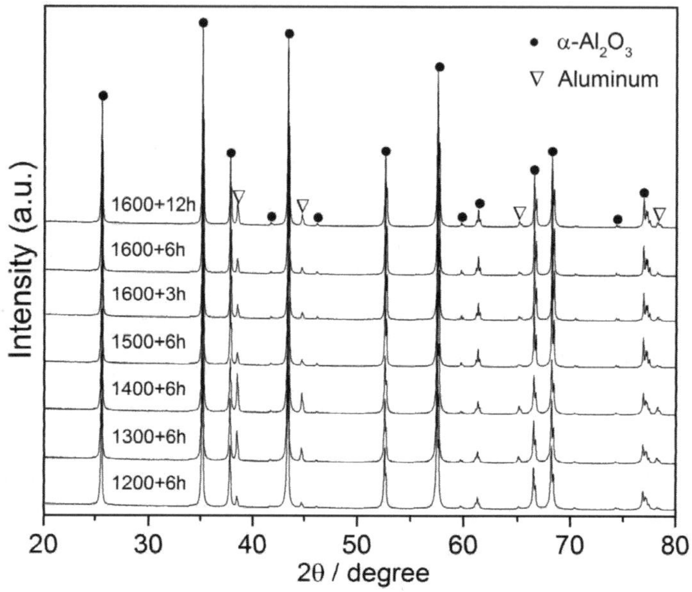

Figure 6. XRD patterns of Al/Al_2O_3 composite foams sintered/oxidized at different temperature and time.

3.3 Compression strength behavior

Representative stress-strain curves in compression of Al/Al_2O_3 composite foams are arranged in Figure 7, and that of plain Al foam and ZTA foam [30] were added for comparison. The curve of the ZTA foam represents typical compression behavior of conventional RPCs. It shows that when compressive stress reaches the critical strength of RPCs, damage occurs rapidly and the strength falls to a low level, which actually implies the catastrophic collapse in deformation mode for such foams. Commonly, crack of struts or through-body cracks in foams can be observed after the testing. With respect to the deformation capacity, it allowed limited strain having a level of only 5% or less. The Al foam, it showed higher ductility, accompanied with continual densification in compression. Its compression strength was so much low that it cannot be identified accurately under the described testing conditions.

Comparing with that of Al and ZTA foam, the as-received Al/Al_2O_3 foams behaved entirely different in compression. The compressive process consisted of three stages, as labelled in the figure. In the first stage the course of all Al/Al_2O_3 foams follows Hook´s law until a tipping point is reached. In contrast to brittle foams (e. g. ZTA) no catastrophic failure follows, instead a negative slope of the course is observed (stage II) until a further tipping point is reached followed by a steep positive slope (stage III). Stage II represents an accumulated damage where the struts collapse in a mixed ductile and brittle mode. Stage III represents a densification, also like in stage II with a ductile and a brittle fraction.

In summary, the failure behavior of these DOC foams in compression mode represents their composite design build of aluminum as a ductile and alumina as a brittle component. The increase of stress until the first tipping point is reached is a consequence of an increased densification by sintering with an increasing temperature, starting at 1400 °C.

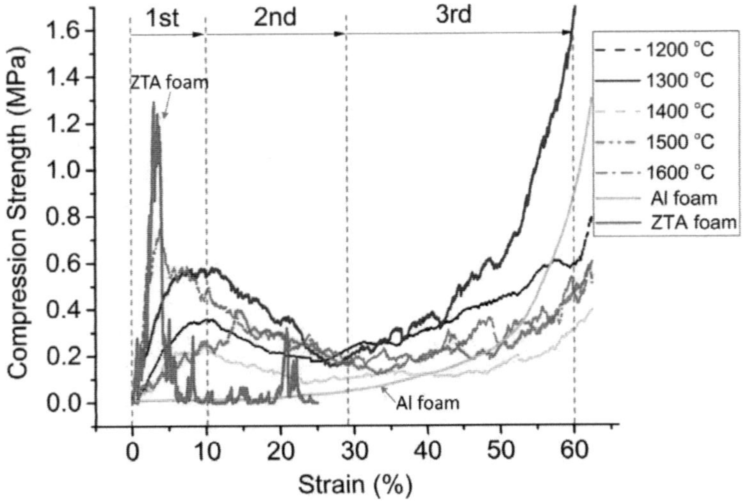

Figure 7. Typical compressive stress-strain curves of Al/Al_2O_3 composite foams with three coating cycles sintered at different temperatures for 6 hours.

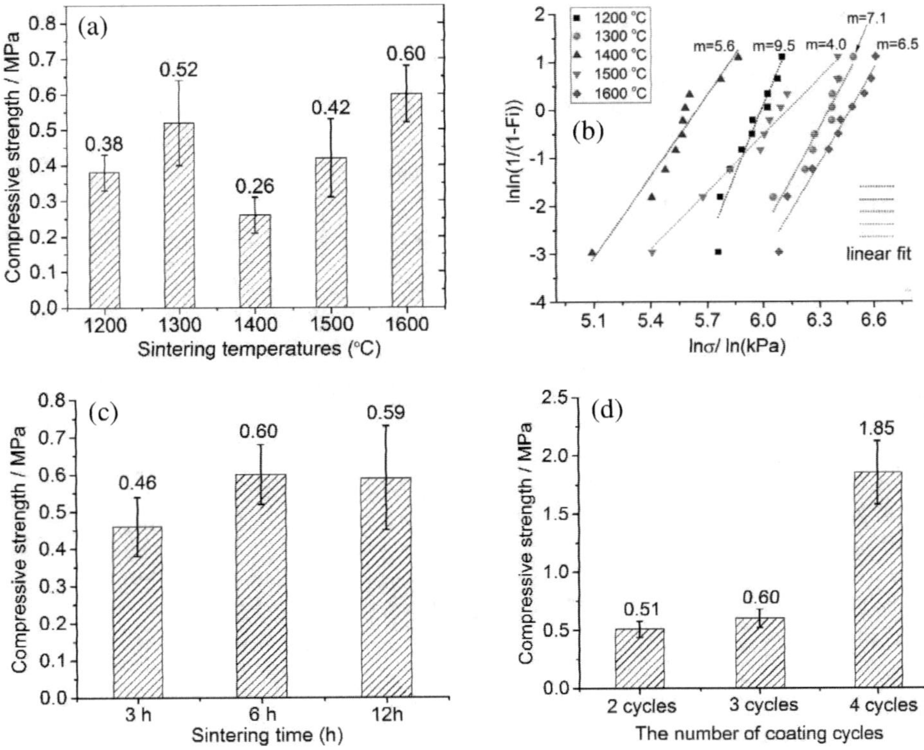

Figure 8. Processing conditions and their influence on compressive strength and corresponding Weibull modulus in Al/Al_2O_3 composite foams: (a) and (b) the influence of the sintering temperature for foams sintered at different temperature for 6 hours; (c) the influence of the compression strength on foams sintered at 1600 °C for different times; (d) the influence of the number of coating cycles for foams sintered at 1600 °C for 6h.

The influence of the sintering conditions on compressive strength of the Al/Al_2O_3 foams is shown in Figure 8: the compressive strength increases from 1200 °C to 1300 °C, drops to 1400 °C and increases with increasing temperature to 1600 °C. A tentative explanation for this course is the following: increasing sintering temperature facilitated the densification of the outer ceramic coating, which is along with increased strut strength [38]. An increasing firing temperature gave rise to surface crack formation, as we had observed by means of μ-CT, and this reduced the strut strength. With increasing temperature the ongoing sintering process reduces strut porosity leading to an increase in strength.

The Weibull modulus in compressive strength also varied with the sintering temperature. The foams processed at 1200 °C and 1300 °C owned relatively larger Weibull moduli. This may be associated with surface morphology of the struts. Totally, though severe surface cracks existed within resulting foams, the Weibull modulus in our foams is still comparable to other studies [7, 30, 34].

The influence of sintering time on compressive strength of the foams as displayed in Figure 8(c) indicates a slight increase in strength with increasing time, which may have influenced the theoretical density of the alumina phase, being higher after longer sintering times. Figure 8(d) shows how the coating thickness, here given as the number of coating cycles, affects the compressive strength. This is related to the thickening of struts under the loss of porosity, see also Figure 4(b).

4 Conclusions

By combining the investment-cast technique with the direct oxidation conversion (DOC) method, novel Al/Al_2O_3 composite foams were prepared. Replacement of commonly polymeric templates as used in the replica processing of ceramic foams by Al foams with dense struts leads to novel-type Al/Al_2O_3 foams with ternary-layered structures in their struts, which consists of an inner layer of Al, a middle layer of α-Al_2O_3 formed by Al oxidation and an outer layer of α-Al_2O_3 ceramic coating. The process parameter sintering temperature played a crucial role, leading to moderate compressive strength when being over 1500 °C. Compared to conventional RPCs, these novel-type composite foams possess a new deformation mode in compression: they show accumulative damages instead of catastrophic collapse, with more deformation capacity and better ductility. The approach of DOC might inspire a new way to manufacture composite ceramic foams, and such composite foams might be the alternatives to other cellular solids in some applications.

Acknowledgements

X Chen acknowledges the Chinese Scholarship Council (CSC) for its financial support.

References

[1] Scheffler M and Colombo P eds 2005 *Cellular ceramics: structure, manufacturing, properties and applications* (Chichester: John Wiley distributor)
[2] Colombo P 2006 *Phil. Trans. Series A, Math., Phy. and Eng. Sci.* **364** 109–24
[3] Studart A R, Gonzenbach U T, Tervoort E and Gauckler L J 2006 *J. Am. Ceram. Soc.* **89** 1771–89
[4] Emmel M and Aneziris C G 2012 *Ceram. Int.* **38** 5165–73
[5] Mujeebu M A, Abdullah M Z, Bakar M A, Mohamad A A and Abdullah M K 2009 *Appl. Energy* **86** 1365–75
[6] Betke U, Klaus M, Eggebrecht J G, Scheffler M and Lieb A 2018 *Microporous and Mesoporous Mater.* **265** 43–56
[7] Schelm K, Schwidder M, Samuel J, Scheffler F and Scheffler M 2017 *Adv. Eng. Mater.* **19** 1700418
[8] Kribus A, Gray Y, Grijnevich M, Mittelman G, Mey C S and Caliot C 2014 *Solar Energy* **110** 463–81
[9] Wu Z, Caliot C, Flamant G and Wang Z 2011 *Solar Energy* **85** 2374–85
[10] Schwartzwalder K and Somers V A 1963 *3090094*
[11] Brown D D and Green J D 1994 *J. Am. Ceram. Soc.* **77** 1467–72
[12] Gibson L J and Ashby M F 2010 *Cellular solids: structure and properties* 2nd edn (Cambridge: Cambridge Univ. Press)

[13] Brezny R and Green D J 1993 *J. Am. Ceram. Soc.* **76** 2185–92
[14] Jun I K, Koh Y H, Song J H, Lee S H and Kim H E 2006 *Mater. Lett.* **60** 2507–10
[15] Pu X P, Jia L P, Zhang D F, Su C H and Liu X J 2007 *J. Am. Ceram. Soc.* **90** 2998–3000
[16] Schelm K, Dammler K, Betke U and Scheffler M 2019 *Adv. Eng. Mater.* **21** 1900635
[17] Zhu X, Jiang D and Tan S 2001 *Mater. Lett.* **51** 363–7
[18] Han F, Zhong Z, Yang Y, Wei W, Zhang F, Xing W and Fan Y 2016 *J. Euro. Ceram. Soc.* **36** 3909–17
[19] Mianxue W and Gary L M 1990 *J. Am. Ceram. Soc.* **73** 3497–9
[20] Paiva A, Sepulveda P and Pandolfelli V 1999 *J. Mater. Sci.* **34** 2641–9
[21] Ahmad R, Ha J H and Song I H 2014 *Ceram. Inter.* **40** 3679–85
[22] Fengzhang R, Gangjun Z, Zhanhon M, Alex V and Baohong T 2014 *J. Ceram. Proc. Res.* **15** 71–5
[23] Herrera A M, Martins A A, Novaes A P and Hotza D 2013 *J. Ceram.* **2013** 1–8
[24] Pu X, Liu X, Qiu F and Huang L 2004 *J. Am. Ceram. Soc.* **87** 1392–4
[25] Voigt C, Jäckel E, Aneziris C G and Hubálková J 2013 *Ceram. Int.* **39** 2415–22
[26] Xinwen Z, Dongliang J, Shouhong T, and Zhaoquan Z 2001 *J. Am. Ceram. Soc.* **84** 1654–6
[27] Xipeng P, Xuejian L, Fagui Q and Liping H 2004 *J. Am. Ceram. Soc.* **87** 1392–4
[28] Yao X, Yang Y, Liu X and Huang Z 2013 *J. Euro. Ceram. Soc.* **33** 2909–14
[29] Jun I K, Kong Y M, Lee S H, Kim H E, Kim H W and Goretta K C 2006 *J. Am. Ceram. Soc.* **0** 2317–2319
[30] Chen X, Betke U, Rannabauer S, Peters P C, Söffker G M and Scheffler M 2017 *Mater.* **10**
[31] Liang X, Li Y, Wang Q, Sang S, Xu Y, Chen Y, Li B and Aneziris C G 2017 *Ceram. Int.* **43** 11197–203
[32] Rannabauer S, Söffker G M, Scheunemann M, Betke U and Scheffler M 2017 *Adv. Eng. Mater.* **19** 1700211
[33] Vogt U F, Gorbar M, Dimopoulos E P, Broenstrup A, Wagner G and Colombo P 2010 *J. Euro. Ceram. Soc.* **30** 3005–11
[34] Betke U, Scheunemann M and Scheffler M 2019 *Mater.* **12**
[35] Liang X, Li Y, Liu J, Sang S, Chen Y, Li B and Aneziris C G 2016 *Ceram. Int.* **42** 13091–7
[36] Newkirk M S, Urquhart A W, Zwicker H R and Breval E 1986 *J. Mater. Res.* **1** 81–929-35
[37] John E H (ed) 1984 *Aluminum Properties and Physical Metallurgy: Properties of pure aluminum* (Geauga, Ohio: Aluminum Association Inc. and ASM International)
[38] Athena T, Christos A and Aggelos M 2001 *J. Euro. Ceram. Soc.* **21** 363–73
[39] Nasch P M and Steinemann S G 1995 *Phy. and Chem. of Liq.* **29** 43–58
[40] Charles J E and Ernest H Z 1959 *J. Am. Ceram. Soc.* **42** 300–5
[41] Jeurgens L P H, Sloof W G, Tichelaar F D and Mittemeijer E J 2002 *J Appl. Phy.* **92** 1649–56
[42] Zhang X, Xie Z, Xu L and Yang W 2009 *Mater. & Des.* **30** 4507–10

IOP Conf. Series: Materials Science and Engineering **882** (2020) 012004 doi:10.1088/1757-899X/882/1/012004

Microstructure and surface investigations of TiAl6V4 and CoCr28Mo6 orthopaedic femoral stems

M Crackau[1], K Harnisch[1], T Baierl[1], P Rosemann[1], C H Lohmann[2], J Bertrand[2] and T Halle[1]

[1] Otto-von-Guericke University Magdeburg, Faculty of Mechanical Engineering, Institute of Materials and Joining Technology, Universitätsplatz 2, 39106 Magdeburg, Germany
[2] Otto-von-Guericke-University Magdeburg, Faculty of Medicine (FME), Department of Orthopaedic Surgery, Leipziger Straße 44, 39120 Magdeburg, Germany

maria.crackau@ovgu.de

Abstract. Total hip arthroplasties (THA) achieve very good clinical results and show annually increasing numbers of implantation. Interactions of the bone with the implants surface are of major importance for a stabile fixation and longevity of the implant. Therefore, manufacturing of the material and the implants as well as their surface properties can have a decisive influence on the functionality of the implant. The aim of the present study is the investigation of two commercially available femur stems with analytical methods. One stem is made of a TiAl6V4 wrought alloy for cementless application and the other one is made of a CoCr28Mo6 cast alloy for cemented fixation. The change of the production-related microstructure within the implant, differences between surface and bulk properties and potential correlations between the production-related changes to predict failures are addressed. Longitudinal cross sections of tested stems were prepared metallographically, investigated using optical and scanning electron microscopy including EDS and EBSD and correlated with micro hardness depth profiles. Due to production and processing, a subsurface layer is formed in both alloys. The TiAl6V4 wrought alloy stem exhibits a homogenous recrystallization microstructure with fine grains of micrometre size. The subsurface layer of the stem is deformed in parts with embedded corundum particles within the depth of 10 μm. Corundum residues were detected on the entire stem surface and partially covered by the applied calcium phosphate spray coating. The CoCr28Mo6 cast alloy stem shows a dendritic microstructure with chromium- and molybdenum-rich interdendritic precipitations and a surface layer of smaller globular grains in the size of 50 μm to 200 μm. The face centred cubic (fcc) crystal structure was predominantly detected in the phase analysis. The brittle hexagonal close packed (hcp) phase was evident at the implants surface. Similar to the TiAl6V4 alloy stem, embedded residues of corundum particles were detected on the implant surface. This study shows different surface integrities for both stems in comparison to the base material. The observed residues from the manufacturing processes are generally well-embedded into the implant surface, however if released they could impair the functionality of the endoprosthesis as the particles might negatively affect the sterilization process or might reduce metal corrosion resistance.

1 Introduction

Total hip arthroplasties (THA) are successful options for the treatment of joint diseases, such as osteoarthritis or fractures due to trauma. As a standard surgery of modern orthopaedics, artificial hip replacement shows excellent clinical outcomes for the patients with increasing numbers of implantation [1]. In Germany, there are approximately 230,000 primary THA implanted annually [2]. All THA consist of a femoral stem component that is inserted into the femur, a modular head and an acetabular cup, as monobloc or with a liner being placed into the acetabulum. The modular systems allow the reconstruction of the centre of rotation with stable anchorage of stem and socket. Standard metals used for femoral stems are cobalt-chromium-molybdenum alloys (Co-28%Cr-6%Mo cast/wrought alloy ISO 5832-4/-12) with high fracture toughness, wear and corrosion resistance [3] or titanium-based alloys (Ti-6%Al-4%V ISO 5832-3, Ti-6%Al-7%Nb ISO 5832-11) with excellent biocompatibility, high corrosion resistance, low density and a comparatively low elastic modulus [4].

Implant-bone fixation of the femoral stem can either be cementless (press-fit), cemented by applying an acrylic bone cement, or hybrid as a combination of both methods. The fixation type predominantly depends on the patient's bone structure, bone quality, age and activity [5–7]. Cementless fixation has the advantages of lower intramedullar pressure, fewer risk of emboli and hemodynamic disturbances [5], but increases the risk for periprosthetic fractures [8]. Cemented fixation is mainly used in older patients or in patients with minor bone quality [9]. It allows an immediate loading after implantation, and is cost-effective [10]. In 2018, more than 75 % of all primary THA in Germany had a complete cementless fixation whereas only 5 % were fully cemented [11]. In cementless application, primary implant stability is based on a rigid press-fit macro-lock and secondary stability by bony apposition and growth, known as osseointegration [12]. The biological fixation requires an appropriate implant design and surface finish, such as porous or roughened surfaces [13, 14]. In addition, osteoconductive surface coatings of calcium phosphates, such as hydroxyapatite (HA), are applied. These coatings promote cell attachment, calcification and adaptive bone remodelling by continuous coating degradation [15, 16]. Titanium-based alloys are favourable for cementless fixation since their elastic modulus of 110 GPa is closer to the elastic modulus of bone thus ensuring a better interface stress transfer and lower stress shielding [17, 18]. In the cemented fixation method, bone cement, usually PMMA (polymethylmethacrylat), forms a bond between the cortical bone and the implant and ensures a firm cement-bone connection. PMMA tolerates compressive loads, but is sensitive to tensile or shear stresses that can cause breakage [19]. Therefore, rounded stem designs with polished surfaces are envisioned to reduce stress peaks and overloads in the PMMA and adjacent bone [17, 20].

The interface between implant, cement and bone tissue is of uppermost importance. A permanently firm, load-bearing connection to the bone is decisive for the long-term functionality of the endoprosthesis [15, 21]. Implant weight-bearing causes inevitable micromotion that is associated with failure of the femoral stem fixation due to cement debonding, fracture, wear generation, corrosion and prevention of osseointegration that can lead to the need of revision surgery [5, 22–24]. Common damage mechanisms initiate at the stem surface and are influenced by multiple factors. Manufacturing of titanium- and cobalt-based alloy stems determines the microstructure and consequently the mechanical properties. Defined surface roughness is achieved by post-machining and finishing. Common finishing techniques for stems are shot peening, grit blasting and vibratory grinding to produce a matte, smooth or roughened surface for improved cemented or cementless fixation [25, 26]. Roughened surface modifications promote the subsequent HA coating adhesion that is commonly deposited in a plasma spray process [15]. All finishing processes affect surface near material layers and can cause microstructural changes, metal distortion, contamination and micro damages that alter the material properties. While some studies report an effect of surface contaminants on early implant failure, production-related changes in material and surface properties of stems are not well known [27, 28].

IOP Conf. Series: Materials Science and Engineering **882** (2020) 012004 doi:10.1088/1757-899X/882/1/012004

This study investigates the effects of stem manufacturing on the microstructure and properties of a TiAl6V4 wrought alloy stem for cementless application and a CoCr28Mo6 cast alloy stem for cemented fixation. In particular, the following questions are addressed:

- How does the production-related microstructure change with component geometry?
- How do the material properties differ between the bulk and the surface?
- Is there any connection between the production-related change in the properties of the implants and known failure mechanisms?

2 Materials and Methods

2.1 Material and sampling

Two commercially available femoral hip stems were investigated. One stem was made of TiAl6V4 wrought alloy (ISO 5832-3) from the company ImplanTec (ImplanTec Deutschland GmbH (today ARTIQO GmbH), Lüdinghausen, Germany) for cementless application. The arch-shaped ANA.NOVA Alpha stem having a rectangular cross-section exhibits a polished neck and the holohedral, bioactive plasma spray coated calcium phosphate coating (BIONIT®). In contrast, the CoCr28Mo6 cast alloy (ISO 5832-4) stem is intended for cemented application from the company Link (Waldemar Link GmbH & Co. KG, Hamburg, Germany). Rounded surfaces medially and laterally serve to protect the cement bed and improve the implant anchoring. The stem has a collar that counteracts the sintering of the stem. Technical data of the investigated femoral stems are summarized in table 1. Length, width and height (L x W x H), CCD-angle (centrum- collum- diaphyseal- angle) and taper design describe geometrical characteristics of the stems.

Table 1. Technical data of the investigated femoral stems.

	TiAl6V4 stem	**CoCr28Mo6 stem**
Manufacturer	ImplanTec Deutschland GmbH	Waldemar Link GmbH & Co. KG
Implant model	modular ANA.NOVA Alpha	modular Lubinus Classic Plus
Material standard	DIN EN ISO 5832-3:2017-03 (TiAl6V4 wrought alloy)	DIN EN ISO 5832-4:2015-12 (CoCr28Mo6 cast alloy)
Size (L x W x H [mm])	150 x 10-38 x 7-19	150 x 5-20 x 5-10
CCD-angle[°]	127	126
Taper design	12/14	12/14
Surface finish	porous calcium phosphate coating	matte

The stems were sectioned perpendicular to the stem axis into 12 samples each with a thickness of 10 mm (figure 1). The TiAl6V4 stem was customized by cut-off grinding with liquid lubrication. For the CoCr28Mo6 cast alloy stem, wire-cut electrical discharge machining (Mitsubishi MV2400S, Mitsubishi Electric, Japan) was used.

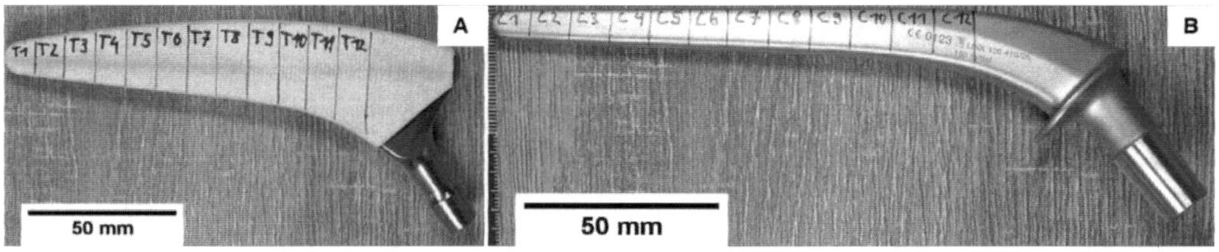

Figure 1. Macro photographs of the femoral stems with marking of the samples. (**A**) Cementless TiAl6V4 alloy stem with calcium phosphate coating and (**B**) CoCr28Mo6 cast alloy stem.

2.2 Surface roughness
Surface roughness of the implant-bone/cement interface was determined with a confocal microscope (µsurf expert, NanoFocus AG, Oberhausen, Germany). The measuring area of 1.57 × 1.57 mm was acquired with a ×10 magnification objective and a vertical resolution of 20 nm. Each stem surface was measured at three different locations (sample T12 and C12, on medial, anterior and lateral side). Besides 3D topography images, the roughness parameters arithmetic mean roughness R_a and average maximum profile height R_z were obtained according to ISO 25178. Based on the obtained data, the Abbott-Firestone curves were calculated and the functional bearing ratio parameters R_k, R_{pk} and R_{vk} were determined according to European project report EUR 15178 EN.

2.3 Metallographic preparation
The samples were hot mounted in a conductive phenol resin with carbon additive (PolyFast, Struers GmbH, Dresden, Germany). TiAl6V4 samples were ground with resin bonded diamond discs (MD-Piano, Struers GmbH, Dresden, Germany) of different grits (125 µm – 10 µm). For the CoCr28Mo6 samples, a manual grinding process was conducted with decreasing grit (200 µm – 8 µm) of abrasive silicon carbide paper under flowing water. Subsequently, a chemical-mechanical polishing process with a mixture of silicon oxide and hydrogen peroxide (35 %) on synthetic fibre for 10 minutes generated a scratch- and deformation-free surface. In order to visualize the microstructure, TiAl6V4 samples were etched in an immersion bath (100 ml deionized water, 5 ml hydrogen peroxid and 2 ml hydrofluoric acid) for 30 seconds. For the CoCr28Mo6 samples, electrochemical etching was applied. The electrolyte was ethanol with 60% perchlorid acid at a voltage of 7 V for 15 seconds (Electrolyte A2 Struers GmbH, Dresden, Germany).

2.4 Optical Microscopy
The optical microscopes Axiovert 200M with the camera system AxisCam MRc5 (Carl Zeiss Jena GmbH, Jena, Germany) and Leica MEF4A (Leica Microsystems GmbH, Wetzlar, Germany) with the camera unit AxioCam 503 were used for optical analyses. Differential interference contrast was applied for a better relief representation of the polished sample surfaces. Three different locations on each sample were examined as shown in figure 2 enabling to compare changes in depth direction (e.g. see A-B) and longitudinal direction of the stem (e.g. see T1-B, T2-B, T3-B).

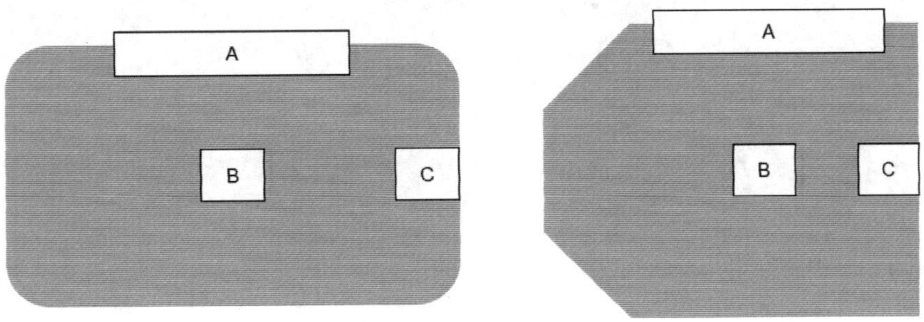

Figure 2. Locations of the optical microscopic examination on cross sections of the (**left**) TiAl6V4 stem and (**right**) CoCr28Mo6 stem.

2.5 Scanning Electron Microscopy
The electron microscopic examination was carried out using a scanning electron microscope (FEI Scios DualBeam, Thermo Fisher Scientific, Waltham, MA, USA) equipped with a TEAM Trident system (energy dispersive X-ray spectroscopy EDS, electron backscatter diffraction EBSD, wavelength dispersive X-ray spectroscopy WDS; EDAX, AMETEK GmbH, Weiterstadt, Germany). Secondary electron (SE) imaging and backscattered electron imaging (BSE) were used to show microstructural and surface features. EDS was used to determine the local chemical composition of different phases and

precipitations. The crystal structures were investigated using EBSD regarding crystallographic orientations, grain sizes and phase distribution.

2.6 Microhardness measurements

Microhardness ($HV_{0.05}$) was measured according to ISO 6507-1 using an automated microhardness tester VH3300 (Buehler, ITW Test & Measurement GmbH, Esslingen am Neckar, Germany).

3 Results and Discussion

3.1 Characterization of TiAl6V4 wrought alloy stem

Figure 3 shows a micrograph of the original surface of the TiAl6V4 stem revealing a strongly inhomogeneous coating. The plasma sprayed surface coating exhibits a characteristic crystalline structure, inherent porosity and cracked, deformed as well as chipped-off areas. Different grey shades in the BSE micrograph indicate different chemical elements according to the atomic number. Fine needle shaped structures correspond to the crystalline calcium phosphate (CaP) coating. Bright features are related to the TiAl6V4 substrate not being covered with CaP. The darker features within the red frame in figure 3 were classified in detail using EDS mapping. The results of the elemental mappings verify the features to consist of aluminum and oxygen.

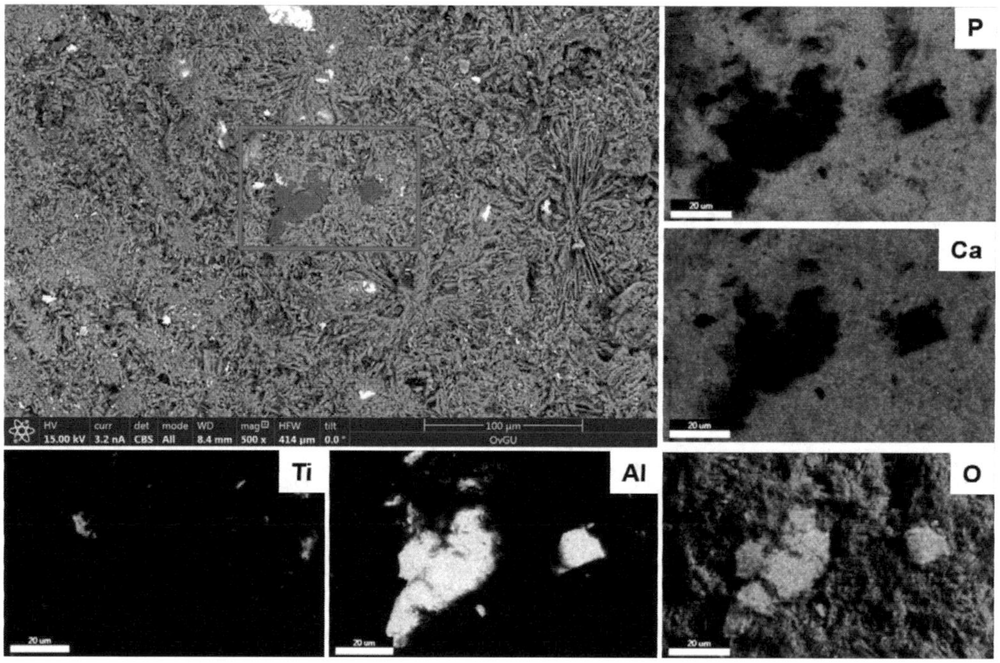

Figure 3. Electron micrograph of the surface structure of the CaP coated TiAl6V4 stem with elemental mappings of titanium, aluminum, oxygen, calcium and phosphor

Presumably, these particles being partly covered with CaP originate from former processing steps. TiAl6V4 implants are commonly grit blasted in order to generate a macro topography that increases the surface area and improves primary bony fixation. Previous studies showed that an insufficient surface cleaning after grit-blasting resulted in the adhesion of Al_2O_3 particles, which are commonly used in the grit blasting process [25–28]. Schuh et al. investigated thermomechanical and pH-dependent cleaning techniques to remove contaminant grit particles and highlighted the relevance of a post-cleaning process for implant application [29, 30].

The inhomogeneous CaP coating, embedded Al_2O_3 particles and the surface structure of the TiAl6V4 substrate influence the resulting roughness of the outer osteoconductive surface. Figure 4 is summarizing of the results of surface roughness measurements.

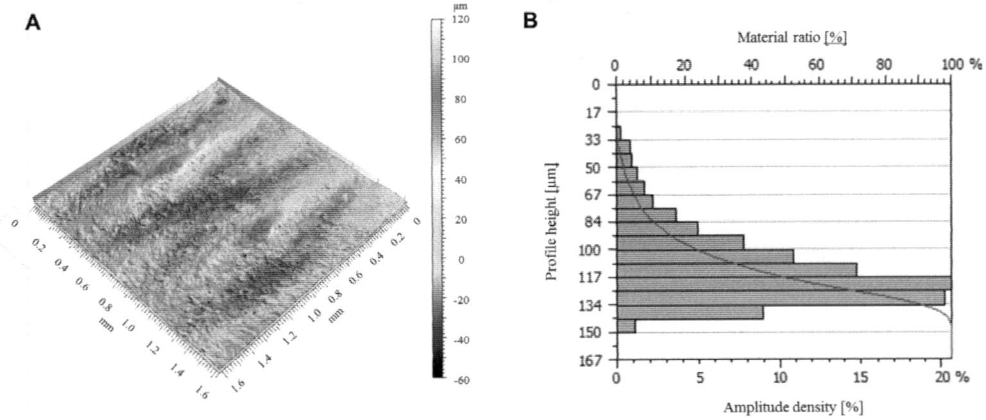

Figure 4. Surface of the TiAl6V4 stem with an osteoconductive CaP coating. (**A**) 3D topographic confocal microscopic image showing macroscopic waviness. (**B**) Bearing area curve of the measured surface area in A.

The macroscopic surface morphology of the TiAl6V4 stem exhibits regular distributed peak ripple of 100 µm height with a spatial distance of 800 µm (macro roughness). The steep drop in the corresponding areal material ratio curve (figure 4B) demonstrates the resulting pronounced peak height above the core roughness. In addition, microroughness is generated by the porous plasma sprayed CaP coating that exhibits an arithmetic mean roughness R_a of 2.77 ± 0.43 µm and roughness height R_z of 16.57 ± 3.45 µm. Various studies reported that complex roughness features in macro, micro and nano length scale exhibit an enhanced osseointegration in comparison to polished or uniform patterned surfaces [26, 31, 32]. The measured roughness parameters for plasma sprayed CaP coatings correspond to other studies that report a wide range of R_a value between 2 µm to 9 µm according to coating thickness and pre-treatment processes [13, 33, 34].

The TiAl6V4 stem exhibits a fine-grained and regular microstructure of a globular α phase and a lamellar α+β phase, as shown in figure 5 A. The average grain size of the homogeneously distributed α phase is smaller than 10 µm. In the SEM image (figure 5B), the lamellae of the β phase are shown and exhibit sub-micrometre thickness. Large α-phase regions often consist of many smaller α grains adjacent to each other and being separated by their orientation. The structure within the α grains that becomes apparent using SEM is based on orientation dependent responses to the etching of the samples.

IOP Conf. Series: Materials Science and Engineering **882** (2020) 012004 doi:10.1088/1757-899X/882/1/012004

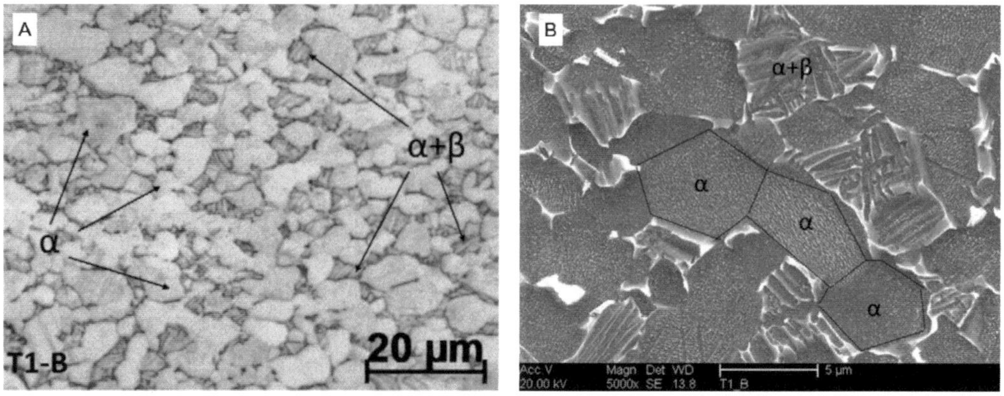

Figure 5. Microstructure of the investigated TiAl6V4 stem. (**A**) Optical micrograph showing two different phases. Light grains represent the globular α phase that is surrounded by a lamellar α+β phase. (**B**) Corresponding SEM image displaying the fine structure of the α+β phase between α grains.

There are no changes in the microstructure in the longitudinal and depth direction (see Appendix figure A 1). The results of the phase analysis by optical greyscale distribution reveal that the majority with 75.3 ± 2.3 % comprises of α phase and 24.7 ± 2.3 % of the lamellar α+β phase. Local EDS measurements revealed that the α phase exhibits a slightly increased concentration of aluminum (8 wt.-%) and a slightly decreased concentration of vanadium (1 wt.-%) in comparison to integral measurements with about 7 wt.-% aluminum and about 4 wt.-% vanadium. In contrast, the lamellar α+β phase shows a slightly increased vanadium concentration (6 wt.-%) and a slightly decreased concentration of titanium (88 wt.-%).

Cross sections of the corresponding stem surface are illustrated in figure 6 confirming the presence of particles within the stem surface and below the CaP coating.

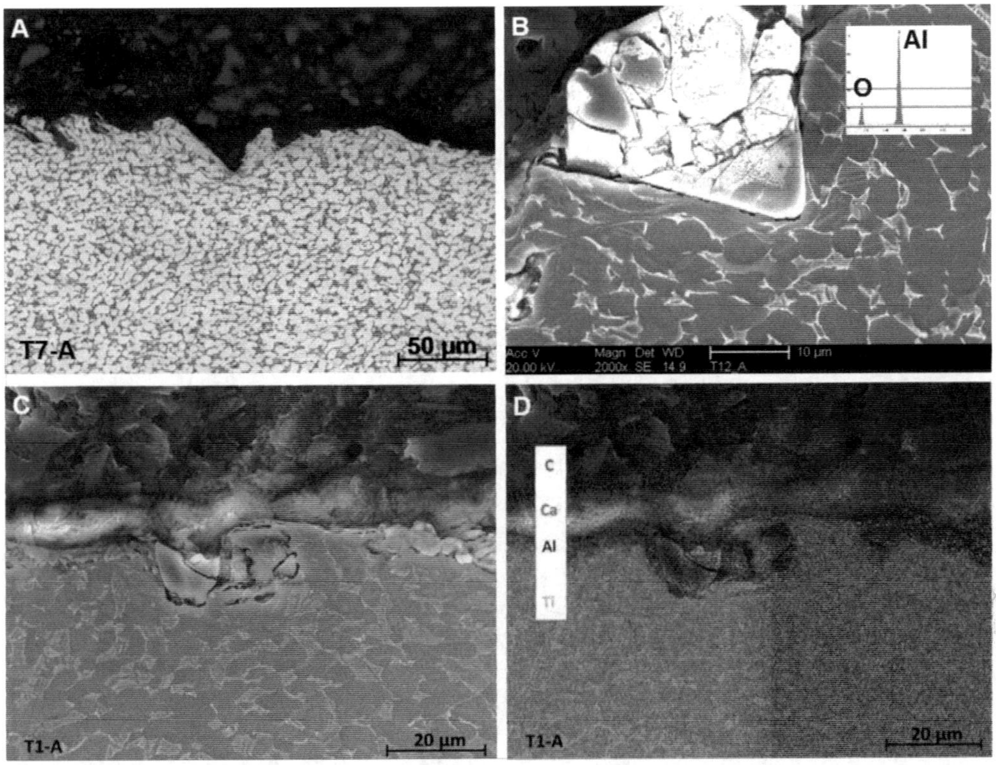

Figure 6. Surface near microstructure of the TiAl6V4 stem (**A**) Micrograph showing the coated surface of the two-phase TiAl6V4 alloy exhibiting a rough surface. (**B**) SEM image displays a fractured particle embedded in the surface and the corresponding EDS spectrum (Al_2O_3, corundum). (**C**) Cross section of the stem surface with porous CaP coating and underlying corundum particles. (**D**) EDS map of (C) with the elemental distribution of carbon (red), calcium (pink), aluminium (blue), titanium (turquoise).

Figure 6 A shows the processing caused imperfections of the stem surface. Blasting induced the generation of a rough inhomogeneous surface with a deformed subsurface within the first 10 µm. Embedded particles of Al_2O_3 that have been detected within the previous surface investigation (see figure 3) can be seen in the cross sections as well. In figure 6 B a fractured particle that is incorporated into the surface is pictured with the corresponding EDS spectrum of the particle showing distinct peaks of aluminum and oxygen. Figures 6 C and 6 D are showing topography and colour-coded elemental mapping (EDS mapping) of a surface near cross section with a high amount of Al_2O_3 particles. The titanium rich TiAl6V4 substrate, several Al_2O_3 particles embedded in the irregular surface structure and a thin layer of CaP can be recognized and separated. Several measurements of Al_2O_3 particles revealed a particle size between 5 µm and 25 µm. The thickness of the CaP coating varies between few microns up to about 15 µm depending on the surface topology and subjacent Al_2O_3 particles. Schuh et al. determined the particle size distribution on different Al_2O_3 blasted titanium hip stems and detected that more than 75 % are smaller than 40 µm [25]. Based on Gburec et al., up to 35 % of the titanium surface can be covered by the blasting grit [35]. The areal amount and size depend on the blasting pressure, grit diameter, process control and cleaning method [25, 26]. The thickness of CaP coating determines the degradation time and consequently the complex dynamic osseointegration process [34]. Recommended layer thickness for the CaP coating is < 50 µm since it provides a good adhesion to the implant, easy resorption and a sufficient strength [34, 36].

IOP Conf. Series: Materials Science and Engineering **882** (2020) 012004 doi:10.1088/1757-899X/882/1/012004

In contrast to the surface characteristics of the investigated TiAl6V4 stem such as topology, roughness and present surface composition, hardness measurements did not indicate major deviations. Figure 7 shows a characteristic microhardness depth profile of the surface near zone of 0.7 mm.

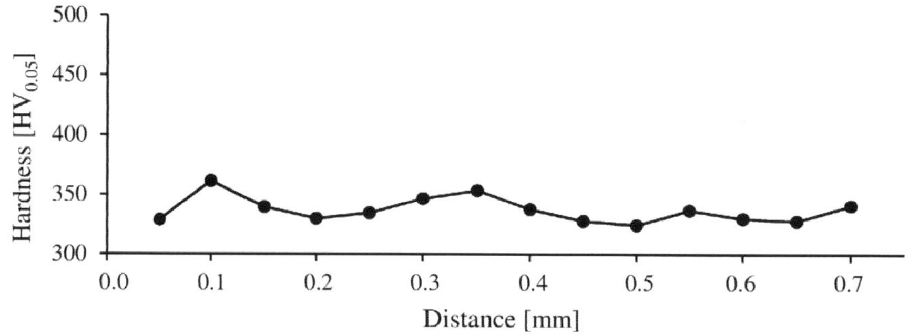

Figure 7. Microhardness depth profile of the TiAl6V4 stem. Distance delineates the perpendicular offset to the surface of the sample.

There is no significant change in hardness with increasing distance to the surface. The mean hardness of the measurement is 337 ± 10.5 $HV_{0.05}$, being in the order of previous general studies of TiAl6V4 wrought alloys [37]. Slight deviations may result from inhomogeneity of the surface and the structure hit by the indenter due to phase composition and grain boundaries. The comparatively low hardness for a distance of 0.05 mm (329 $HV_{0.05}$) may be due to an edge effect.

3.2 Characterization of CoCr28Mo6 cast alloy stem

The matte surface of the CoCr28Mo6 cast alloy stem shows a macroscopic inhomogeneity. A SEM BSE image illustrates a characteristic surface area in figure 8 A.

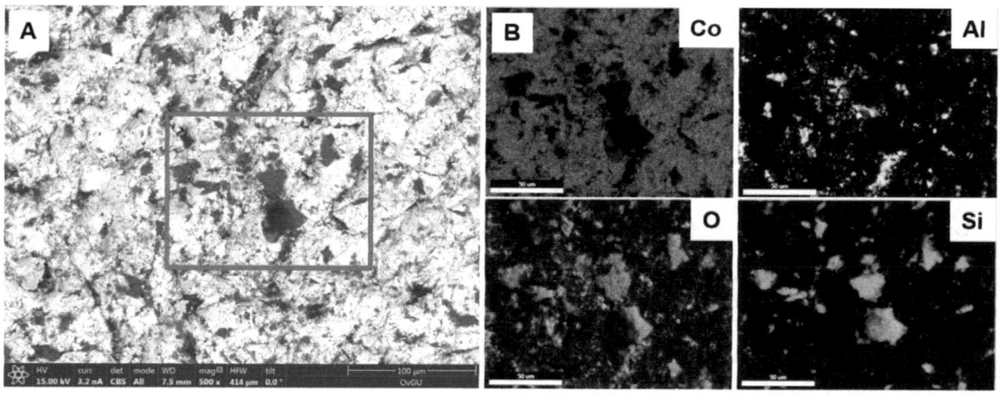

Figure 8. Surface of the CoCr28Mo6 stem. (**A**) BSE image of the surface. (**B**) EDS mapping of cobalt, aluminium, oxygen and silicon.

The fissured surface exhibits particulate inclusions, deformation, scratches and cracks. Different grey scale values in the BSE image confirm variations in the chemical composition. Based on EDS, aluminium and silicon rich particles are present and embedded in the CoCr28Mo6 alloy surface. Based on previous publications, we assume that these particles are residues from the surface finishing applying corundum grit and glass bead blasting to generate a matte surface [26, 30]. Ricci et al. detected similar residues of silicon- and aluminium-rich particles on explanted cemented cobalt-chromium hip implants [27].

IOP Conf. Series: Materials Science and Engineering **882** (2020) 012004 doi:10.1088/1757-899X/882/1/012004

The results of the measurements of surface roughness are presented in figure 9.

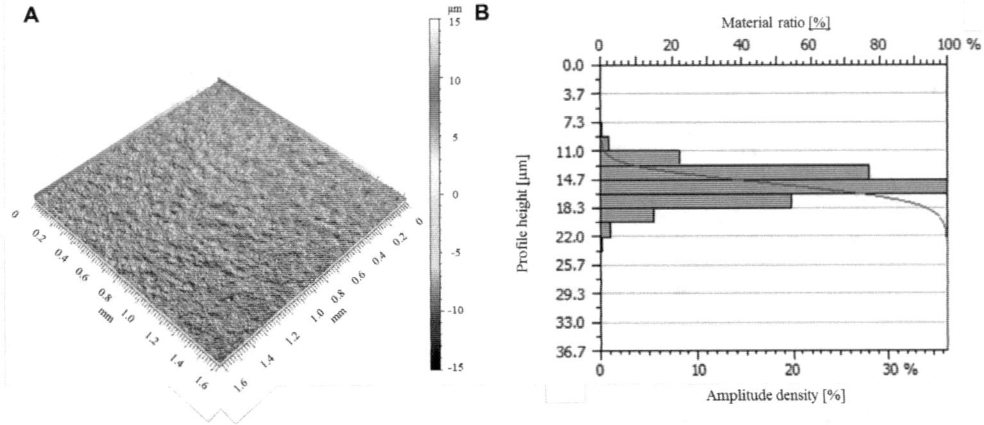

Figure 9. Confocal microscopic measurements of the CoCr28Mo6 alloy stem. (**A**) 3D topographic image of the matt surface. (**B**) Bearing area curve of the measured surface area in A.

The matte stem surface is characterized by an irregular surface texture as shown in **figure 9 A**. The corresponding material bearing ratio curve with a low curve inclination represents a pronounced core roughness of the surface that allows a homogenous load distribution. The determined micro-roughness R_a value of 1.12 ± 0.17 µm and the mean value of roughness height R_z with 7.18 ± 1.1 µm are in accordance to cementless matte surface roughness of other manufacturers [20]. For cementless fixation surface finish and roughness determine the critical implant-cement bond and thus affect mechanical stress transfer, magnitude of post-operative stem subsidence and possible cement mantle damage leading to PMMA and metal wear causing osteolysis [24, 38–40]. Verdonschot et al. emphasize that polished cemented stem surfaces (R_a <0.04 µm) are favourable in comparison to unpolished, matte surfaces since the amount of wear debris due to micromotion is reduced [41].

Figure 10 is illustrating the microstructure of the CoCr28Mo6 cast alloy stem. A coarse dendritic microstructure with a considerable amount of interdendritic precipitations can be identified in optical and electron micrographs.

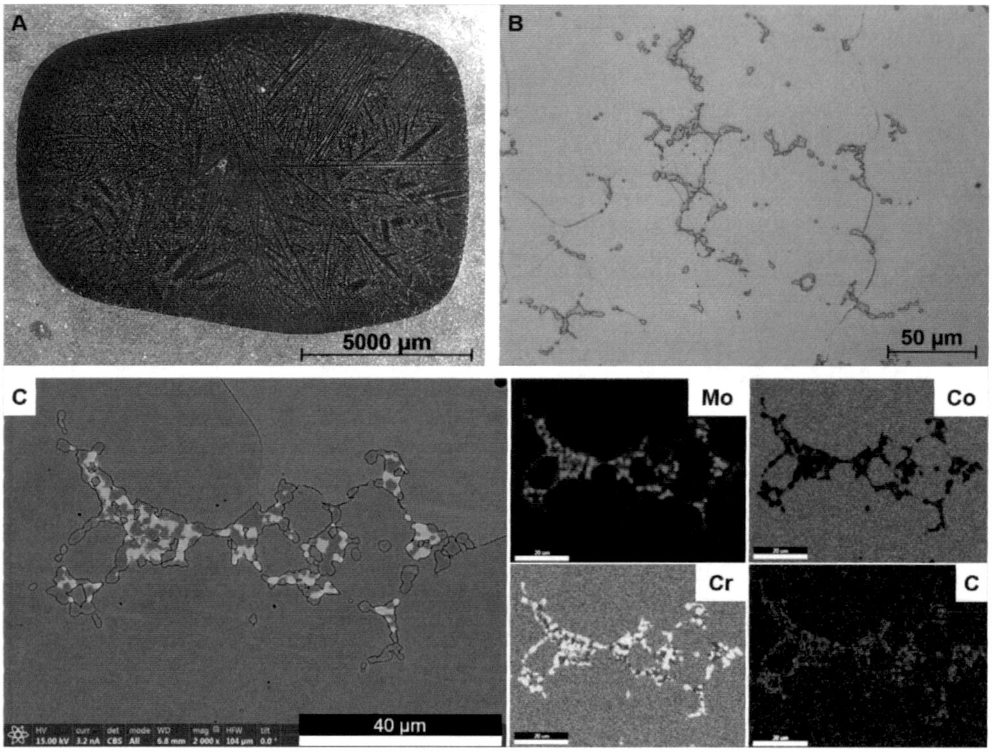

Figure 10. Top: Microstructure of the CoCr28Mo6 cast alloy stem. (**A**) Overview of a cross section showing a coarse dendritic structure. (**B**) Micrograph illustrating segmented interdendritic precipitations. (**C**) SEM BSE image indicating the segmented structure of interdendritic precipitations. EDS mapping of the area shown in (**C**) revealing the structure of the interdendritic precipitations consisting of either fractions rich in chromium and carbon or increased concentration of molybdenum.

Dendrites with a length of a few millimetres can be recognized over the complete cross-sectional area (figure 10 A). Numerous homogenously distributed precipitations within the interdendritic regions and at grain boundaries are visible at higher magnifications (figure 10 B). The segmented structure of the precipitations is shown in the electron micrograph using the BSE contrast in figure 10 C. A detailed analysis of the precipitations has been performed using EDS mapping and the results are shown in the bottom line of figure 10. Fractions of the precipitations that appear in darker grey values in BSE image are rich in molybdenum, lighter segments are rich in chromium. The entire precipitations are characterized by a lower cobalt concentration. Based on the determined characteristic composition, it is assumed that the chromium and carbon rich phase represents carbides. Grains within the base material exhibit large grain sizes due to the dendritic structure. In surface near areas, grain size is decreased considerably from few millimetres in the centre to about 150 µm at the surface (figure 11 A), respectively.

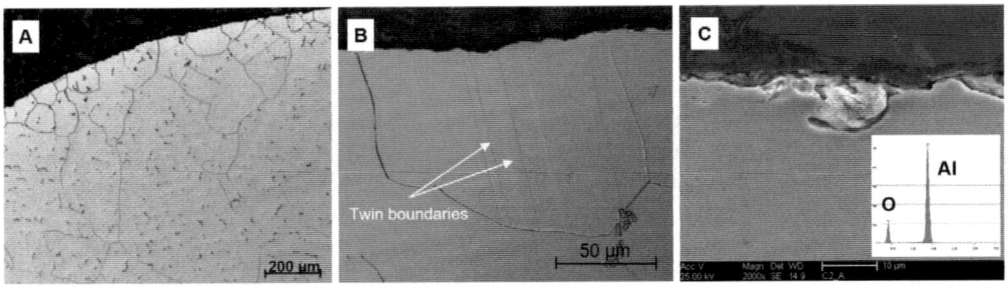

Figure 11. Microstructure of the CoCr28Mo6 cast alloy stem. (**A**) Surface area exhibiting finer grain size. (**B**) Twin boundaries within surface near grains. (**C**) Corundum particles embedded in the surface of the stem and the corresponding EDS spectrum.

Depth-related reduction of the grain size, as shown in figure 11 A, might be a result of the faster cooling and solidification with more nuclei at the crucible interface. The grain size reduction along the component contour was observed in all cross sections. There are no further changes in the dendritic microstructure in the longitudinal direction (see Appendix figure A 2).

Twin boundaries are a common feature in CoCr28Mo6 cast alloys due to a low stacking fault energy [42]. Figure 11 B illustrates the present twin boundaries in the stem investigated within this study. Similar to the TiAl6V4 stem, the surface of the CoCr28Mo6 stem exhibits irregularities as shown in figure 11 C. An Al_2O_3 (corundum) particle presumably originating from a previous blasting or grinding in the manufacturing process is embedded in the deformed surface. Clarification of the present crystal structure was provided using EBSD. Figure 12 A-C illustrate the central area of a cross section, figure 12 D-F show a surface near area.

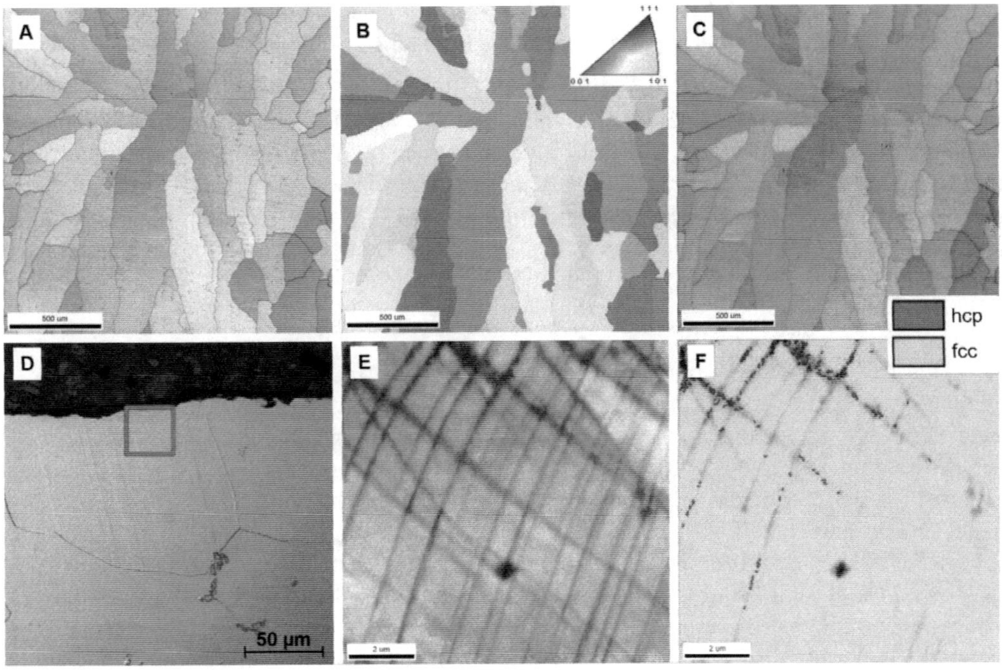

Figure 12. EBSD results of the CoCr28Mo6 stem from central (**A-C**) and surface area (**D-F**): (**A**) Image Quality map (IQ). (**B**) Inverse Pole Figure (IPF map) indicating crystallographic orientations. (**C**) Phase map assigning crystal structure. (**D**) SEM image of the surface with a red frame highlighting the area for EBSD in (**E**) and (**F**). (**E**) IQ map showing deformation marks (thin lines). (**F**) Phase map revealing hcp phase being present along the determined lines and surrounded by the fcc matrix.

Grains within the material do not show a preferred orientation as proven by IPF map in figure 12 B. EBSD measurements revealed that the face centred cubic (fcc) structure is present in the investigated stem, see figure 12 C. This correlates with common studies that characterized CoCr28Mo6 cast alloys [43–45]. Small amounts of hexagonal close packed (hcp) phase can only be located in surface near areas in the vicinity of deformation marks (thin lines) indicating a strain-induced phase transformation during manufacturing process of the stem. Figure 12 D-F are visualizing these features. The thin lines exist in hcp structure (red) while the matrix is present in fcc structure (green). Plastic deformation of cobalt base alloys leads to work hardening due to twin formation or strain-induced phase transformation from fcc to hcp (known as ε-martensite) [42]. The volume fraction of deformation-induced formation hcp phase is limited to 0.4 to 0.5 and causes an increase of hardness, yield and tensile strength as well as improved wear properties [46, 47].

The corresponding microhardness depth profile of a cross section of the CoCr28Mo6 alloy stem is illustrated in figure 13.

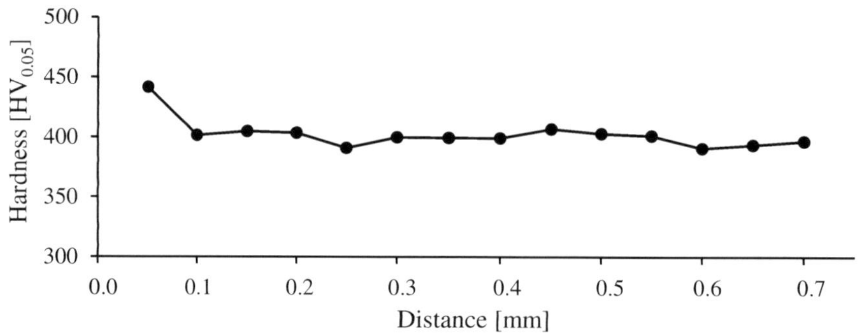

Figure 13. Microhardness depth profile of the CoCr28Mo6 stem. Distance delineates the perpendicular offset to the surface of the sample.

Besides a 10 % increase of hardness in the outer surface layer (distance of 0.5 mm) there is a uniform profile of the hardness with increasing distance to the surface. The mean hardness of the measurement illustrated in figure 13 is 402 ± 12.3 $HV_{0.05}$. The increased hardness (442 $HV_{0.05}$) at the surface can be attributed to the observed lower grain size according to the Hall-Petch equation [48]. In addition, hcp phase was detected that is associated with a strain-induced phase transformation due to stem processing and results in an increased hardness.

4 Conclusion

The aim of this study was the examination of the microstructure and basic properties of a TiAl6V4 stem for cementless application and a CoCr28Mo6 stem for cemented application. The TiAl6V4 wrought alloy exhibits a homogenous two-phase microstructure and constant hardness in comparison to a characteristic cast microstructure with interdendritic precipitations and an increased surface hardness possessed by the CoCr28Mo6 cast alloy. Residues from the manufacturing process were detected in the surface and sub-surface on both new prostheses. Corundum blasting was applied to roughen the surface for a better CaP adhesion and as surface finishing after investment casting thereby causing areal grit inclusions of Al_2O_3 particles. Individual observations suggest that there could be a possible negative impact of residual particles on femoral stems on implant longevity. Böhler et al. observed adverse local tissue reactions in periprosthetic tissue with presence of corundum particles in early retrieved titanium based hip implants [28]. Silicon and aluminium particles were also detected on explanted hip stems of titanium and cobalt based alloys examined by Ricci et al. [27]. If corundum particles are not cleaned from the surfaces after the blasting process and liberated from the surfaces, these hard particles might migrate and could act abrasive [26, 29, 30, 40]. It was shown that adjoining cracks and crevices around embedded particles deteriorate the metal corrosion resistance and promote bacterial adhesion [27, 49].

Despite these individual investigations, the long-term results of corundum blasted stems with ultrasonic post-cleaning (e.g. SPII stem prostheses) verify good clinical results for more than 30 years [50].

A stable femoral stem fixation is particularly relevant since aseptic loosening of the stem component represents one of the main reasons for revision surgery [11, 51]. Micromotion caused by cyclic implant loading are inevitable due to differences in stiffness between stem metal, cement, cancellous and cortical bone. Therefore, surface and material properties should be adjusted. In order to minimize the risk of critical crevices around embedded particles, it is suggested to enhance the finishing process of cemented and cementless stems with regard to the avoidance of residual particles by ensuring good adhesion of the CaP coating. In particular, blasting with metallic particles instead of abrasive grit followed by pickling is a suitable method to generate a contaminant free roughened surface. In addition, the metal's potential of a targeted adjustment of the subsurface microstructure that aims at an increased wear resistance should be exploited. In this context defined work hardening (e.g. deep rolling) or thermomechanical processing (e.g. solution annealing, quenching) should be envisaged. Research in the area of postprocessing after blasting and customized surface treatments in order to develop a contamination free, wear resistant implant surface offers great potential for improvement of resulting properties.

Acknowledgements
The authors Maria Crackau and Karsten Harnisch contributed equally to this work.

The authors would like to thank Michael Reppin and Oliver Michael for their excellent technical assistance and valuable support. This work was in part conducted within the context of the International Graduate School MEMoRIAL at Otto von Guericke University (OVGU) Magdeburg, Germany, kindly supported by the European Structural and Investment Funds (ESF) under the program "Sachsen-Anhalt WISSENSCHAFT Internationalisierung“ (project grant no. ZS/2016/08/80646).

Finally, the authors would like to thank the German Research Foundation (DFG) for the funding of the used DualBeam FEI Scios within the framework of the funding of large-scale facilities according to GG§91.

Appendix

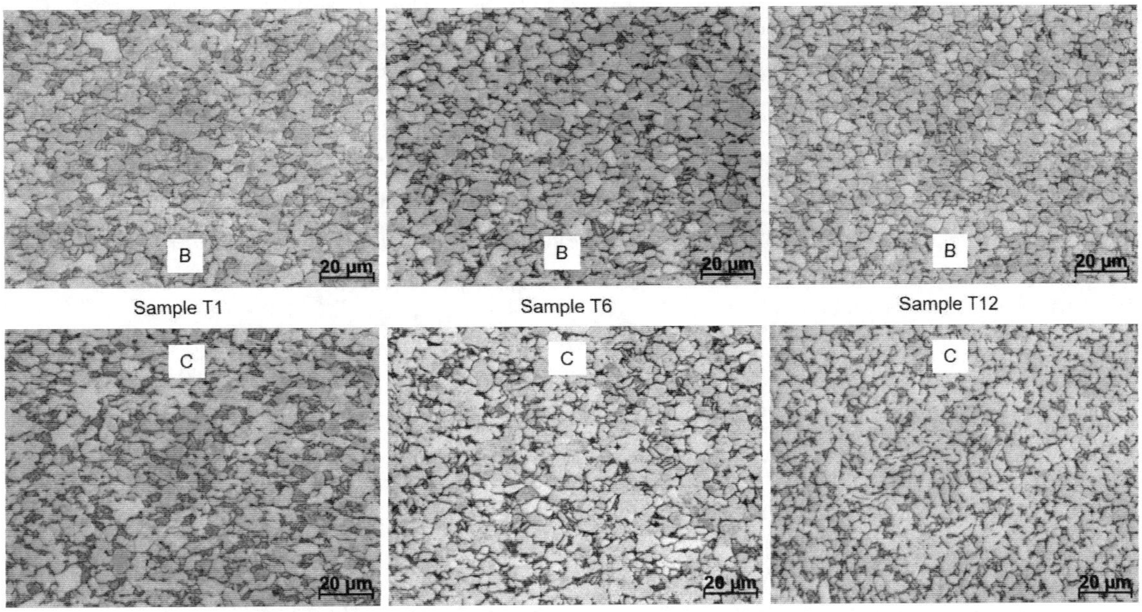

Figure A 1. Comparison of the microstructure of samples from the central (B) and edge positions (C) of various sections from the TiAl6V4 stem (location see figure 1 and figure 2).

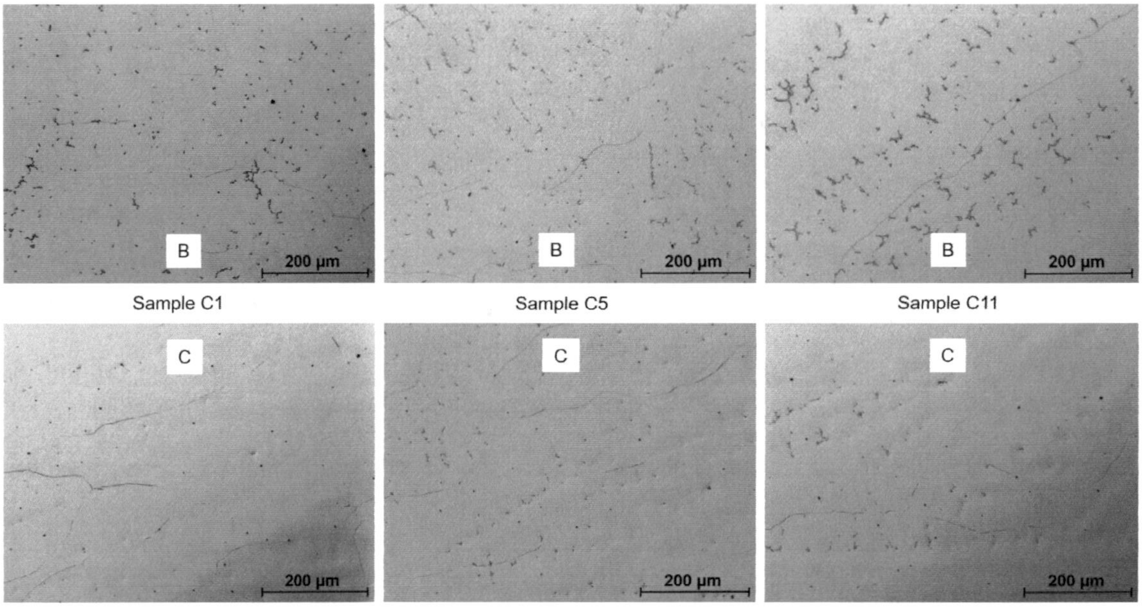

Figure A 2. Comparison of the microstructure of samples from central (B) and edge position (C) of various sections from the CoCr28Mo6 stem (location see figure 1 and figure 2).

References

[1] Pilz V, Hanstein T and Skripitz R 2018 *Acta Orthop.*

[2] Statistisches Bundesamt 2019 *Die 20 häufigsten Operationen insgesamt (OPS 5): Vollstationär behandelte Patientinnen und Patienten in Krankenhäuser 2018* https://www.destatis.de/DE/Themen/Gesellschaft-Umwelt/Gesundheit/Krankenhaeuser/Tabellen/drg-operationen-insgesamt.html

[3] Kaivosoja E, Tiainen V M, Takakubo Y, Rajchel B, Sobiecki J, Konttinen Y T and Takagi M 2012 *Woodhead Publ. Ser. Biomater.* 178–218

[4] Davis J R (ed) 2006 *Handbook of materials for medical devices* 4th ed (Materials Park, Ohio: ASM International)

[5] Karuppal R 2016 *J. Orthop.* **13** 190–2

[6] Bleß H-H and Kip M 2018 *White Paper on Joint Replacement* (Berlin, Heidelberg: Springer)

[7] Konan S, Abdel M P and Haddad F S 2019 *Bone Joint Res.* **8** 604–7

[8] Abdel M P, Houdek M T, Watts C D, Lewallen D G and Berry D J 2016 *Bone Joint J.* **98-B** 468–74

[9] Moskal J T, Capps S G and Scanelli J A 2016 *Arthroplast. Today* **2** 211–8

[10] Abdulkarim A, Ellanti P, Motterlini N, Fahey T and O'Byrne J M 2013 *Orthop. Rev.* **5** e8

[11] Alexander Grimberg, Volkmar Jansson, Oliver Melsheimer, Arnd Steinbrück 2019 0–66

[12] Albrektsson T, Brånemark P I, Hansson H A and Lindström J 1981 *Acta Orthop. Scand.* **52** 155–70

[13] Wong M, Eulenberger J, Schenk R and Hunziker E 1995 *J. Biomed. Mater. Res.* **29** 1567–75

[14] Jones C P and Kelley S S 2001 *Curr. Opin. Orthop.* **12** 52–6

[15] Herrera A, Mateo J, Gil-Albarova J, Lobo-Escolar A, Ibarz E, Gabarre S, Más Y and Gracia L 2015 *Biomed. Res. Int.* **2015**

[16] Kienapfel H, Sprey C, Wilke A and Griss P 1999 *J. Arthroplasty* **14** 355–68

[17] Claes L, Kirschner P, Perka C and Rudert M 2012 *AE-Manual der Endoprothetik* (Berlin, Heidelberg: Springer)

[18] Head W C, Bauk D J and Emerson R H 1995 *Clin. Orthop. Relat. Res.* 85–90
[19] Kuehn K-D, Ege W and Gopp U 2005 *Orthop. Clin. N. Am.* **36** 29-39, v-vi
[20] Scheerlinck T and Casteleyn P-P 2006 *J. Bone Joint Surg. BR* **88** 1409–18
[21] Pilliar R M, Lee J M and Maniatopoulos C 1986 *Clin. Orthop. Relat. Res.* 108–13
[22] Singh S and Harsha A P 2016 *J. Inst. Eng. India Ser. D* **97** 113–20
[23] Emerson R H, Head W C, Emerson C B, Rosenfeldt W and Higgins L L 2002 *J. Arthroplasty* **17** 584–91
[24] Ramaniraka N A, Rakotomanana L R and Leyvraz P-F 2000 *J. Bone Joint Surg. BR* **82-B** 297–303
[25] Schuh A, Holzwarth U, Kachler W, Göske J and Zeiler G 2004 *Orthopäde* **33** 905–10
[26] Darvell B W, Samman N, Luk W K, Clark RKF and Tideman H 1995 *J. Dent.* **23** 319–22
[27] Ricci J L, Kummer F J, Alexander H and Casar R S 1992 *J. Appl. Biomater.* **3** 225–30
[28] Böhler M, Kanz F, Schwarz B, Steffan I, Walter A, Plenk H and Knahr K 2002 *J. Bone Joint Surg. BR* **84** 128–36
[29] Schuh A, Uter W, Holzwarth U, Kachler W, Goske J, Raab B and Knetsch T 2005 *Biomed. Te.-Biomed. Eng.* **50** 404–7
[30] Schuh A, Uter W, Holzwarth U, Kachler W, Göske J and Müller T 2005 *Zentralbl. Chir.* **130** 346–52
[31] Gittens R A, Olivares-Navarrete R, Schwartz Z and Boyan B D 2014 *Acta Biomater.* **10** 3363–71
[32] Novaes A B, Souza S L S de, Barros R R M de, Pereira K K Y, Iezzi G and Piattelli A 2010 *Braz. Dent. J.* **21** 471–81
[33] Zweymüller K A, Lintner F K and Semlitsch M F 1988 *Clin. Orthop. Relat. Res.* 195–206
[34] Heimann R B 2016 *J. Therm. Spray. Tech.* **25** 827–50
[35] Gbureck U, Masten A, Probst J and Thull R 2003 *Mat. Sci. Eng. C* **23** 461–5
[36] Wintermantel E and Ha S-W (eds) 2009 Medizintechnik (Berlin, Heidelberg: Springer)
[37] Wank A, Wielage B, Podlesak H, Matthes K-J and Kolbe G 2005 *J. Therm. Spray. Tech.* **14** 134–40
[38] Gardiner R C and Hozack W J 1994 *J. Bone Joint. Surg. BR* **76** 49–52
[39] Huo M H, Salvati E A, Lieberman J R, Betts F and Bansal M 1992 *Clin. Orthop. Relat. Res.* 157–68
[40] Howell JR, Blunt LA, Doyle C, Hooper RM, Lee AJC and Ling RSM 2004 *J. Arthroplasty* **19** 88–101
[41] Verdonschot N and Huiskes R 1998 *Biomaterials* **19** 1773–9
[42] Huang P and López HF 1999 *Mater. Lett.* **39** 244–8
[43] Giacchi J V, Morando C N, Fornaro O and Palacio H A 2011 *Mater. Charact.* **62** 53–61
[44] Vander Sande J B, Coke J R and Wulff J 1976 *Metall. Mater. Trans. A* **7** 389–97
[45] Rajan K 1983 *J. Mater. Sci.* **18** 257–64
[46] Kashani H, Laridjani M S, Amadeh A, Khodagholi M and Ahmadzadeh S 2008 *Mater. Sci. Eng. A* **478** 38–42
[47] Mani A, Salinas-Rodriguez and Lopez H F 2011 *Mater. Sci. Eng. A* **528** 3037–43
[48] Sylwestrowicz W and Hall E O 1951 *Proc. Phys. Soc. B* **64** 495–502
[49] Avsec K *et al* 2019 *Coatings* **9** 868
[50] Kärrholm J, Lindahl H, Malchau H, Mohaddes M, Nemes S, Rogmark C and Rolfson O 2017 *The Swedish Hip Arthroplasty Register Annual Report 2016*
[51] Aust. Orthopaedic Assoc. National Joint Replacement Registry 2017 *Hip, Knee & Shoulder Arthroplasty Annual Report 2017*

IOP Conf. Series: Materials Science and Engineering **882** (2020) 012005 doi:10.1088/1757-899X/882/1/012005

Preceramic polymer-derived ceramic foams with lamellar strut porosity

K Dammler, K Schelm, D Kniep, G Hasemann and M Scheffler

Otto-von-Guericke-University Magdeburg, Faculty of Mechanical Engineering, Institute for Materials and Joining Technology, Universitätsplatz 2, 39106 Magdeburg, Germany

Abstract. Open porous replica foams with additional lamellar strut pores were prepared by preceramic polymer processing combined with a freeze-drying process. Polymer foam templates were coated with an aqueous polymethylsilesquioxane system followed by freezing at - 20 °C to generate hierarchical porosity within the foam struts. The cellular structure was preserved during freeze drying, cross-linking, template removal and pyrolysis, and additional strut porosity was formed. The size of the lamellar strut pores was found to depend on the solid load of the preceramic slurry and its pretreatment procedure. The cellular materials obtained in this work possess additional lamellar strut pore channels with a thickness of approximately 40 μm and a total foam porosity of > 90 %.

1 Introduction

Open porous ceramic foams are frequently used as supports for heterogeneous catalysts, filters for liquid metal filtration, heat exchangers and other applications.[1–5] Within these applications the combination of mechanical strength and an increased surface area is advantageous and favored.[1]

With the freeze-casting technique a systematic positioning of pores into suspension-generated materials is possible. Typical for the production of porous ceramics are water based slurries primarily consisting of ceramic particles (in the nanometer to micrometer range), processing additives such as viscosity modifiers and deagglomerating agents and water.[6] Ceramic slurries based on other liquids than water are also suitable for freeze-drying processing. Specifically with regard to preceramic polymer systems, organic solvents such as cyclohexane, *tert*-butyl alcohol or camphene are used, as several preceramic polymers are not soluble in water.[7] During freeze casting/freeze drying processing the solvent solidifies, is removed by a freeze drying process (sublimation) and leaves behind pores instead of frozen material, within the sample.[8] Ice crystals from frozen water leave behind lamellar pore channels after sublimation within the structure, separated by ceramic lamellae. As the superior objective of this study was the accessibility of hollow struts in replica foam samples and to enlarge the foam surface, water was chosen as dispersant in this work; replica foams are typically manufactured by coating a polyurethane (PU) foam template with a ceramic slurry and subsequent thermal processing with a) template burnout and b) sintering or pyrolysis. The modification of lamellar pore channel structures was realized by varying the solid content or the freezing temperature. The pore channel formation is a result of nucleation and growth of ice crystals located in the suspension and has already been investigated extensively.[9–14]

Most research related to freeze drying was carried out with bulk ceramic materials [1, 3] or combined with direct foaming processes.[7, 15]

The combination of the freeze drying process with the replica (impregnation of a polyurethane sponge as a template) method [16] was described in the literature,[17] but its focus was on alumina based porous ceramics. Besides alumina, preceramic polymers may be used for the manufacturing of porous ceramics. In this context these are mostly Si-based oligomers [18] and they are suitable for the replica technique.[19] Due to a polymer–to–ceramic conversion, preceramic polymers can be transferred into Si-based ceramics with a high oxidation and creep resistance up to high temperatures.[18]
The combination of freeze-drying and sponge replication technique with preceramic polymer based slurries is presented in this paper with a focus on the pretreatment of the starting preceramic polymer powders.

Characterization was carried out with respect to the coating process, the micro- and macrostructural features and the mechanical strength of the samples.

2 Experimental Section

The workflow of sample preparation is illustrated in *Figure 1* showing the combination of the replica process and the freeze drying process.

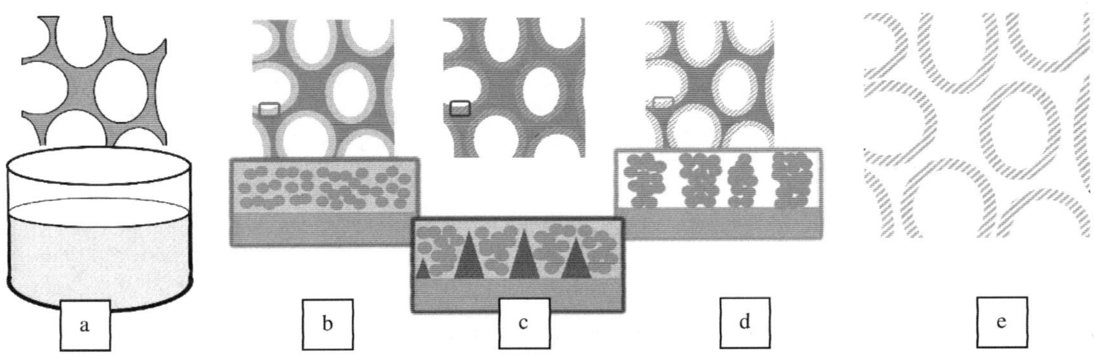

Figure 1. Scheme of the workflow and basic processes: Replica technique combined with freeze drying: (a) Impregnation of a PU template with a ceramic slurry; (b) Freezing of coated PU foam; (c) growth of ice crystals; (d) Sublimation of ice crystals with reduced temperature and pressure; (e) PU template burnout and sintering of the ceramic foam with pores of freeze drying.

2.1 Powder processing

As a preceramic polymer powder commercially available polymethylsilesquioxane (SILRES®MK; d_{50}: 25 µm; information from the manufactures datasheet; Wacker Silicone AG, Munich, Germany) was used. The powder was grinded with a ball mill (PM 400, Retsch GmbH, Haan, Germany). Two different milling process modifications were applied: dry milling, whereby the powder was placed in the Al_2O_3 ceramic containers with Si_3N_4 milling balls (d = 10 mm); and wet-milling, whereby water was added to the mixing containers. For the wetting of the hydrophobic polymer powder with water a surfactant was added (Alboclean n24-5, Albon-Chemie Dr. Ludwig-E. Gminder KG, Metzingen, Germany). Milling was performed for 30 min with 250 rpm. The preparation of the coating suspensions was performed with the grinded, dried and sieved ($d_{particles}$ < 20 µm) SILRES®MK powders.

2.2 Slurry preparation

The preceramic polymer-processing route requires a crosslinking step to obtain a stable polymer network before pyrolysis. Therefore, aluminum acetylacetonate (Merck Schuchardt OHG, Germany) was added as crosslinking catalyst.

To increase the viscosity of the preceramic slurry, a suitable amount of methylcellulose (Carl Roth GmbH & Co. KG, 160 000 g mol^{-1}, Karlsruhe, Germany) was added to distilled water before adding it to the mixing container (5.3 wt. % of water). Different solid contents of the preceramic powder (20, 30 and 40 vol. %) were used within this testing series. A planetary centrifugal mixer (THINKY Mixer ARE-250, THINKY Corp. Tokyo, Japan) was used for homogenization of the slurry (mixing mode: 15 min, 2000 rpm) and for degassing of the slurry (degassing mode: 10 min, 2200 rpm). The composition of the slurries with a different solid content is shown in *Table 1.*

Table 1. Composition of investigated slurries.

Slurry	*20 vol. % solid*		*30 vol. % solid*		*40 vol. % solid*	
Component/Share	(vol. %)	(g)	(vol. %)	(g)	(vol. %)	(g)
SILRES®MK	18.1	8.7	27.3	8.7	36.8	10.9
Water+Cellulose	81.7	36.1	72.3	21.2	62.7	17.1
Tenside	0.2	0.1	0.3	0.1	0.3	0.1
Catalyst	0.1	0.1	0.1	0.1	0.2	0.1

2.3 Coating of the PU templates

Open cell polyurethane foam templates (S30P20R, Koepp-Schaum GmbH, Oestrich Winkel, Germany) with a volume of (15*15*20) mm³ and a pore count of 20 ppi (pores per linear inch) were squeezed and immersed into the suspension, withdrawn and excess suspension was removed manually by facile squeezing. Coated foams were weighed to ensure almost the same load of all templates (~1.5 g per foam piece). After coating of approximately 10 foam pieces they were put into the freezing compartment of a laboratory fridge at approximately - 20 °C and kept there for at least 24 hours to ensure complete freezing of all samples.

2.4 Freeze drying and heat treatment

After the freezing process was completed, the samples were put into a freeze dryer (Gefriertrockenschrank BETA 1-16 LMC 2, Martin Christ Gefriertrocknungsanlagen GmbH, Osterode, Germany) at - 50 °C and 0.1 mbar for 24 hours. The last step of sample processing was the heat treatment consisting of crosslinking and pyrolysis. Crosslinking of the dried preceramic polymer slurry was performed with a heating rate of 1 K min^{-1} at a maximum temperature of 180 °C with a dwelling time of 3 hours in air (KU 40/04/A, THERMCONCEPT Dr. Fischer GmbH, Bremen, Germany). In this temperature range the PU template is still intact, so it is pyrolyzed together with the preceramic polymer in the next heating step. Pyrolysis took place in a tube furnace (THERMCONCEPT Dr. Fischer GmbH, Bremen, Germany) with a heating rate of 2 K min^{-1} in nitrogen (nitrogen flow rate set to 15 l min^{-1}) for 3 hours at a maximum temperature of 800 °C.

3 Characterization

The measurement of particle size distribution was performed with dried, sieved powders after ball-milling (wet- and dry-milled) in air with the light scattering method using a Mastersizer 2000 (Malvern Instruments Ltd., Malvern, United Kingdom). For the determination of the particle size the Fraunhofer model [20] was used for particle sizes distribution analysis for a particle size range from 10 µm to 20 µm for the wet-milled powders and of 100 µm for the dry-milled powders. The particle size of the unmilled SILRES®MK powder was extracted from the supplier´s data sheet.

For the characterization of the rheological behavior of the slurries a viscometer (Anton Paar GmbH, Graz, Austria) with plate-plate configuration (d_{Plate} = 50 mm) was used and the distance between the parallel plates was set to 1 mm. The shear rate was varied between 0.1 s^{-1} and 100 s^{-1}, the corresponding shear stress values were recorded with the software package Rheoplus V. 3.40 (Anton Paar GmbH, Graz, Austria). During each measurement 31 data points were recorded, the measuring duration for each point was logarithmically reduced from 60 s to 1 s. The rheological data – shear stress τ as a function of the shear rate γ – were fitted with the Power - Law model (equation 1) as a shear thinning flow behavior was observed for these slurries:[21, 22]

$$\tau = K\dot{\gamma}^{n}. \quad (1)$$

The parameter *K is* defined as consistency coefficient of the slurry and the parameter *n* defines the variation to Newtonian flow behavior of the fluid. For *n* equal to 1 the fluid possesses Newtonian flow behavior (shear stress is proportional to the shear rate), $n < 1$ describes a shear-thinning flow (the shear stress reduces significantly for increased shear rates) and $n > 1$ means a shear-thickening behavior (the shear stress increases strongly with an increase of the shear rate).[21]

To determine the bulk density of the strut material after the heat treatment, the volume of the solid was measured with a helium pycnometer (AccuPyc 1330, Micromeritics GmbH, Unterschleißheim, Germany). For this, pieces of pyrolyzed foams were ball-milled with a planetary ball mill (Retsch PM400, Haan, Germany). The powder was filled into the cylindrical measuring chamber of the pycnometer. The measuring program was carried out for five times. After determination of the sample weight (Cubis, Sartorius, Göttingen, Germany) the density was calculated using equation (2):

$$\rho_{\mathrm{bulk}} = \frac{m}{V}. \quad (2)$$

Scanning electron micrographs were taken from sample pieces mounted on the SEM sample carriers with electrical and electrically conductive glue (Plano Leit C, Plano GmbH, Wetzlar, Germany). The samples were sputtered in a vacuum-camber with gold to guarantee electric conductivity of the samples. The accelerating voltage of in the SEM (SEM, XL30 ESEM-FEG, FEI Company, Hilsboro, OR, USA) was set to 10 kV, specifically to avoid melting of the preceramic polymer before pyrolysis, when green foams were analyzed.

Cell size and porosity measurements were carried out with a microcomputer tomograph Nanotom S (GE Sensing & Inspection Technologies, Wunstorf, Germany). For the determination of the pore size distributions generated with freeze drying the incremental pore volumes were plotted against the mean pore diameter and their maxima were determined with a Gaussian curve fit. For scanning in a µ-CT, pieces of reticulated foams after pyrolysis (to determine the lamellar thickness of the strut pores small pieces of foams were used) were fixed separately on the sample holder of the µ-CT. The voltage was 60 kV, the tube current was set to 160 µA, and the distance of the sample holder to the X-ray tube and detector was adjusted to obtain a voxel size of 10 µm³ for scanning of the complete samples (resp. 7 µm³ for foam pieces). The reconstruction of the datasets was performed with the software package Phoenix Datos |X 2.0 (GE Sensing & Inspection Technologies, Wunstorf, Germany). With the software CT Analyzer 1.18 (Skyscan/Bruker microCT, Kontich, Belgium) different structural parameters of the foams were calculated in a 3D mode. The thickness of ceramic lamellae and the width of lamellar pore channels were determined with the software CT Analyzer 1.18, too. For more details of the calculation procedure (e.g. thresholding and binarization) please refer to Betke et al.[23]

The compressive strength of the pyrolyzed foam samples was determined with the mechanical test machine TIRATEST 2820 (TIRA GmbH, Schalkau, Germany). Six samples of each series were measured. To ensure a uniform pressure distribution to the planar faces, a cardboard was placed underneath and over the foams and the compression stamp. The traverse-speed was set to 2 mm min^{-1}. The compressive strength was calculated by scaling the maximum force to the loaded area of the sample.

IOP Conf. Series: Materials Science and Engineering **882** (2020) 012005 doi:10.1088/1757-899X/882/1/012005

4 Results and discussion

4.1 Rheological behavior of the coating slurries

The shear stress of the three slurries prepared with wet-milled SILRES®MK powder are presented as a function of the shear rate, see **Figure 2**. Since the flow curves of the slurries prepared with the dry-milled powder show comparable results, they are not shown here. The viscosity increases with increasing solid content, whereby for 20 vol. % and 30 vol. % of solid load the viscosity values are almost identical, **Figure 2**. This behavior is related to the increasing amount of particles, which results in an increase of the shear resistance and gets significant for a solid load above 30 vol. % of the coating slurry.[23–25]

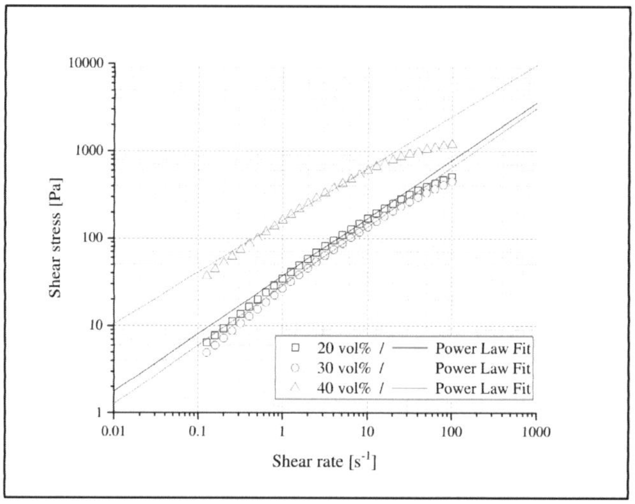

Figure 2. Flow curves of preceramic slurries (wet-milled powder) with three different solid contents.

A shear thinning behavior of the slurries is advantageous for the replica method [25–27], as an almost ideal impregnation (penetration of the slurry into the template foam cells and windows) is possible at higher shear rates due to a lower viscosity by movement the template within the coating slurry. Right after finishing the coating process a draining off of the slurry is prevented due to a viscosity increase by reducing the shear stress to zero. In this coating system the viscosity is very low for a given shear rate in comparison to standard suspensions suitable for the replica technique (shear rate of 1 s^{-1} shear stress is in the range of 100 Pa, referring to Voigt et al. [27]).[25, 28]

No yield strength was found when the slurries were characterized. Both, a low viscosity and the lack in yield strength, is critical for template coating. However, draining off of the coating was prevented by a fast transfer of the samples into the freezer.

The parameters K and n describing the flow behavior as determined by the Power-Law model are listed in *Table 2*. The Power-Law model was chosen because of the slurry shows a shear-thinning behavior without yield strength. Further information can be found in the literature.[29–33]

Table 2. Calculated flow parameters from the power law model: consistency factor K and flow index n.

Sample/Parameter	*K/Pas^n^*	*n*
20 vol. %	37.15	0.72
30 vol. %	28.64	0.67
40 vol. %	161.34	0.59

The Power-Law model serves as a mathematical description of the shear thinning or shear thickening process of suspensions having non - Newtonian flow behavior.[21, 22] The flow index *n* decreases for an increasing solid content of the slurry and indicates that the fluid has a shear thinning flow behavior. The consistency factor *K* describes in which extend the suspension "works" against the shear stress and represents the shear stress at a shear rate of 1.0 s^{-1}. The more the suspension consists of particles, the harder it is to move it with the plate, what is one reason for the highest *K*-value for the 40 vol. %-slurry. The experimental data fit well to the Power-Law model for a particular range of the shear rate, as shown in **Figure 2**.

4.2 Microstructure of the foams
Heat treatment
Since ceramic processing based on preceramic polymers is connected to cross-linking and pyrolysis, the SEM micrographs in **Figure 3** show the microstructures of dried, cross-linked and pyrolyzed samples with respect to the lamellar pore channels. All four samples were prepared from the wet-milled powder and a slurry with a solid content of 30 vol. %.

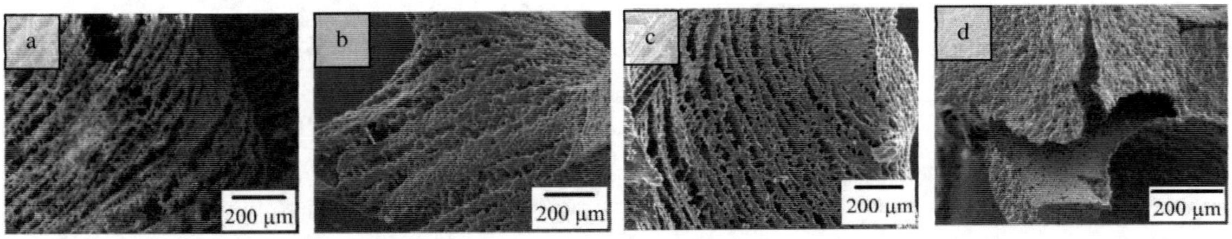

Figure 3. SEM-micrographs of (a) freeze-dried, (b) freeze-dried and cross-linked and (c) freeze-dried, cross-linked and pyrolyzed samples manufactured with a 30 vol. % solid-loaded slurry with the wet-milled preceramic polymer powder, (d) sample dried under ambient conditions.

SEM images of the pyrolyzed foams show lamellar pore channels inside of the foam struts that are a result of freezing and freeze-drying. The freeze-formed pores are parallel aligned, and it was found that this type of alignment is independent on the solid load.

Replica foams without freeze-drying process and the same slurry do not have any lamellar pore channels, **Figure 3** *(d)*. Therefore we state, that the development of lamellar structures is a consequence of freeze processing consisting of freezing and freeze drying. While freezing leads to the formation of water-ice crystals at ambient pressure, drying leads to the removal of these ice crystals by sublimation. These two process steps are not part of processing in the foam shown in **Figure 3** *(d)*. The struts show smaller non lamellar pores which are a result of the pyrolysis process, whereby the cross-linked preceramic material undergoes a mass loss of more than 20 vol. % after pyrolysis at 800 °C. These pores could be closed if pyrolysis temperature is increased or fillers are used. More information dealing with the thermal rearrangement and gas release in polymer pyrolysis can be found elsewhere.[34–43]

Milling process
Milling of the preceramic polymer powder, with addition of water led to smaller particles [44] with a more uniform shape and rounded edges, compared to the sharp edges of dry-milled powders. **Figure 4** shows (a) the particle size distribution of the wet-milled and the dry-milled powder, (b-d) the microstructure of the untreated, dry-milled and wet-milled powder.

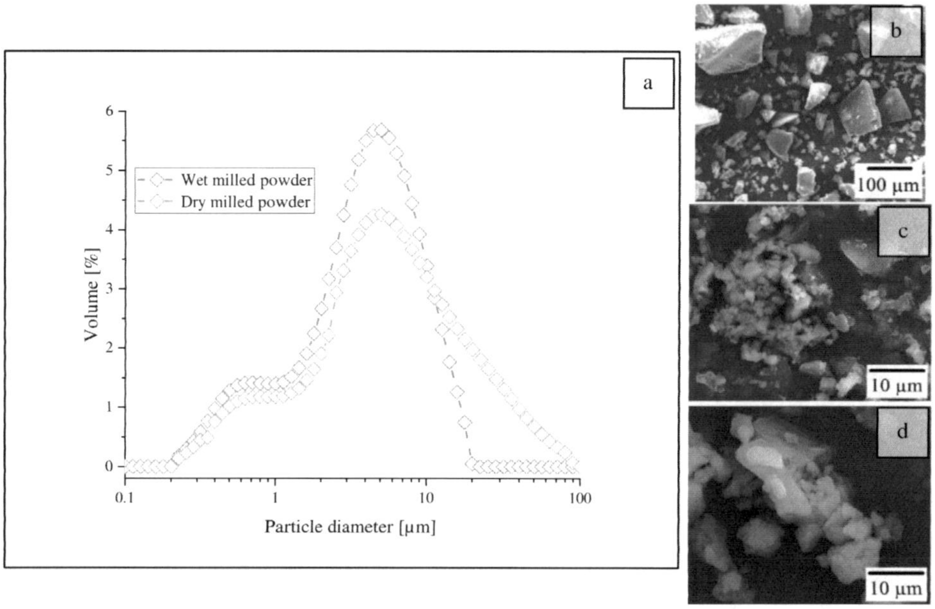

Figure 4. (a): Particle size distribution of dry and wet milled SILRES®MK powder; (b - d): SEM-micrographs of powder samples: (b) untreated; (c) dry milled; (d) wet milled.

Porosity measurements and determination of pore size distributions were performed by microcomputer tomography. The lower the solid content inside of the coating slurry was, the higher was the porosity and the width of the lamellar pore channels *Table 3*, **Figure 6**. Prior to the calculation of porosity, the bulk density of the pyrolyzed samples was measured with helium pycnometry. The value the bulk density is 2.259 g cm^{-3}. Samples prepared of dry- and wet-milled powders show the same bulk density, both sample series prepared with SILRES®MK powder. The foam porosity was calculated with the following equation (3):

$$P_{total} = 1 - \frac{\rho_{geometric}}{\rho_{bulk}} * 100\%. \tag{3}$$

whereby the geometric density is calculated by dividing the weight of the sample to its volume:

$$\rho_{geometric} = \frac{m_{sample}}{V_{sample}}. \tag{4}$$

Based on the values listed in *Table 3* the open strut porosity of the samples was determined with equation (5):

$$P_{Strut} = 1 - \frac{\frac{m_{piece\ in\ CT}}{V_{piece\ in\ CT}}}{\rho_{bulk}} \tag{5}$$

Table 3. Geometric values of samples, volume of pyrolyzed sample piece determined with µCT, calculated strut porosity and width of ceramic lamellae and lamellar pore channels.

Parameter/ Sample	$\rho_{geometric}$	P_{total}	$m_{piece\ in\ CT}$	$V_{piece\ in\ CT}$	P_{Strut}	*Width of ceramic lamellae*	*Width of lamellar pore channels*
Unit	(g/cm³)	(%)	(g)	(cm³)	(%)	(µm)	(µm)
30 vol. % dry	0.151	93.30	0.0386	0.138	87.65	57.62	55.31
30 vol. % wet	0.148	93.44	0.0574	0.197	87.12	53.92	51.17
40 vol. % wet	0.204	90.96	0.0193	0.045	81.11	33.72	42.34

The strut porosity decreases with an increase of solid content, for wet- and dry-milled powder the strut porosity is almost identical and in general the porosity of samples being in the range from 91 % to more than 93 % is very high.

To investigate the pore size distribution, the thickness of the lamellar pore channels and the ceramic lamellae within the struts, defined pieces of foam sample were investigated with µ-CT and the data was evaluated with the CTAnalyzer software. The results are shown in **Figure 5** and *Table 3*.

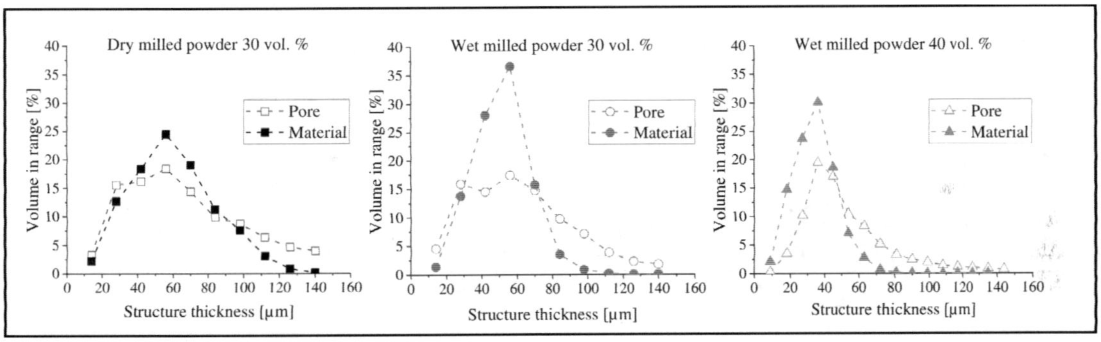

Figure 5. Width of ceramic lamellae (■/●/▲) and lamellar pore channels (□/○/△) determined by µ-CT data with structure separation and structure thickness function (3D analysis).

The width of ceramic lamellae increases with an increased solid content in the preceramic slurry, and the width of lamellar pore channels was found to behave inversely. The width change of the lamellar pore channels and the ceramic lamellae is illustrated in **Figure 6**, exemplified with samples from the wet-milled powder with a (a) 20, (b) 30 and (c) 40 vol. % solid content of the coating slurry.

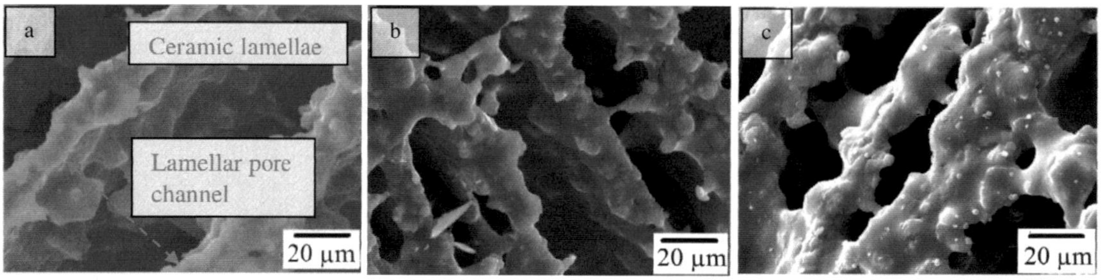

Figure 6. SEM-micrographs of samples from wet-milled powder after pyrolysis with a slurry-solid content of (a) 20 vol. %, (b) 30 vol. % and (c) 40 vol. %

In summary it can be stated that the freezing and freeze-drying process leads to the formation of lamellar pore channels separated by dense ceramic lamellae. The width of lamellar pore channels decrease and the width of ceramic lamellae increases with an increasing solid content in the coating slurry. Both effects are in accordance to the literature, found for bulk materials processes via freezing and freeze –drying [13, 14, 17, 45] Furthermore, an increased number and thickness of dendritic-like features (**Figure 7**) connecting the lamellar pore channels was found with an increasing solid content of the coating slurries, in this work from 20 vol. % to 40 vol. %.

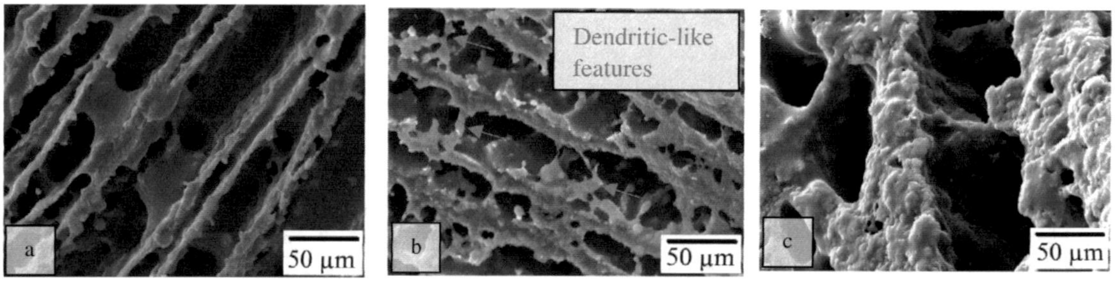

Figure 7. SEM-micrographs of freeze-dried; cross-linked and pyrolyzed samples from (a) dry-milled powder with 30 vol. % solid content; wet-milled powder with a slurry solid content of (b) 30 vol. % and (c) 40 vol. %

Compressive strength

The compressive strength correlates with the solid content of the preceramic coating suspension. The higher the solid content, the more stable are the foams. This is due to the increase of the ceramic lamellae thickness, the decreasing width of lamellar pore channels and the increased number and thickness of dendritic-like features between lamellar pore channels, **Figure 6**, **Figure 7**. Hence, the solid content determines the porosity of the pyrolyzed samples.[46] The extent of porosity acts contrary to the aim of highly mechanical stable foams, but our goal was to maximize the porosity, but anyway, the samples have to have a stability high enough to handle them. Therefore, mechanical strength suffered from high share of porosity among produced foams. Additionally the separate measurements induce variant compressive strength values, which is basically due to the inherently fracturable foam microstructure. During compression, only a small number of struts has to retain the total weight force operated to the foam sample. As the load of individual struts is not implicitly homogeneous within a sample series, collapse may appear at diversify values.[47]

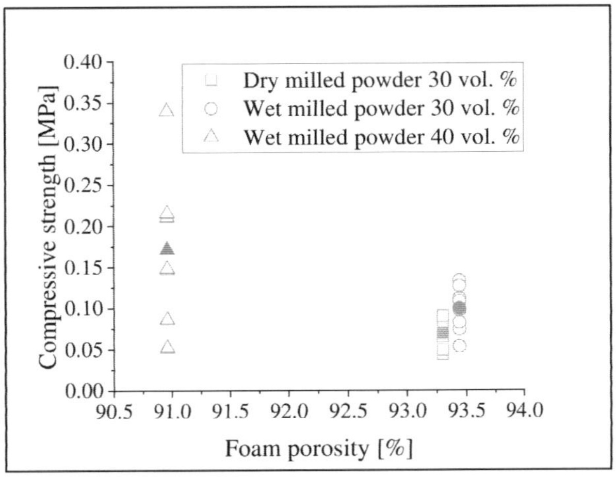

Figure 8. Compressive strength plotted against the total foam porosity of the investigated samples, filled symbols represent average values

Precise numerical figures are not given here as numerous affecting conditions concerning the measuring system are existent and therefrom the numerical figures are not significant.[48] Despite this, there is a visible trend: as the solid content increases, the compressive strength of the samples increases. **Figure 8** shows the compressive strength of the produced foams to the foam porosity.[5, 14]

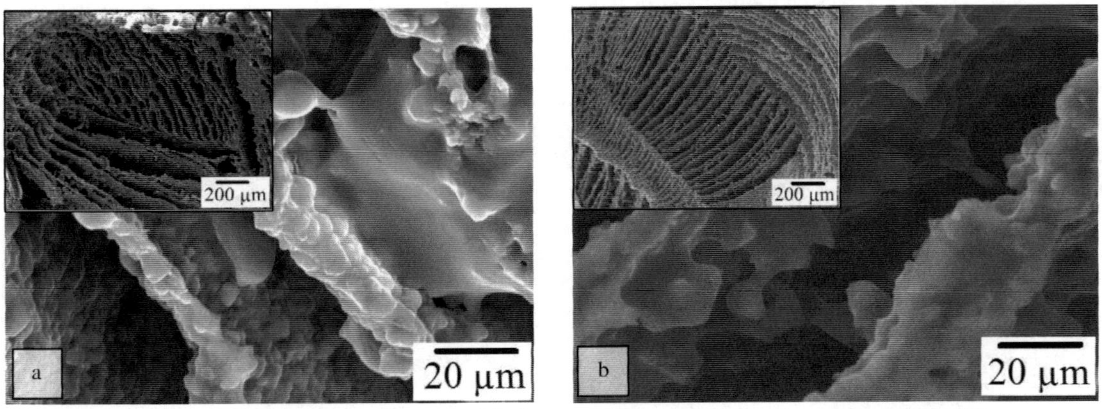

Figure 9. SEM-micrographs of freeze-dried; cross-linked and pyrolyzed samples with a slurry-solid content of 20 vol. % from (a) dry-milled powder and (b) wet-milled powder; both inserts show microstructure with a smaller magnification to illustrate the parallel alignment of ceramic lamellae and lamellar pore channels.

Furthermore, the powder milling process influenced the mechanical stability of the samples. Foams produced with wet-milled powder consisted of smaller particles without sharp edges and a more uniform shape, which leads to more consistent ceramic lamellae, **Figure 4**, **Figure 7**, *Figure 9*.

5 Conclusion

Freeze-dried replica foams from water based preceramic polymer slurries were successfully prepared. The solid load in the coating slurries was varying from 20 vol. % and 40 vol. %, a solid load of 30 vol. % and more lead to mechanically stable polymer derived ceramic foams. The freeze-dried PDC foams possess a low compressive strength, which was assigned due to the parallel alignment of lamellar shaped pores within the pore struts, *Figure 9*.

Acknowledgements

This work was carried out in the frame of a scholarship from the *International Graduate School MEMoRIAL*, which is funded by the *European Structural Investment Funds* under the code and number *MEMoRIAL* 64 1003 2013, which grateful acknowledged. The authors wish thank to Diethard Kürschner for his help with the freeze-dryer, Michael Schwidder for viscosity measurements and Andreas Schlinkert for particle size distribution measurements; all persons are from the Faculty for Process Engineering of the Otto-von-Guericke-University, Magdeburg. We also thank Ulf Betke for his kind replay to our questions.

Conflict of interest

The authors state no conflict of interest.

References

[1] Fukasawa T, Deng Z-Y, Ando M, Ohji T and Goto Y 2001 *Journal of Materials Science* **36** 2523–7
[2] Zaman E and Keleş Ö 2014 *Acta Phys. Pol. A* **125** 445–8
[3] Soltani N, Simon U, Bahrami A, Wang X, Selve S, Epping J D, Pech-Canul M I, Bekheet M F and Gurlo A 2017 *Journal of the European Ceramic Society* **37** 4809–20
[4] Naviroj M, Voorhees P W and Faber K T 2017 *J. Mater. Res.* **32** 3372–82
[5] Deville S 2008 *Advanced engineering materials* 155–69
[6] Deville S 2017 *Freezing Colloids: Observations, Principles, Control, and Use* (Cham, Switzerland: Springer International Publishing)
[7] Naviroj M, Miller S M, Colombo P and Faber K T 2015 *Journal of the European Ceramic Society* **35** 2225–32
[8] Deville S, Saiz E, Nalla R K and Tomsia P 2006 *Science (New York, N.Y.)* **311** 515–8
[9] Koch D, Andresen L, Schmedders T and Grathwohl G 2003 *Journal of Sol-Gel Science and Technology* **26** 149–52
[10] Deville S, Maire E, Lasalle A, Bogner A, Gauthier C, Leloup J and Guizard C 2010 *Journal of the American Ceramic Society* **93** 2507–10
[11] Munch E, Franco J, Deville S, Hunger P, Saiz E and Tomsia A P 2008 *Journal of The Minerals, Metals & Materials Society (TMS)* **60** 54–8
[12] Deville S 2013 *J. Mater. Res.* **28** 2202–19
[13] Deville S, Saiz E and Tomsia A P 2006 *Biomaterials* **27** 5480–9
[14] Sylvain Deville, Eduardo Saiz, Antoni Tomsia 2007 *Acta Materialia* **55** 1965–74
[15] Chino Y and Dunand D C 2008 *Acta Materialia* **56** 105–13
[16] Schwartzwalder K and Somers Holly and Arthur V. 1963 Method of makinkg porous ceramic articles *3,090,094*
[17] Schelm K, Fey T, Dammler K, Betke U and Scheffler M 2019 *Adv. Eng. Mater.* **10**
[18] Colombo P, Mera G, Riedel R and Sorarù G D 2010 *Journal of the American Ceramic Society* **73** 1805-1837
[19] Jana P, Zera E and Sorarù G D 2017 *Materials & Design* **116** 278–86
[20] Cyr M and Tagnit-Hamou A 2001 *Materials and Structures* **34** 342–50

IOP Conf. Series: Materials Science and Engineering **882** (2020) 012005 doi:10.1088/1757-899X/882/1/012005

[21] George K E 2009 Non-Newtonian fluid mechanics and polymer rheology *Advances in polymer processing: From macro- to nano-scales* ed S Thomas (Cambridge: Woodhead Publishing Ltd) pp 13–46
[22] Hanswalter Giesekus 1994 *Phänomenologische Rheologie: Ein Einführung* 1st edn (New York: Springer-Verlag Berlin Heidelberg GmbH)
[23] Betke U, Dalicho S, Rannabauer S, Lieb A, Scheffler F and Scheffler M 2017 *Adv. Eng. Mater.* **19**
[24] Adebowale A-R A and Sanni L O 2013 *Journal of food science and technology* **50** 573–8
[25] Mouazer R, Mullens S, Thijs I, Luyten J and Buekenhoudt A 2005 *Adv. Eng. Mater.* **7** 1124–8
[26] Brown D D and Green D J 1994 *Journal of the American Ceramic Society* **77** 1467–72
[27] Voigt C, Aneziris C G and Hubálková J 2015 *J. Am. Ceram. Soc.* **98** 1460–3
[28] ZHAO H-s, LIU Z-g, YANG Y, LIU X-x, ZHANG K-h and LI Z-q 2011 *Transactions of Nonferrous Metals Society of China* **21** 1329–34
[29] Ostwald W 1925 *Kolloid-Zeitschrift* **36** 99–117
[30] Rha C (ed) 1975 *Theory, Determination and Control of physical Properties of Food Materials* (Dodrecht, Holland: D. Reidel Publishing Company)
[31] George K E 2009 Non-Newtonian fluid mechanics and polymer rheology *Advances in polymer processing: From macro- to nano-scales* ed S Thomas (Cambridge: Woodhead Publishing Ltd) pp 13–46
[32] Krishnan J M, Deshpande A P and Kumar P B S 2010 *Rheology of Complex Fluids* (New York, NY: Springer New York)
[33] Rao M A 2014 *Rheology of Fluid, Semisolid, and Solid Foods* (Boston, MA: Springer US)
[34] Ma J, Shi L, Shi Y, Luo S and Xu J 2001 *Journal of Applied Polymer Science* **85** 1077–86
[35] Černý M, Chlup Z, Strachota A, Halasová M, Rýglová Š, Schweigstillová J, Svítilová J and Havelcová M 2015 *Ceramics International* **41** 6237–47
[36] Bois L, Maquet J, Babonneau F, Mutin H and Bahloul D 1994 *Chemical Materials* **6** 796–802
[37] Bahloul-Hourlier D, Latournerie J and Dempsey P 2005 *Journal of the European Ceramic Society* **25** 979–85
[38] Belot V, Corriu R J P, Leclercq D, Mutin P H and Vioux A 1992 *Journal of Polymer Science: Part A: Polymer Chemistry* **30** 613–23
[39] Bahloul D, Pereira M and Goursat P 1995 *Journal of the American Ceramic Society* **76** 1156–62
[40] Parcianello G, Bernardo E and Colombo P 2013 *Ceramics International* **39** 8893–9
[41] Mutin P H 1999 *Journal of Sol-Gel Science and Technology* **14** 27–38
[42] Peuckert M, Vaahs T and Brück M 1990 *Advanced Materials* **9** 398–404
[43] Belot V, Corriu R J P, Leclercq D, Mutin P H and Vioux A 1992 *Journal of Non-Crystalline Solids* 52–5
[44] Rajkhowa R, Wang L and Wang X 2008 *Powder Technology* **185** 87–95
[45] Preiss A, Su B, Collins S and Simpson D 2012 *Journal of the European Ceramic Society* **32** 1575–83
[46] Munch E, Saiz E, Tomsia A P and Deville S 2009 *Journal of the American Ceramic Society* **92** 1534–9
[47] Carstens S, Dammler K, Scheffler M and Enke D 2019 *Adv. Eng. Mater.* **21**
[48] Schelm K, Abreu Morales E and Scheffler M 2019 *Materials (Basel, Switzerland)* **12**

IOP Conf. Series: Materials Science and Engineering **882** (2020) 012006 doi:10.1088/1757-899X/882/1/012006

Improvement of the martensitic stainless steel X46Cr13 by Q&P heat treatment

S Dieck[1], M Ecke[2], T Halle[2] and P Rosemann[2]

[1] DeltaSigma Analytics GmbH, Mittagstraße 16p, 39124, Magdeburg, Germany
[2] Otto-von-Guericke-University Magdeburg, Faculty of Mechanical Engineering, Institute of Materials and Joining Technology, Universitätsplatz 2, 39106 Magdeburg, Germany

sebastian.dieck@deltasigma.de

Abstract. The innovative heat treatment concept of "Quenching and Partitioning" (Q&P) enables the production of high-strength martensitic steels with high ductility. Residual austenite is stabilized in the partitioning process by carbon diffusion. This enables the strain-induced phase transformation from austenite to martensite and increases the ductility without reducing the strength. So far, little research has been done on the application of the Q&P process on martensitic stainless steels and their related corrosion resistance. The present work supplements the first investigations by Yuang and Raabe with important material-technical parameters as well as the achievable corrosion resistance. The Q&P heat treatment offers enhanced mechanical properties and a higher corrosion resistance in comparison to the conventional quenching and tempering heat treatment.

1 Introduction

The usability of all products depends on the properties of the used materials. Whenever a further development of technical applications exceeds the application limits of the materials used, an alternative material concept is required. For example, the increase in efficiency of aircraft engines due to the increase in the combustion temperature necessitates a substitution of the nickel-based alloys currently used as a turbine material [1]. However, new developments are not always necessary. The targeted optimization of the properties of existing material concepts is often sufficient, which can be ensured by adapting the manufacturing parameters and the applied heat treatment.

The so-called "Quenching and Partitioning" (Q&P) is a new heat treatment concept for high-strength steels with increased ductility [2, 3]. Figure 1 presents the Q&P heat treatment process. The first step is the austenitization which aims on the homogenous distribution of alloying elements within the austenite. The subsequent quenching is stopped at a specific temperature between the martensite start (M_s) and finish temperature (M_f) to target a defined proportion of retained austenite (RA) in the martensite. The final partitioning at low tempering temperatures induces carbon diffusion from martensite into the RA, which is favoured due to the higher carbon solubility of the face-centred cubic crystal structure [4, 5]. This carbon enrichment further reduces the M_s temperature and stabilizes the RA [6]. A special feature of this stabilized RA is the property of transforming strain-induced into martensite under mechanical load. The effect, referred to as TRIP (transformation induced plasticity), increases strength and deformability of the steel [7, 8]. While previous studies on the Q&P process mostly focus on low-alloy steels, there are current studies on corrosion-resistant, martensitic steels like X46Cr13. Q&P heat treatment offers several advantages for this type of steel. On the one hand, the heat treatment process is simplified because M_f is below room temperature (RT) [9]. This eliminates the need for interrupted quenching in the first part of the Q&P process, which is necessary for low-alloy steels. On the other hand, with optimal heat treatment parameters, very high strengths (~ 1,800 MPa) with good elongation at fracture (20 %) can be achieved [11] and may be combined with high corrosion resistance.

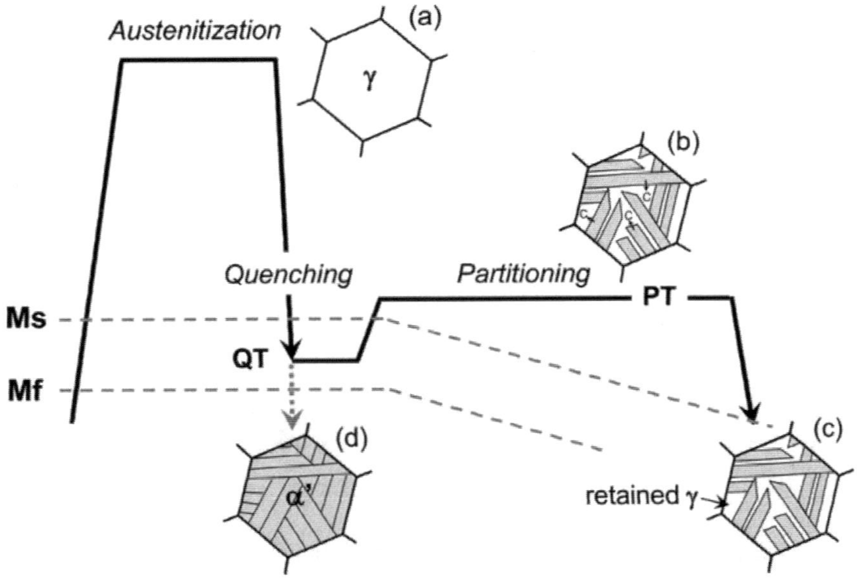

Figure 1. Temperature-time curve and structural change in Q&P heat treatment [10].

Furthermore, carbon accumulation occurs at large and small angle grain boundaries of the martensite. This enables the local transformation of martensite back into so-called "reverted" austenite (revA) [11], see Figure 2.

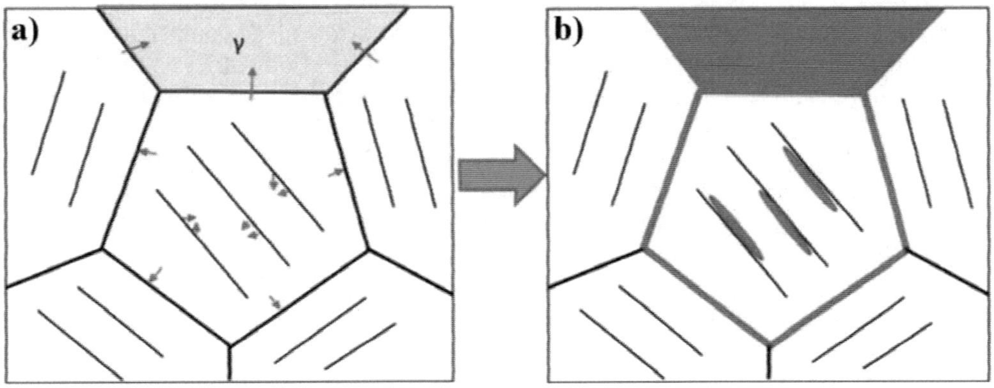

Figure 2. a) Diffusion of carbon during partitioning (red arrows), b) resulting microstructure (martensite = white; stabilized austenite = red) [9].

Due to the high chromium content, finely disperse carbides are formed during partitioning, which further increase the strength by particle hardening [9, 10]. The present work is intended to contribute to the characterization of the property profile (structure, strength, deformability and corrosion resistance) of the stainless steel X46Cr13 depending on the duration of the partitioning.

2 Material and methods
The used material was the martensitic stainless steel X46Cr13 (1.4034, AISI 420). The chemical composition of the cold-rolled and then soft annealed initial state is given in Table 1.

Table 1. Chemical composition of the martensitic stainless steel X46Cr13 in wt.-%.

Element	Fe	C	Cr	Mn	Si	Ni	Mo	N	P	S
Content	bal.	0,45	13,95	0,65	0,34	0,12	0,02	0,02	0,012	<0,001

The selection of heat treatment is based on the studies by Yuan [11], but were partially adjusted. The samples were austenitized for 15 min at 1150 °C in order to dissolve large carbides (≈ 10 µm) from the soft annealed initial state. The samples were quenched to room temperature in water after austenitization. The partitioning was carried out at 400 °C (T_P) and the partitioning time (t_P) was varied in order to investigate the time-dependent phase formation of revA and its influence on the properties. Three main states were created:

- not partitioned — quenched reference condition
- 5 min partitioned — short-term partitioning
- 30 min partitioned — long-term partitioning

The mechanical behavior was determined by tensile and compression tests (according to DIN 50125 and DIN 50106). The phases in the microstructure were quantified by X-ray diffraction (XRD) and electron backscatter diffraction (EBSD) with a scan step size of 100 nm. Furthermore, the structure was visualized using optical microscopy (OM). For these investigations, the samples were first ground and polished. The XRD analysis was carried out on sample surfaces additionally etched with V2A. For the EBSD examinations, the samples were polished with OPS to completely remove preparation-related surface deformation. The light microscopic evaluation was carried out on Beraha II etched surfaces.

The corrosion resistance of the stainless steel X46Cr13 is also significantly influenced by the heat treatment [12]. To demonstrate the advantages of Q&P heat treatment compared to conventional hardening and tempering, electrochemical potentiodynamic reactivation (EPR) and the KorroPad test were used. The EPR method allows to characterizes the passivation and reactivation behavior of the different heat treatment conditions [12]. A 0.1 M H_2SO_4 solution and a classic three-electrode arrangement, consisting of working electrode (sample), Ag/AgCl reference electrode (+198 mV_{NHE}) and platinum counter electrode, were used. The sample was polarized with 2 mV/s from -500 mV to +100 $mV_{Ag/AgCl}$ to form the passive layer (passivation). The direction of polarization was reversed at +100 $mV_{Ag/AgCl}$ and polarized back to -500 $mV_{Ag/AgCl}$ (reactivation) in order to induce the dissolution of chromium-depleted zones. The influence of the heat treatment on the pitting corrosion resistance was examined with the KorroPad test [13, 14]. This enables very quick detection of corrosion-prone microstructure conditions on martensitic stainless steels [15, 16]. All samples were ground with 180 grit SiC paper and then passivated for 24 h at more than 95 % relative humidity to form a stable passive layer. The KorroPads with 0.1 M NaCl and 1 mM $K_3[Fe(CN)_6]$ were applied on the passivated surface for 15 min and compared visually. Blue color indications are the result of pronounced pitting corrosion.

3 Results and discussion

3.1 Mechanical properties
The influence of the partitioning time on the mechanical properties of the martensitic stainless steel X46Cr13 is shown in Figure 3. The quenched reference state exhibits brittle fracture during the tensile tests before the yield strength is reached. This behavior is typical for brittle materials, which include

X46Cr13 in the hardened state. Even with a partitioning time of 5 minutes, an enormous increase in strength and elongation can be seen, while the fracture occurs in the plastic deformation area. By increasing the partitioning time to 30 min, the deformability is furthermore increased. A tensile strength of nearly 1.700 MPa and a total elongation of 12 % are achieved. It should be noted that a deviating behavior in the elastic range was detected for the partitioned states, which contradicts the literature [17] and is attributed to the experimental setup.

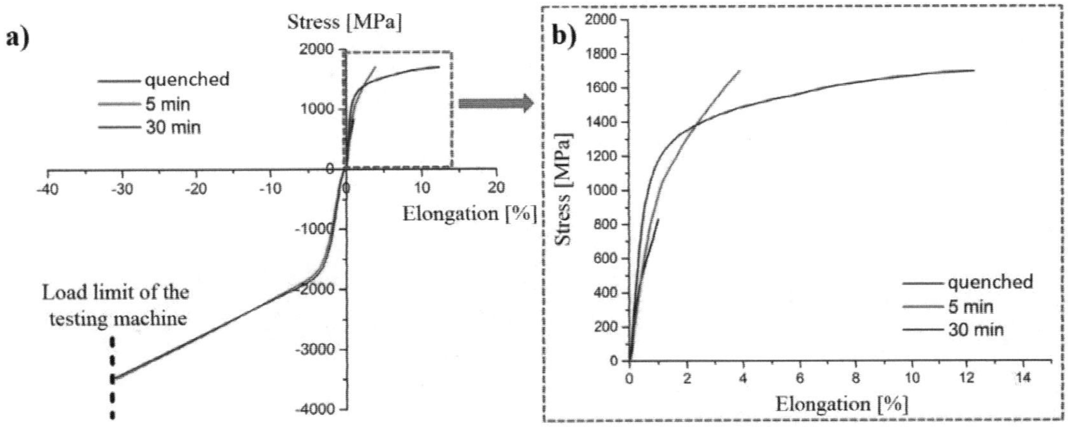

Figure 3. Results of the mechanical testing: a) tensile and compression test, b) tensile test.

In contrast to the tensile tests, the compression tests show identical behavior for all states. No statement can be made about the maximum strength and the maximum deformability, since the tests had to be stopped when the load limit of the testing machine was reached, while no sample failure occurred. However, it can be seen that yield point, maximum strength and deformability exceed the characteristic values of the tensile tests. Based on the assumption that plastic deformation is only caused by shear stresses, the compressive and tensile strengths of materials should be identical [18].

However, this is not the case with high-strength steels and especially with martensite. It has been proven experimentally that high-strength steels can withstand higher pressure loads. This behavior is referred to the SD (strength differential effect) [19, 20]. The causes of the different mechanical parameters have not yet been fully clarified. It is assumed that several influencing factors are involved, but there is no consensus on the specific contribution.

Accordingly, it is conceivable that microcracking limits the deformability under tensile load. Other possible influencing factors are the Bauschinger effect, interactions between dislocations and interstitially embedded atoms, and the residual stresses that depend on the direction of mechanical load [19–21]. For the steel 100Cr6, Ellermann and Scholtes have determined, that the conversion of residual austenite to martensite by the TRIP effect is suppressed under pressure loads [21].

This last aspect in particular would explain the higher deformability of the partly austenitic samples. In addition, the SD effect in martensitic steels, of a maximum of 10 %, cannot be explained with the results of the present work. SINGH et al. also report that the SD effect decreases with increasing tempering temperature [19]. Further investigations are required to explain the different mechanical behavior of Q&P heat treatment under tensile and compressive loads.

Figure 4 presents the fractured surface of the tensile specimens imaged by SEM. The proportion of ductile fracture increases with the partitioning time, which can be seen from the typical honeycomb structure. This observation is in line with the increase in the total elongation in the tensile test with increased partitioning time (Figure 3). Furthermore, all samples show a terrace-shaped fracture surface and the crack growth is preferably transverse to the tensile direction. The cause of this behavior is determined by the microstructure analysis in section 3.2. in more detail.

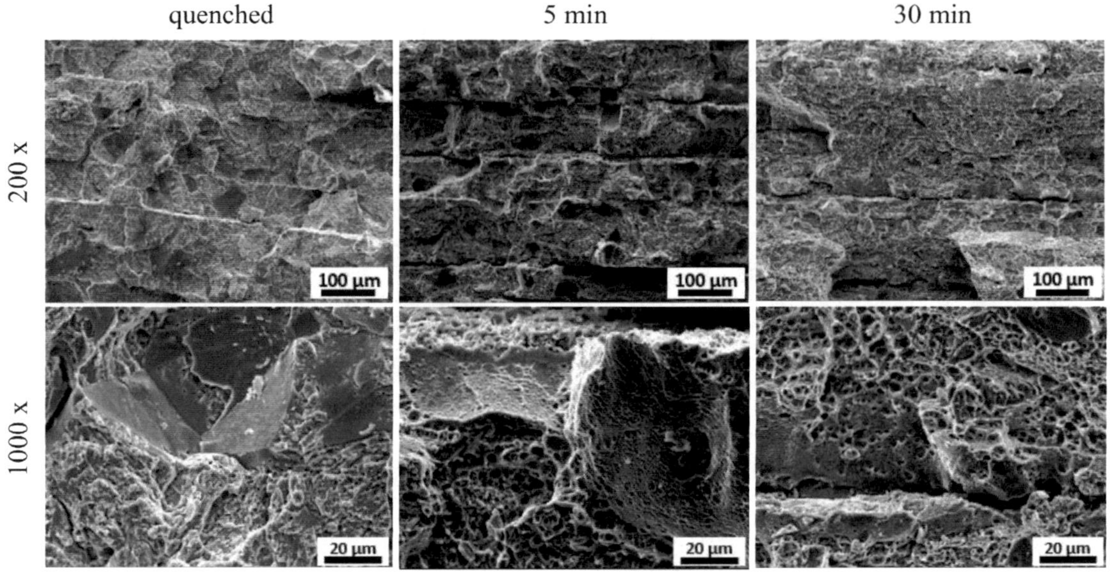

Figure 4. Fracture surfaces of the tensile specimens after Q&P heat treatment.

3.2 Phase and microstructure analysis

The influence of the Q&P heat treatment on the phase components was determined by X-ray analysis and the corresponding results are presented in Figure 5. The quenched state consists mainly of strongly distorted α-Fe (martensite), which can be seen from the broad α-Fe peak which is shifted slightly relative to the reference position. In addition, large fractions of austenite were detected.

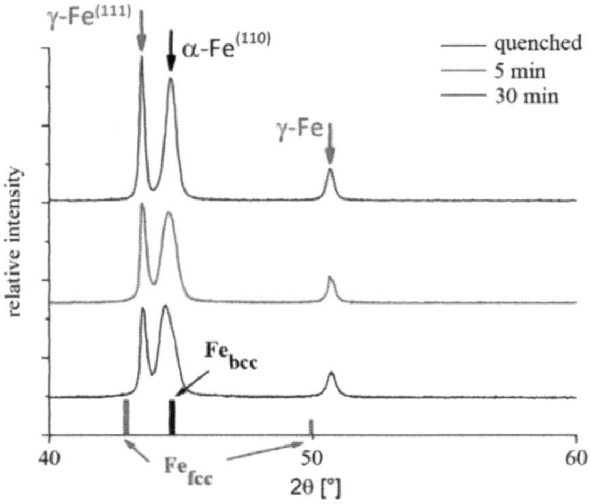

Figure 5. X-ray diffraction diagram after Q&P heat treatment in the 2θ range from 40 ° to 60 °.

The width of the α-Fe peak decreases with increasing partitioning time, which is related to the decrease of residual stresses in the martensite. This is caused by the diffusion-related decrease of the carbon content in the martensite. The peak intensity of the austenitic fcc phase increases with longer partitioning. The cause is the time-dependent austenite reversion due to the partitioning, which increases the austenite content. The results of EBSD presented in Figure 6 confirm the XRD results. The structure of the quenched state consists of martensite and islands of residual austenite.

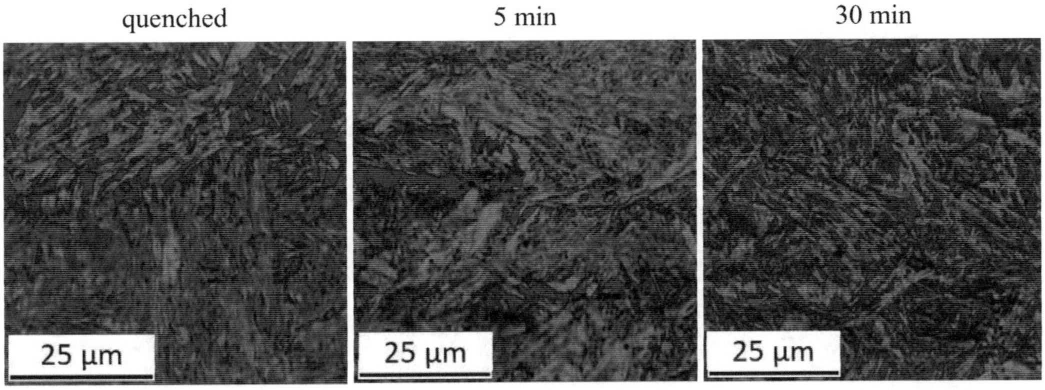

Figure 6. EBSD images after Q&P heat treatment (martensite/bcc: red; austenite/fcc: blue).

5 min partitioning exhibits more dispersed austenite areas between the martensite. These phases grow with increasing partitioning time, which can be seen at 30 min partitioning time. Reverted austenite forms at the sub-grain boundaries of martensite. According to YUAN et al. the reverted austenite also forms like a film at the boundaries of the primary austenite grains [11]. However, this only has an extent of a few nanometers and cannot be detected by EBSD. For additional evaluation, OM images of the microstructures are finally presented in Figure 7. A pronounced line like structure was detected for all states, which partially extends over the entire cross-section of the sample.

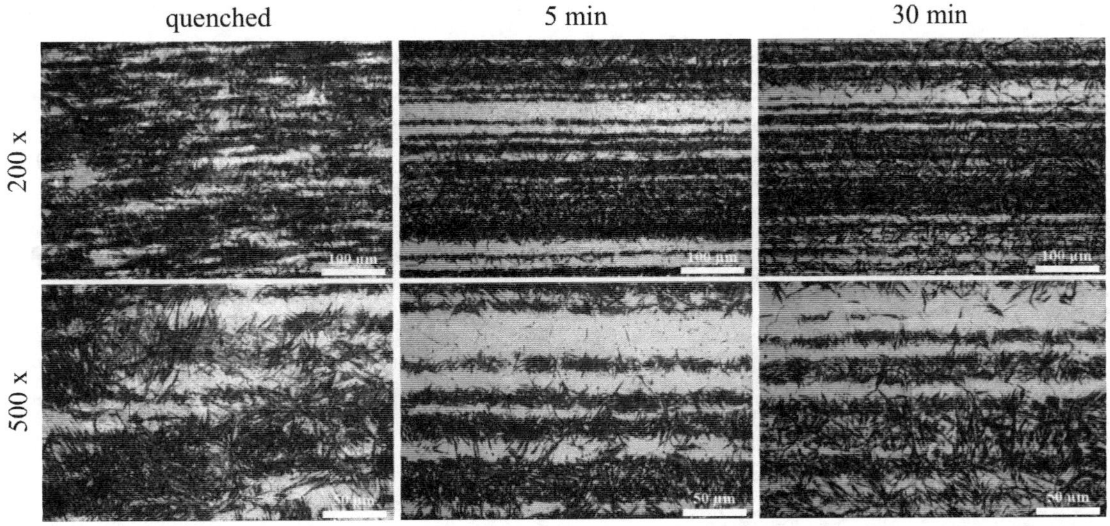

Figure 7. OM after Q&P heat treatment etched with Beraha II (martensite: black, austenite: white).

The white lines are primary austenite and the comparison with the images of the terraced fracture surfaces in Figure 4 suggests that the crack propagation takes place preferentially in the austenitic lines. SEM images in BSE contrast in combination with EDX analyses also show that there is a further phase within the austenite lines, see Figure 8. EDX analyses show that these are chromium-rich carbides within the austenitic lines. Their poor deformation properties cause the preferred crack growth during tensile load along the carbides. The large carbides are already contained in the soft-annealed starting material and thus originate from the hot- and cold-rolling during sheet metal production. These are the result of too low rolling temperatures, which lead to the formation of carbide lines along the grain boundaries of the stretched grains. The subsequent soft annealing leads to further growth of this carbides and cause the formation of large carbide agglomerations.

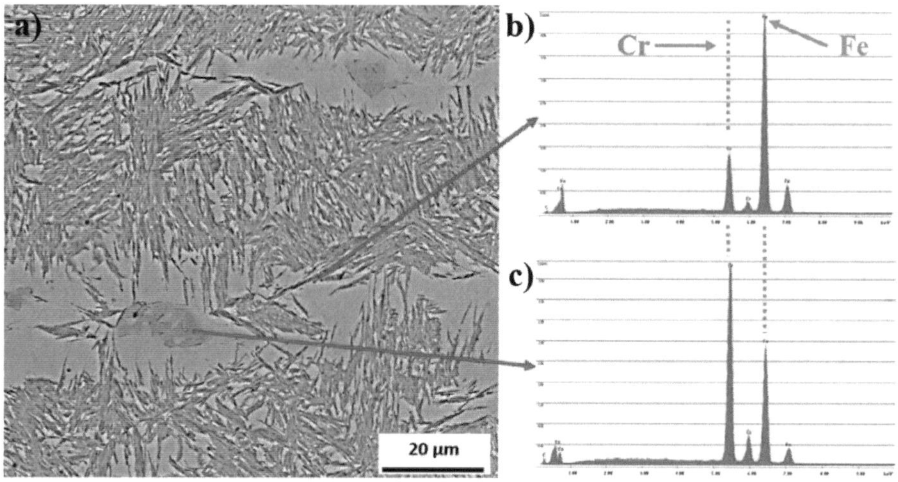

Figure 8. SEM analysis of the austenite line of a 5 min partitioned sample - a) acquisition with BSE contrast and EDX spot analysis b) the austenite, c) the darker phase (carbide).

These agglomerates were not completely dissolved by austenitizing at 1150 °C for 15 min in the first step of Q&P heat treatment. Obviously, the homogeneous distribution of all alloy elements within the austenitic matrix was not achieved. The partly dissolved carbides result in a local enrichment of the surrounding matrix with C and Cr. The residual austenite preferably remains around the carbides due to the locally decreased M_f temperature.

3.3 Electrochemical investigations of the corrosion resistance

Figure 9 shows the results of the electrochemical potentiodynamic reactivation as a direct comparison of tempering (a) with Q&P heat treatment (b). Conventional hardening at 1050 °C for 15 min followed by water quenching achieves a corrosion-resistant state, since chromium is homogeneously distributed in the martensitic matrix, although a large number of carbides are still present in the structure. Conventional tempering results in the formation of sub-microscopic, chromium-rich carbides and induce chromium depletion at a low tempering temperature of 400 °C. This can be seen from the increased passivation and reactivation current densities.

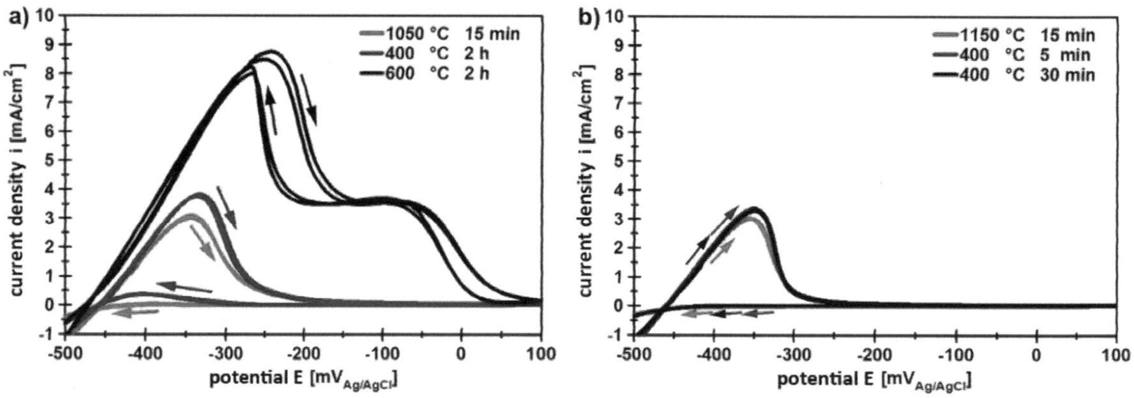

Figure 9. Results of the EPR procedure as a comparison of a) remuneration and b) Q&P.

If the tempering temperature is increased to the usually used tempering temperature of 600 °C, the passivation behavior is severely disturbed, since the matrix is strongly depleted by progressive carbide

IOP Conf. Series: Materials Science and Engineering **882** (2020) 012006 doi:10.1088/1757-899X/882/1/012006

formation and carbide growth. This condition is no longer able to form a stable passive layer and exhibits a detrimental corrosion resistance.

In contrast, the Q&P heat treatment, even after partitioning at 400 °C for 30 min, no chromium depletion can be detected using the EPR method and a stable passive layer is formed in all conditions. This is due to the lower temperature and shorter duration during partitioning which avoid carbide formation and chromium depletion. The differences in passive layer stability demonstrated with the EPR method have a very strong influence on the pitting corrosion resistance. This is illustrated by the results of the KorroPad test in Figure 10.

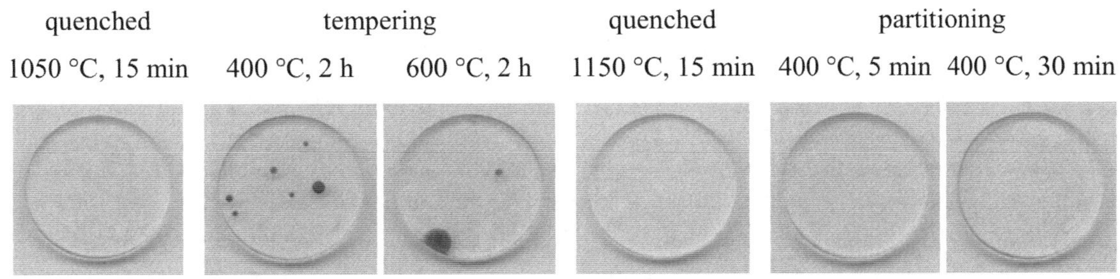

Figure 10. Results of the KorroPad test as a comparison of remuneration and Q&P.

Conventional hardening results in high pitting corrosion resistance, but tempering at 400 °C and 600 °C causes a strong susceptibility to pitting corrosion, which can be seen from the large blue indications that appear when individual pits grow during the KorroPad test. The portioned states, on the other hand, have a high pitting corrosion resistance. From a corrosion point of view, Q&P heat treatment has clear advantages compared conventional tempering.

4 Conclusions

The Q&P heat treatment was tested on the martensitic stainless steel X46Cr13. The focus of the investigations was the influence of the partitioning on mechanical properties, microstructure and corrosion resistance. The following findings were achieved:

1. The quenched state suffers from brittle fracture before the yield strength is reached. A tensile strength of ≈ 1.700 MPa and a total elongation of ≈ 12 % were achieved by partitioning up to 30 min and a pronounced SD effect was detected.
2. The improvement of the mechanical properties is based on the mechanisms of austenite stabilization and reversion. Both are the consequence of time-dependent diffusion of carbon during partitioning.
3. The corrosion resistance after Q&P heat treatment is significantly higher in comparison to the conventional tempered condition, since the diffusion-induced formation of austenite does not cause chromium depletion during partitioning.

Furthermore, it was found that austenitizing (1150 °C, 15 min) must be classified as insufficient, since the large, chromium-rich carbides of the soft-annealed initial state were not dissolved. Consequently, there are line areas of residual austenite around remaining carbides within the microstructure. The carbon diffusion during partitioning does not completely stabilize them. It is assumed that optimization of the soft annealing initial microstructure may further increase the tensile strength and total elongation. This requires the loosening of all carbides and a homogeneous distribution of the alloy elements during austenitizing as an optimized starting point for partitioning.

Acknowledgments
The authors would like to thank the Graduiertenkolleg 1554 for financial support. The authors thank Sebastian Fritsch from the Chemnitz University of Technology for performing the pressure tests and the Department 7.6 Corrosion and Corrosion Protection of the Federal Institute for Materials Research and Testing for the provision of KorroPads.

Furthermore, the authors would like to thank the German Research Foundation (DFG) for the funding of the used DualBeam FEI Scios within the framework of the funding of large-scale facilities according to GG§91.

References

[1] Dieck S, Baumann T, Hasemann G, Rannabauer S and Krüger M 2014 *15. Sommerkurs Werkstoffe und Fügen* (Magdeburg: University Press) 143
[2] Edmonds DV, He K, Rizzo FC, Cooman BC, Matlock DK and Speer JG 2006 *Materials Science and Engineering* A **438** 25
[3] Wang L and Speer JG 2013 *Metallography, Microstructure, and Analysis* **2** 268
[4] Speer J, Matlock DK, Cooman BC and Schroth JG 2003 *Acta Materialia* **51** 2611
[5] Clarke AJ, Speer JG, Miller MK, Hackenberg RE, Edmonds DV, Matlock DK, Rizzo FC, Clarke KD and Moor E 2008 *Acta Materialia* **56** 16
[6] Santofimia MJ, Zhao L and Sietsma J 2011 *Metallurgical and Materials Transactions* A **42** 3620
[7] Arlazarov A, Bouaziz O, Masse JP and Kegel F 2015 *Materials Science and Engineering* A **620** 293
[8] Arlazarov A, Ollat M, Masse JP and Bouzat M 2016 *Materials Science and Engineering* A **661** 79
[9] Dieck S, Rosemann P, Kromm, A and Halle T 2017 *IOP Conference Series: Materials Science and Engineering* **181** 12034
[10] Tsuchiyama T, Tobata J, Tao T, Nakada N and Takaki S 2012 *Materials Science and Engineering* A **532** 585
[11] Yuan L, Ponge D, Wittig J, Choi P, Jiménez JA and Raabe D 2012 *Acta Materialia* **60** 2790
[12] Rosemann P 2017 Einfluss der Wärmebehandlung auf die Korrosionsbeständigkeit martensitischer nichtrostender Stähle (Magdeburg: Shaker Verlag)
[13] Burkert A, Burkert A, Lehmann J, Müller T and Bohlmann T 2014 *final report of the research project 17136 N*
[14] Lehmann J, Burkert A and Steinhoff UM 2012 *german patent* DE 10 2010 037 775 A1
[15] Burkert A, Klapper HS and Lehmann J, *Materials and Corrosion* **64** 675
[16] Lehmann J, Burkert A and Mietz J 2016 *Materials and Corrosion* **67** 84
[17] Yuan L 2012 Nanoscale austenite reversion through partitioning, segregation and kinetic freezing in martensitic stainless steels (Aachen: Shaker)
[18] Bridgman PW 1964 *Collection of experimental papers (*Harvard: University Press)
[19] Singh AP, Padmanabhan KA, Pandey GN and Murty GMD 2000 *Journal of Materials Science* **35** 1379
[20] Rauch GC and Leslie WC 1972 *Metallurgical and Materials Transactions* B **3** 377
[21] Ellermann A and Scholtes B 2015 *Materials Science and Engineering* A **620** 262

IOP Conf. Series: Materials Science and Engineering **882** (2020) 012007 doi:10.1088/1757-899X/882/1/012007

On the effect of aluminum and chromium on the deformation twinning of body-centered cubic iron

M Ecke, O Michael and T Halle

Otto-von-Guericke-University Magdeburg, Faculty of Mechanical Engineering, Institute of Materials Science and Joining Technology, Universitätsplatz 2, 39106 Magdeburg, Germany

martin.ecke@ovgu.de

Abstract. This paper reports on investigations on the effect of aluminum and chromium on the formation of deformation twins in body-centered cubic iron. For this purpose, impact loading experiments ($\dot{\varepsilon} = 10^4\ s^{-1}$) were carried out at room temperature on samples of iron and the alloys Fe-10at.-% Al and Fe-10at.-% Cr. Furthermore, the grain size was varied in order to consider the effect of the alloying elements for both fine-grained and coarse-grained microstructures. The investigations on twin formation were carried out by electron backscatter diffraction (EBSD) and by quantitative image processing using optical microscopy. The observations show a comparable twin portion in iron and Fe 10at.-% Cr after impact loading for both grain sizes. In the alloy Fe-10at.-% Al the portion of twins is significantly increased for fine-grain and coarse-grain state. In order to clarify the possible reason, X-ray diffraction analyses (XRD) were additionally carried out. These show a shift in the reflexes which can be attributed to an elastic distortion of the lattice. This lattice distortion due to the formation of solid solution correlates very well with the determined twin fractions.

1 Introduction

Deformation twinning is an important mechanism of plastic deformation and is of great meaning for many metals in everyday technical life. In hexagonal closed packed (hdp) magnesium and magnesium alloys, the formation of twins is fundamental for the technical processability of the material [1]. In face-centered cubic (fcc), high-manganese TWIP steels (TWinning Induced Plasticity), deformation twinning enables an extremely high ductility of 95% with tensile strength above 1.000 MPa [2–4]. This makes such steels of particular interest in components for automotive engineering.

Twinning can also be observed in body-centered cubic (bcc) iron materials. This is mainly the case at low temperatures below the room temperature and at impact loads with strain rates above $\dot{\varepsilon} - 10^0\ s^{-1}$ [5,6]. For temperatures around room temperature and quasi-static loads, dislocation-based slip is the dominant deformation mechanism in bcc iron. The decreasing temperature or increasing strain rate leads to an increase in the shear stress required to activate twin formation or atomic slip. However, both effects have a clearly different dependence on temperature. While the required stress for dislocation-based slip increases with decreasing temperature, the stress for twin formation remains almost constant with decreasing temperature. The result is more hindered slip at low temperatures and thus the preferred deformation due to twin formation. From investigations of MEYERS ET AL. an overview of this correlation can be seen in Figure 1 [6]. Dislocation-based slip has no linear progression over temperature, so twinning is preferred below a critical temperature.

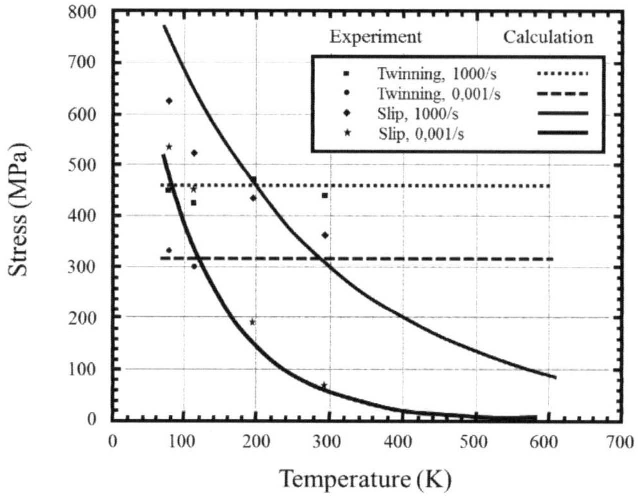

Figure 1. Critical stresses for the activation of dislocation-based slip and deformation twinning as a function of temperature and strain rate. The diagram shows a transition point from slip to twin formation below a critical temperature. The figure is based on calculations for iron with a grain size of 100 µm performed by MEYERS ET AL. [6].

Since NEUMANN'S discovery of deformation twins in iron meteorites in 1848, a large number of investigations on twin formation in metals have been carried out [6–11]. On the one hand, these studies focus on the effects of different influencing variables such as temperature and strain rate but also grain size. The grain size as an internal influencing variable is of particular importance. Several studies have already shown that an increase in grain size also has a favorable effect on twin formation. Similar to the temperature and strain rate, a grain size change also has a different effect on the preferred mechanism [12]. A schematic overview is shown in Figure 2. Therefore, the effect of grain size should also be considered in this study.

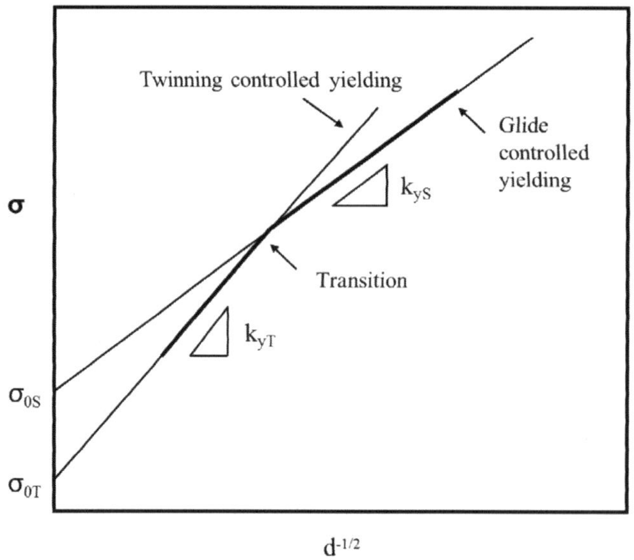

Figure 2. Schematic representation of the course of the formation stress for twin formation as well as displacement based sliding as a function of the grain size [12].

On the other hand, investigations on the nucleation mechanism of twinning are of great interest [13–22]. The theories available include both dislocation-based processes and mechanisms without dislocation involvement. The experimental proof of these formation mechanisms is currently in the focus of research. In addition to the use of experimental methods such as TEM and STEM, computer-based investigation methods are playing an increasingly important role in describing atomic motion processes during impact loading [23,24].

In the context of this study, investigations on the influence of alloying elements on the formation of deformation twins in bcc iron are presented. Specifically, the influence of aluminum and chromium on the formation of twins in bcc iron is investigated and compared with the twin formation in unalloyed iron. Similar to the influence of manganese in the already mentioned fcc TWIP steels, the effect of Al and Cr in the iron lattice will be investigated. Possible effects on twin formation are the influence on the stacking fault energy and an elastic lattice distortion. With respect to the portion of twins in the three materials, these effects will be considered and discussed.

2 Materials and methods

Iron (99.8% Fe) and iron alloys of Fe-10at.-% Al and Fe-10at.-% Cr are used for this study. In the following, the alloys subsequently abbreviated as Fe-10Al and Fe-10Cr. The experiments are carried out on the as-received state (cold-rolled) and on a heat-treated state. The respective grain size before and after heat treatment can be taken from Table 1. The grain size was determined using the line intersection method according to DIN EN ISO 643 and is compared with EBSD investigations. A more precise adjustment of the grain size was not possible for process-technical reasons, so that the scatter in the coarse grain sizes can be justified.

Table 1. Overview of grain sizes and heat treatment parameters for the three investigated materials Fe, Fe-10Al and Fe-10Cr

Grain size	**Fe**	**Fe-10Al**	**Fe-10Cr**
fine-grained	*44 μm*	*48 μm*	*30 μm*
coarse-grained	*116 μm (12 h @ 1150°C)*	*140 μm (1 h @ 950°C)*	*201 μm (10 h @ 1150°C)*

A mechanical test set-up is used to apply an impact load, which accelerates a hardened bolt to 50 m/s and applies a single load to the sample. The sample geometry is 6.0 mm x 6.0 mm (diameter x height). The bolt accelerates the samples, which are caught in a bullet trap 2.0 m away. The resulting plastic deformation is less than 1 % and can therefore be neglected for further considerations. A schematic representation of the experimental set-up shows Figure 3.

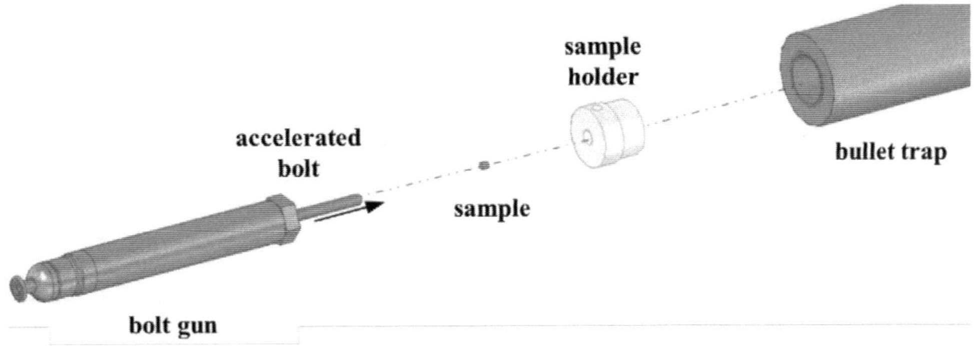

Figure 3. Schematic illustration of the experimental set-up of the set-up for impact loads.

In order to characterize the microstructure, metallographic preparations are required first. The metallographic preparation is carried out by hot mounting, grinding with SiC paper (up to grain size 2400) and polishing with diamond suspension (3 µm and 1 µm). To contrast the microstructure for subsequent optical microscopic examinations, a V2A-Beize is used for iron and Nital for the two iron alloys Fe-10Al and Fe-10Cr. For the EBSD analyses, the diamond polishing is not followed by contrasting, but by fine polishing with oxide particle suspension. Here, different suspensions are used depending on the material. An OP-S suspension is suitable for iron and Fe-10Cr, an OP-U suspension for Fe-10Al.

EBSD analyses are carried out for a random determination of the grain size as a comparison to the line section measurements as well as for the characterization of the crystallographic orientation conditions after the impact loading. With the help of EBSD, crystallographic properties can be determined which are used to reveal the preferred twin systems. This is possible by using the coincidence site lattice (CSL), which characterizes the crystallographic relationship between differently oriented areas. This method is particularly well suited for the study because of the misorientation between matrix and twin. The expected coincidence for the twin system in iron is the CSL-$\Sigma3$ [25].

Furthermore, possible plastic deformations before and after loading as well as second phases weren't observed. The EBSD examinations were performed on a DualBeam FEI SCIOS with AMETEK-EDAX analysis. The selected scan areas have a size between 100 x 100 µm and 1000 x 1000 µm. The scan step size varies between 100 and 500 nm. In this work, the results are presented as color-coded figure of the inverse pole figure IPF.

EBSD measurements on the initial state show that the microstructure is single-phase and free of deformation. Deformation twins or a second phase are also not present. An overview of the crystallographic orientation conditions in the initial state shows Figure 4.

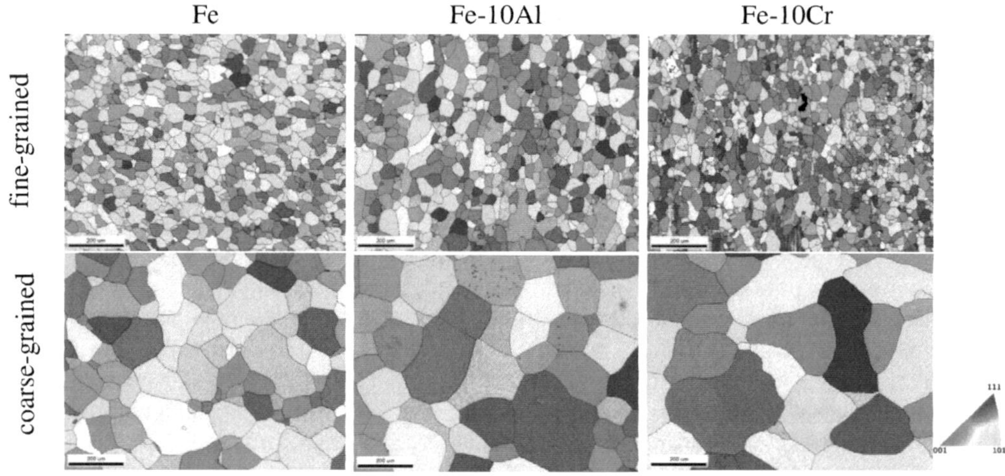

Figure 4. EBSD study of the three investigated materials Fe, Fe-10Al and Fe-10Cr in the initial state and the heat treated state.

The quantification of the twin portion (indicated as area fraction) is carried out using metallographic practices. Using optical microscopic images, the area fraction after impact loading are determined by image processing. The determination of the area fraction of twins is also possible using EBSD measurements. However, the section to be viewed is much smaller than in the optical microscopic section. For a representative and statistical evaluation, a high number of EBSD measurements per sample are necessary.

Binarized dark field images are used for quantification. These offer the advantage that grain boundaries and twin boundaries as well as microstructural artifacts and impurities are easier to detect. By software-based determination the portion of twins in several samples were calculated. This procedure is carried out for areas without and with deformation twins for each sample. The difference in detected signal (no microstructural feature = white, detected microstructural feature = black) between the two

IOP Conf. Series: Materials Science and Engineering **882** (2020) 012007 doi:10.1088/1757-899X/882/1/012007

parts of information (area fraction with twins A_{twins} and area fraction without twins $A_{no\ twins}$) shown in Figure 5 automatically results in the pure portion of deformation twins.

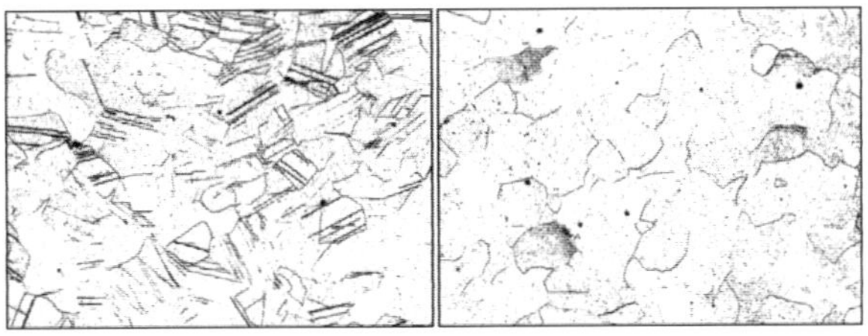

Figure 5. Procedure of quantitative measurement of twinning portion A_{twins} (left) und $A_{no\ twins}$ (right) using binarized dark field images.

X-Ray Diffraction (XRD) analyses are used to determine the structure and phase characteristics of the produced alloys Fe-10Al and Fe-10Cr. Thus it can be ensured that the investigated materials are present in a single-phase, body-centered cubic lattice.

Furthermore, information about the influence of the alloying elements on the iron lattice can be derived from the XRD results. This results are discussed in the results section in connection with the optical microscopic analyses. The sample preparation of the XRD measurements is analogous to the metallographic preparation up to the deformation-free polished state. The used radiation for the XRD analysis is Co-$K_{\alpha 1}$. The diffraction diagrams show a normalized representation of the intensities.

3 Results and Discussion

The results of the impact loaded materials (Fe, Fe-10Al and Fe-10Cr) shown below, reveal the formation of deformation twins in the microstructure in all three materials and both grain sizes. First of all, the EBSD investigations show that the formed twins have an orientation difference to the initial crystal (matrix) of 59 ° and are thus a twin boundary of the type CSL-Σ3. From this it can be deduced that <111> {112} is the preferred twin systems in all three materials. There is no change in the twin system by alloying Al and Cr. As an example, Figure 6 shows EBSD overview measurements of the three materials with fine-grained and coarse-grained microstructures. The random distribution of twins can be seen. A preferred crystallographic orientation in which twinning takes place is not observed.

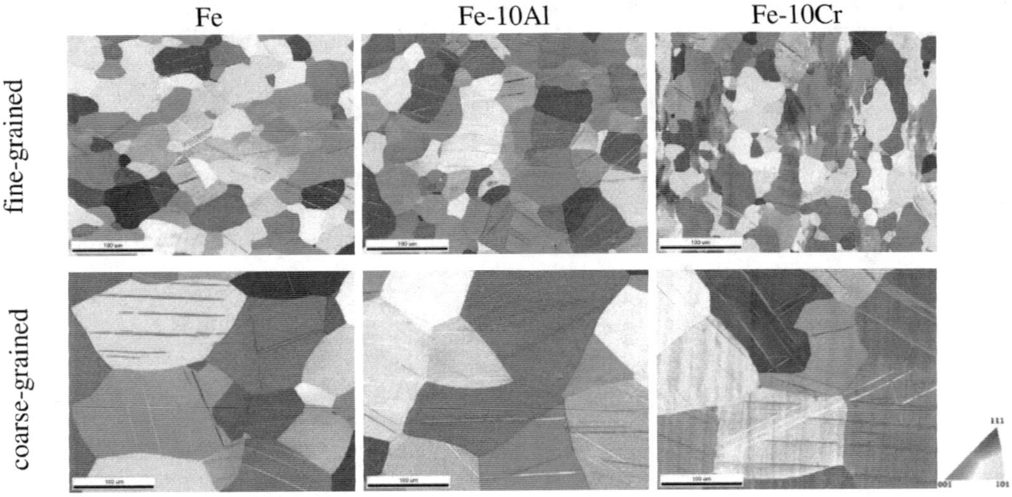

Figure 6. EBSD study of the three investigated materials Fe, Fe-10Al and Fe-10Cr with fine-grained and coarse-grained microstructure after impact loading.

Table 2 presents the results of the measured twin amount after impact loading. The determined portion of twins for iron with fine-grained microstructure is 3.1 %. For the alloy Fe-10Cr the portion is 3.4 % on average and thus in a similar range. In comparison, the determined portion of twins in the material Fe-10Al is 7.9 % on average. For the coarse-grained microstructure, the investigations show an average portion in iron of 6.5 % and for Fe-10Cr of 7.5 %. For Fe-10Al the average portion is 9.1 %. For both grain size it can be observed that the alloy Fe-10Al has a higher portion of twins. For Fe-10Cr a preferred effect on the formation of twins can also be observed, but this is much less pronounced.

In addition, there is the superimposition of the grain size effect. As already mentioned, an increasing grain size leads to preferred twinning. This effect can also be observed in the comparison of fine and coarse-grained microstructure. In the fine-grained state, the average grain sizes determined are 44 µm for iron, 48 µm for Fe-10Al and 30 µm for Fe-10Cr. The differences in grain size are small and therefore exert little influence on the formation of twins. A different situation exists for the coarse-grained microstructure due to the deviation in grain sizes. The grain sizes of iron (116 µm), Fe-10Al (140 µm) and Fe-10Cr (201 µm) have a significantly higher absolute deviation. It is therefore assumed that with similar grain sizes, e.g. around 200 µm, iron and Fe-10Cr would be closer together (~ 7.0 %) and the twin portion in Fe-10Al would increase to ~10.0 % or higher.

Table 2. Overview of the twin portion with standard deviation in Fe, Fe-10Al and Fe-10Cr after impact loading depending on the grain size

Grain size	Fe	Fe-10Al	Fe-10Cr
fine-grained	*3.1 ± 0.8*	*7.9 ± 1.0*	*3.4 ± 1.0*
coarse-grained	*6.5 ± 2.6*	*9.1 ± 0.9*	*7.5 ± 2.7*

The XRD investigation were done on the initial material with the fine-grained microstructure. A section of the obtained diffraction diagram is shown in Figure 7. The overlaid diffraction diagram shows the (011) reflexes for Fe, Fe-10Al and Fe-10Cr in the angular range 50°...2θ...55°. The measurements reveal the influence of the alloying elements in the iron lattice. A shift in the position of the reflexes are detected, which can be explained by the elastic distortion of the lattice. The lattice parameter of the (011) reflexe in iron is 202.68 nm. With 10 at.-% Cr this shifts to 203.05 nm, for 10 at.-% Al to 203.85 nm. The increase of the lattice parameter correlates very well with the determined twin fractions in both the fine-grained and coarse-grained state. The XRD investigations thus provide a good explanation for the observed twin formation behavior.

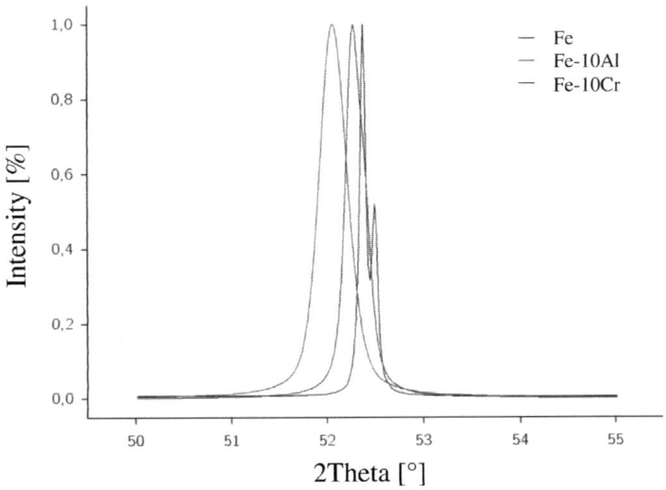

Figure 7. Diffraction diagram from XRD investigations on the three materials Fe, Fe-10Al and Fe-10Cr in the unloaded initial state. The normalized intensity for the (011) reflex in the angular range 50°...2θ...55° is shown. It reveals a peak shift resulting from an elastic lattice distortion due to the addition of Al and Cr

The XRD measurements confirm an influence of the alloying elements Al and Cr on the iron lattice and provide an explanation for the observed twin formation. Which effect is caused by the alloying elements on the stacking fault energy cannot be fully answered by this study.

On the one hand, there are methodological challenges in such nanoscopic scale investigations. The direct determination of the stacking fault energy by means of TEM investigations as well as the indirect determination from XRD measurements should be the aim in future studies.

Furthermore, the question of the existence of stacking faults in bcc iron is not entirely solved. VITEK was one of the first to deal with the existence of stable stacking faults in the bcc lattice [26,27]. The work suggests that stable intrinsic stacking faults in the bcc lattice are not likely. Even the formation of a single-layer stacking fault would not bring any energetic advantages. This fact also includes the preferred (112) plane for twin formation. However, extrinsic stacking faults are not excluded, which are formed as a result of dislocation dissociation and are stable. A generalized stacking fault energy (GSF energy) has therefore been defined to describe dislocation dissociation. This approach has been developed in a large number of computer-aided investigations using various methods [28]. In principle, the question of the existence of stacking faults in bcc iron is not completely clarified and requires a large number of further investigations, also with regard to their influence on the formation of twins in bcc iron.

4 Summary

In this study the influence of Al and Cr on the formation of twins in bcc iron is reported. Samples of the three materials Fe, Fe-10 at.-% Al and Fe-10 at.-% Cr were impact loaded at a strain rate of 10^4 s^{-1} at room temperature. Furthermore two different grain sizes were considered instead of. Fine-grained in the range from 30 μm to 48 μm and coarse-grained in the range from 116 μm to 201 μm. The investigations by EBSD first show that in all materials and grain size states the misorientation between twin and crystal matrix is ~59°. It can be concluded from this that the preferred twin system is the <111> {112}. The optical microscopic analyses show that for both states, Fe-10Al has the largest portion of twins after impact loading. For fine-grained state the mean portion is 7.9 % and for the coarse-grained 9.1 %. For the other two materials, the percentages are lower. For fine-grained microstructure, the mean portion is

IOP Conf. Series: Materials Science and Engineering **882** (2020) 012007 doi:10.1088/1757-899X/882/1/012007

comparable and is 3.1 % for Fe and 3.4 % for Fe-10Cr. In the coarse-grained microstructure, the mean portion are 6.5 % for Fe and 7.5 % for Fe-10Cr. However, it can be assumed that the difference in the coarse-grained state is an influence which must be taken into account. Suppose, that the grain sizes are identical (deviation below 10 µm), it can therefore be assumed that Fe and Fe-10Cr have a tendency to form similar twin portions and Fe-10Al has a much stronger tendency to form twins. XRD investigations show a shift of the peak positions in the diffraction diagram. An example is the shown reflex (011). The increase of the lattice parameter from 202.68 nm (Fe) to 203.05 nm (Fe-10Cr) or to 203.85 nm (Fe-10Al) is in very good agreement with the shown twin portions. For a validation of the effect, further experiments with identical grain sizes are useful. However, basic manufacturing and process engineering approaches must be revised in the preliminary field.

Furthermore, an expansion of the alloying elements and the alloy contents is considered useful in order to investigate the effect for other alloys. In principle, the experimental investigation of possible stacking fault formations by TEM and XRD studies is also considered useful for future studies. In addition, molecular dynamic simulations could provide further information about the effect of alloy atoms on twin nucleation and twin growth.

Acknowledgments
The authors would like to thank the Methodisch Diagnostisches Zentrum Werkstoffprüfung e.V. (MDZWP e.V.) for financial support. Furthermore, the authors would like to thank the Deutsche Forschungsgemeinschaft (DFG) for the funding of the used DualBeam FEI Scios within the framework of the funding of large-scale facilities according to GG§91.

References

[1] Martin Krauß 2006 Zur thermischen Ermüdung der Magnesiumbasislegierungen AZ31 und AZ91 *Forschungsberichte aus dem Institut für Werkstofftechnik und Metallische Werkstoffe der Universität Kassel Band 8* (Kassel: University Press)
[2] Frommeyer G, Brüx U and Neumann P 2003 Supra-Ductile and High-Strength Manganese-TRIP/TWIP Steels for High Energy Absorption Purposes. *ISIJ Int.* **43** p 438
[3] Grässel O, Krüger L, Frommeyer G and Meyer L W 2000 High strength Fe-Mn-(Al, Si) TRIP/TWIP steels development - properties - application *Int. J. Plast.* **16** p 1391
[4] Gutierrez-Urrutia I, Zaefferer S and Raabe D 2010 The effect of grain size and grain orientation on deformation twinning in a Fe–22wt.% Mn–0.6wt.% C TWIP steel *Mater. Sci. Eng. A* **527** p 3552
[5] Hertzberg R W, Vinci R P and Hertzberg J L 2012 *Deformation and Fracture Mechanics of Engineering Materials, 5th Edition* (Wiley)
[6] Meyers M A, Vöhringer O and Lubarda V A 2001 The onset of twinning in metals: a constitutive description *Acta Mater.* **49** p 4025
[7] Foley F B and Howell S P 1922 Neumann Bands as Evidence of Action of Explosives upon Metal *Trans. Am. Inst. Min. Metall. Eng.* **1186** p 1
[8] Ecke M, Wilke M, Heyse H, Wendt U, Krüger M, Schwarz F and Krüger L 2014 Characterization of deformation twins in meteorites and dynamically loaded ferrous materials by means of electron backscatter diffraction, EBSD *Prakt. Metallogr. Metallogr.* **51** p 765
[9] Christian J W and Mahajan S 1995 Deformation Twinning *Prog. Mater. Sci.* **39** p 1
[10] Mahajan S 2002 Deformation Twinning *Encycl. Mater. Sci. Technol.* p 1
[11] Guo Y, Wang Y and Zhao D 2007 Atomistic simulation of stress-induced phase transformation and recrystallization at the crack tip in bcc iron *Acta Mater.* **55** p 401
[12] Barnett M R 2008 A rationale for the strong dependence of mechanical twinning on grain size *Scr. Mater.* **59** p 696
[13] Hirth; J P and Lothe J 1972 The influence of grain boundaries on mechanical properties *Metall. Trans.* **3** p 3047
[14] Cottrell A H and Bilby B A 1951 A mechanism for the growth of deformation twins in crystals

IOP Conf. Series: Materials Science and Engineering **882** (2020) 012007 doi:10.1088/1757-899X/882/1/012007

Philos. Mag. **42** p 573
[15] Sleeswyk A W 1963 ½<111> screw dislocations and the nucleation of {112}<111> twins in the b.c.c. lattice *Philos. Mag.* **8** p 1467
[16] Sleeswyk A W 1974 Perfect dislocation pole models for twinning in the f.c.c. and b.c.c lattices *Philos. Mag.* **29** p 407
[17] Lagerlöf K P D 1993 On deformation twinning in b.c.c. metals *Acta Metall. Mater.* **41** p 2143
[18] Priestner R and Leslie W C 1965 Nucleation of deformation twins at slip plane intersections in B.C.C. metals *Philos. Mag.* **11** p 895
[19] Ogawa K 1965 Edge dislocations dissociated in {112} planes and twinning mechanism of b.c.c. metals *Philos. Mag.* **11** p 217
[20] Price P B 1961 Nucleation and groth of twins in dislocation-free zinc crystals *Proc. R. Soc. A Math. Phys. Sci.* **260** p 251
[21] R.E. Reed-Hill 1964 *Deformation Twinning : proceedings of a conference sponsored by the Metallurgical Society*
[22] Christian J W 1968 Twinning : Crystallography *Encycl. Mater. Sci. Technol.* p 9423
[23] Ecke M, Michael O, Wilke M, Hütter S, Krüger M and Halle T 2020 Deformation Twinning in bcc Iron - Experimental Investigation of Twin Formation Assisted by Molecular Dynamics Simulation *Plasticity, Damage and Fracture in Advanced Materials* (Springer) p 43
[24] Ojha A, Sehitoglu H, Patriarca L and Maier H J 2014 Twin migration in Fe-based bcc crystals: Theory and experiments *Philos. Mag.* **94** p 1816
[25] Hull D and Bacon D J 2011 *Introduction to Dislocations* (Elsevier)
[26] Vítek V 1970 On the stability of stacking faults in B.C.C. crystals *Philos. Mag.* **21** p 1275
[27] Vítek V 1968 Intrinsic stacking faults in body-centred cubic crystals *Philos. Mag.* **18** p 773
[28] Yan J-A, Wang C-Y and Wang S-Y 2004 Generalized-stacking-fault energy and dislocation properties in bcc Fe: A first-principles study *Phys. Rev. B* **70** 174105

IOP Conf. Series: Materials Science and Engineering **882** (2020) 012008 doi:10.1088/1757-899X/882/1/012008

Investigations on metallurgical and mechanical properties of resistant spot welded dissimilar joints between a high manganese steel and a low alloyed steel grade

C Frohwein[1], M Otto[2], M Ecke[2], S Jüttner[2] and T Halle[2]

[1] Volkswagen AG, 38440, Wolfsburg, Germany

[2] Otto-von-Guericke-University Magdeburg, Faculty of Mechanical Engineering, Institute of Materials and Joining Technology, Universitätsplatz 2, 39106 Magdeburg, Germany

chris.frohwein@volkswagen.de

Abstract. The present investigation gives a brief overview into the correlation of the amount of molten ferrite and metallurgical as well as mechanical properties of resistance spot welded austenitic-ferritic joints. The fusion rate describes the ferritic quantity in the weld metal and provides a reliable reference to the integrity of dissimilar joints. Metallurgical effects due to varied fusion rates are shown by element distribution, micro hardness measurement and phase analysis. In addition, this work states specific fusion rates which lead to a degradation of the mechanical properties and can be used as a basis for application guidelines. This can be seen in certain tensile tests with single and multiple spot specimens.

1 Introduction

The use of new lightweight materials for the body in white structure is an important aspect regarding the weight reduction targets of the automotive industry. In order to reduce emissions and manufacturing costs the development of high strength steels is an important factor. The implementation of these materials in future electro mobility concepts proves to be even more significant. High manganese steels with their full austenitic structure offer a special property profile better than conventional steel grades. The TWIP Effect (Twinning Induced Plasticity) allows for extraordinary ductility combined with tensile strength higher than 1.000 MPa [1].

Contrary to conventional austenitic FeCrNi based steels FeMn alloys contain high amounts of carbon, silicon, aluminum and of course manganese for stabilizing the austenitic phase [2]. The weldability of these grades has become an important aspect for the processing sector. Especially heterogeneous spot welded joints between austenitic and ferritic sheets represent a great challenge for the automotive production. It is well known that the chemical composition of the base material determines the stacking fault energy Γ (SFE) and thus ultimately the hardening mechanism [3]. This aspect should therefore also be transferable to the common weld metal. Previous investigations have shown the presence of brittle martensitic phases which often lead to an early failure [4]. Consequently, it is necessary to estimate the influence of this embrittlement on the spot integrity of dissimilar joints.

As illustrated in figure 1, dissimilar spot welds of austenitic and ferritic steels show a deviating behavior in terms of nugget formation. Due to the higher thermal conductivity of low alloyed grades with their body centered cubic lattice (bcc) nuggets grow predominantly in direction of the austenitic steel and form asymmetrical weld metals as well as heat affected zones. This goes also along with unequal crystallization rates within the nugget.

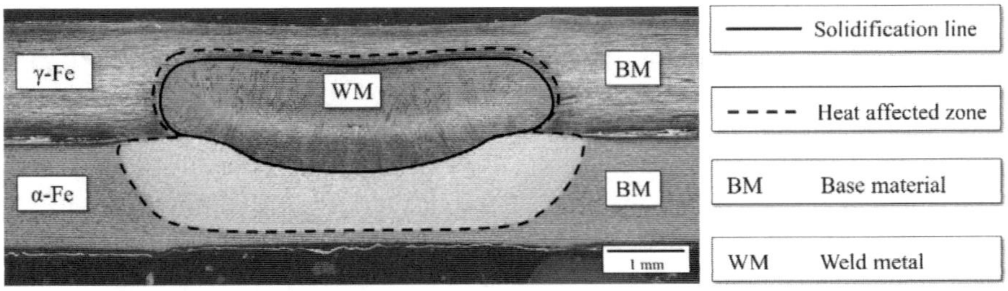

Figure 1. Cross section of an austenitic-ferritic spot welded joint.

2 Materials and experimental setup

The used austenitic base material is an uncoated innovative FeMn steel with a tensile strength of $R_m \geq 950$ MPa and an elongation of $A_{80} \geq 41$ %. The ferritic joining partner is a widespread zinc coated micro alloyed steel. The tensile strength is $R_m \geq 410$ MPa and the elongation is $A_{80} \geq 23$ %. Both sheets are used in a thickness of 1.5 mm. The chemical composition is presented in table 1.

Table 1. Chemical composition of test materials [in wt.%].

Steel	C	Mn	Al	Si	Cr	Ni	Ti	Nb	V	Fe & other
FeMn	0.64	17.0	1.8	0.05	0.01	0.007	0.075	0.0002	0.002	bal.
Ferritic	0.06	0.53	0.037	0.222	0.088	0.046	0.003	0.014	0.002	bal.

To characterize the metallurgical properties specimens according to SEP 1220-2 [5] are used. For the welding current range and the element distribution the two coupons of dimension 45 mm x 45 mm with 40 mm overlap are joined by one single spot without shunt. Micro hardness measurement, phase analysis and destructive testing of mechanical properties is performed under quasi-static load with shear tensile and cross tensile specimens according to SEP 1220-2 [5] as well. In addition a multiple spot specimen is designed as illustrated in figure 2. This certain geometry should provide an impression over the load transfer capacity between the spot welds. In all cases the external load is applied by a static material testing machine Zwick Z100 with a testing speed of 1/10 mm/s.

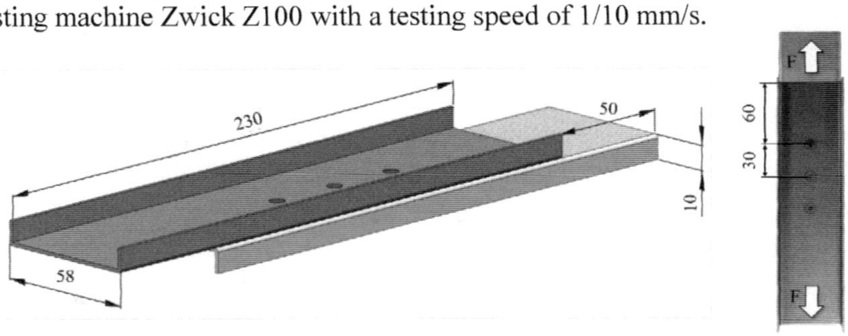

Figure 2. Multiple spot specimen (dimensions in mm).

For the element distribution the Electron Probe Microanalyzer (EPMA) JEOL JXA 880 L is used. The analysis is performed in a scan field of 900 x 900 µm with a step interval of 3 µm. The scan field is located at the notch tip of the spot weld between the sheets. This area is usually exposed to a distinctive stress under shear load [6].

The micro hardness measurement (HV 0.1) by UCI method (Ultra Sonic Impedance) is performed in the same area as the element distribution and done by the automated hardness scanner BAQ UT 200. The step interval is 100 µm.

For the phase analysis by EBSD (Electron Backscatter Diffraction) the DualBeam FEI Scios equipped with AMETEK-EDAX analytic is applied. The acceleration voltage is 20 kV and for the step size a distance of 200 nm is chosen.

3 Definition and adjustment of fusion rates

This investigation into dissimilar γ-α-joints defines the fusion rate Φ as the quantity of molten ferrite referring to the entire weld metal. According to equation (1), the determination of Φ is made in a polished cross section by forming the area ratios of molten ferrite and austenite as illustrated below in figure 3.

$$\Phi = \frac{A_{Ferrite}}{A_{Ferrite} + A_{Austenite}} \tag{1}$$

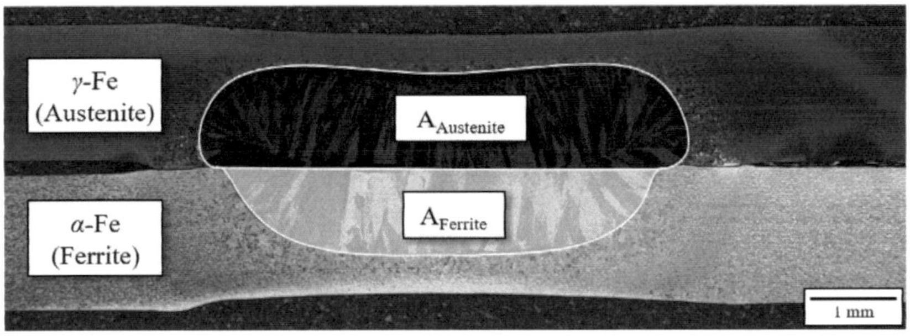

Figure 3. Fusion rate determination by area ratios in a cross section of a dissimilar γ-α-joint.

For the following investigations two fusion rates are considered. The adjustment is made by applying commercial CrNi process tapes in a thickness of 0.15 mm between the electrode cap and the sheet metal as shown in figure 4. In the first case one tape is applied on the austenitic side of the joint and in the second case two tapes are applied on the ferritic side. With this procedure two certain fusion rates of $\Phi = 13 \pm 3$ % and $\Phi = 50 \pm 3$ % are achieved. The welding parameters are shown in table 2.

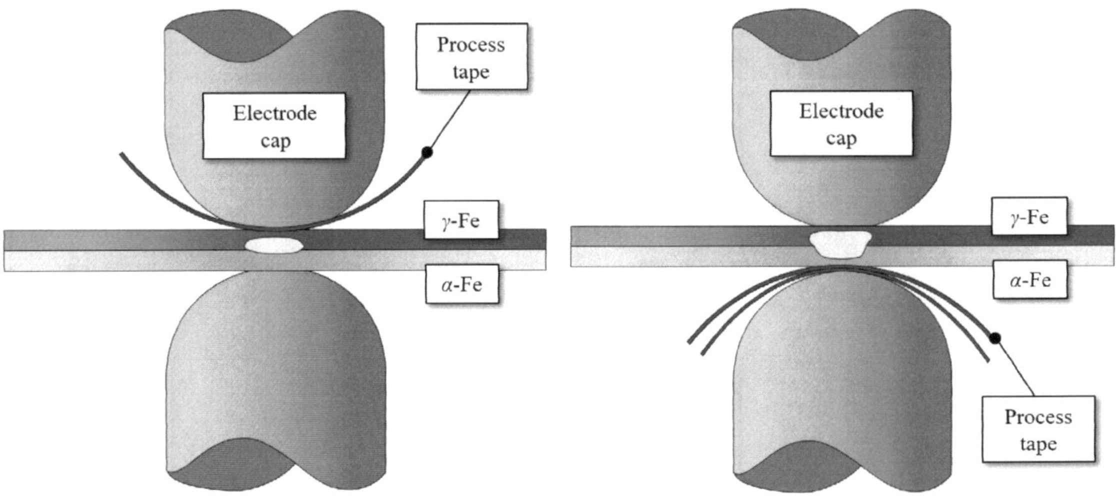

Figure 4. Adjustment of fusion rates with process tapes (left: $\Phi = 13$ %, right: $\Phi = 50$ %).

In order to evaluate an influence of the fusion rate on the mechanical properties of resistance spot welds the welding parameters are set in a certain way to obtain a uniform nugget diameter of 5 ± 0.1 mm. This ensures a comparability and provides more information about the joint properties.

Table 2. Welding parameters according to fusion rate.

Φ [%]	Electrode Force [kN]	Squeeze Time [ms]	Prepulse Time [ms]	Prepulse Current [kA]	Pause Time [ms]	Welding Time [ms]	Weld Current [kA]	Hold Time [ms]
13	4	500	200	4	200	400	5,8	300
50	4	500	200	4	200	400	5,0	300

4 Metallurgical properties

During welding the chemical composition of dissimilar metal welds in the common fusion zone must be expected to differ from that of the base materials. In the present study two certain fusion rates of $\Phi = 13$ % and $\Phi = 50$ % are considered. Therefore, it is necessary to first characterize the distribution of the main alloying elements and then measure the resulting micro hardness level in areas which determine the integrity of the joint. For a better understanding of microstructural mechanism which lead to an early failure of the joint a phase analysis is carried out. The hardness measurement as well as the phase analysis is done under as-welded conditions and in certain prestrained states.

4.1 Element distribution

Figure 5 shows a quantitative course of Fe, Mn and Al in sheet thickness direction in the center of the nugget at a maximum fusion rate of $\Phi = 50$ %. At the boundaries γ-Fe (I) → weld metal (II) and weld metal (II) → α-Fe (III) increases in Fe concentration occur as expected. The Mn content of the weld metal is significantly reduced due to the fusion process. Within the weld metal (II), however, no significant gradient in sheet thickness direction can be detected on a macroscopic level which is representative for an even course. Only the detected cavity in the center influences the course. For Al there is no difference in concentration at the selected resolution, neither at the transitions nor within the weld metal.

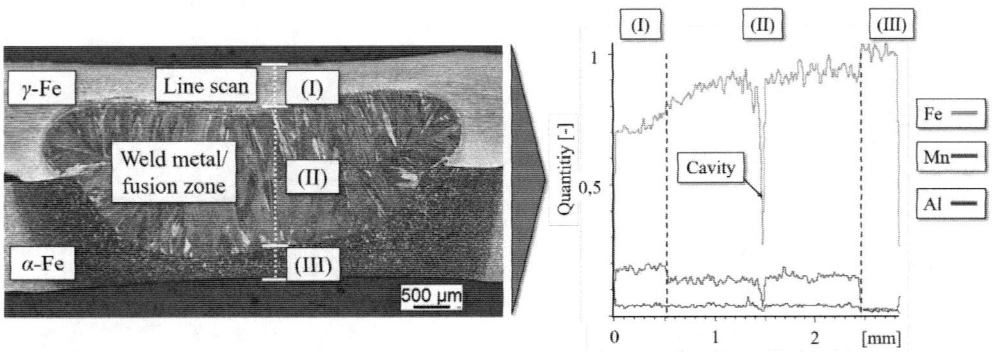

Figure 5. Cross section with EDX line scan in the center of the nugget ($\Phi = 50$ %).

To assess the element distribution at the microstructural level, electron probe microanalysis (EPMA) is carried out in discrete areas of the nugget as illustrated in figure 6. The analysis shows the distribution of the elements Fe, Mn and Al determined for a minimum fusion rate of $\Phi = 13$ % (left) and a maximum of $\Phi = 50$ % (right).

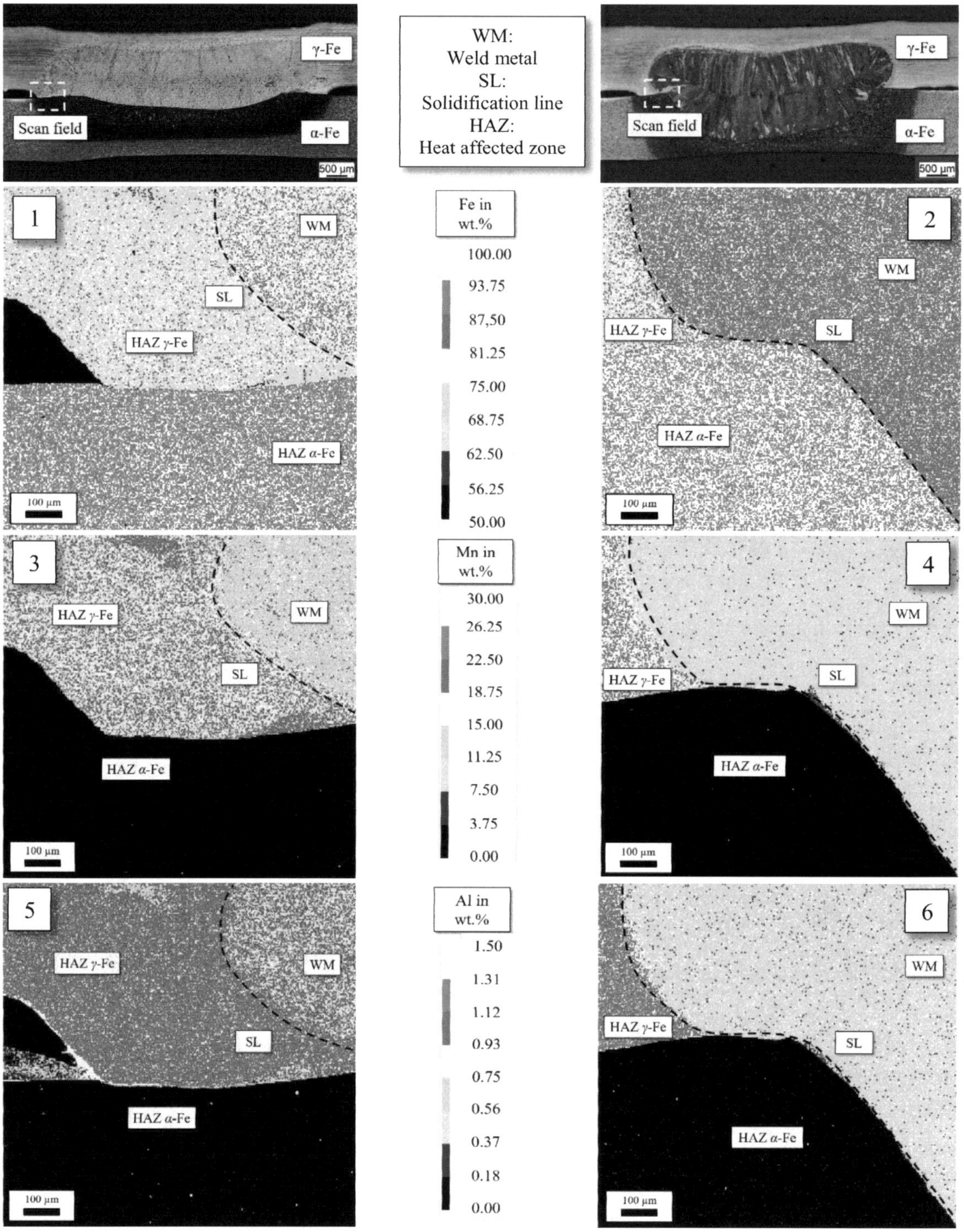

Figure 6. Distribution of alloying elements Fe (1 + 2), Mn (3 + 4) and Al (5 + 6) at the notch tip of the spot weld (Micro analysis; left: Φ = 13 %; right: Φ = 50 %).

The analysis is carried out in the notch tip of the joint and shows parts of the base materials, the heat affected zones and the weld metal. According to [6], the area shown in the figure represents the most stressed area for resistance spot welds and is therefore of particular interest.

It is obvious that the chemical compositions of the weld metals differ significantly from those of the base materials, which were displayed in table 1. Part 1 and 2 of figure 6 show the Fe distribution for the areas of heat affected zone (HAZ) and weld metal (WM) of the γ-α-joint separated by the solidification line (SL). While in part 1 at $\Phi = 13$ % the Fe content in the HAZ of the FeMn steel (HAZ γ-Fe) is about 80 % on average, values of 87 % can be detected in the weld metal. In part 2 at $\Phi = 50$ % the Fe content in the weld metal is even above 90 %. This is accompanied by a change in the proportions of the main alloying elements Mn, Al and C. With an increasing fusion rate the concentration of the main alloying elements is reduced in the weld metal.

In the analysis of Mn clear differences between the two fusion rates can be seen. The average Mn content of the austenite stabilizing element in the base material is about 17 wt.%. When comparing the Mn contents of the weld metals (WM) from part 3 ($\Phi = 13$ %) and part 4 ($\Phi = 50$ %) a significant difference in concentrations become apparent as a result of the increasing fusion rate. While large areas in the weld metal (WM) are between 11 - 18 wt.% of Mn at a low fusion rate the Mn content decreases to values between 7 - 11 wt.% and in some cases even below this at a high fusion rate.

According to the segregation behavior of Mn figure 7 shows certain areas of Mn concentration. At a high fusion rate areas of Mn enrichment and depletion lie directly next to each other. This observation is confirmed by the investigations of [7] and [8] in which Mn follows the solidification structure. In interdendritic spaces Mn is enriched while in the dendrite itself Mn depletion occurs.

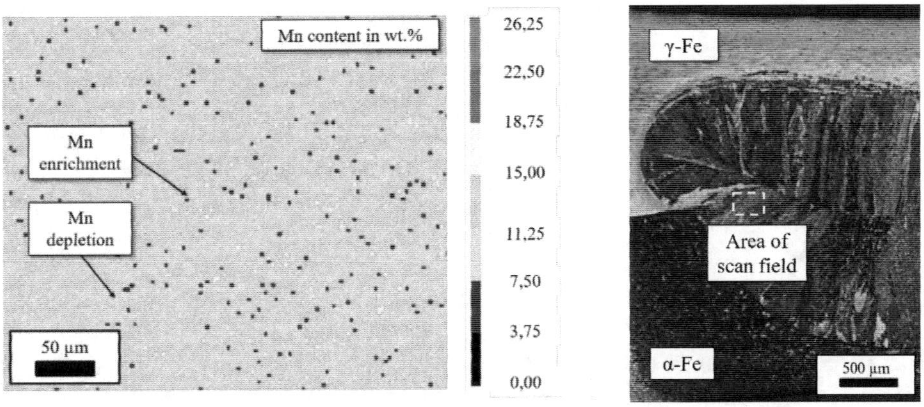

Figure 7. Enrichment and depletion of Mn in the weld metal (Micro analysis at $\Phi = 50$ %).

These differences in the concentration of Mn on microstructural level can lead to different hardening mechanisms in the respective areas. An important aspect in this context is the SFE which is dependent on the alloying elements. In the literature are different values for the SFE in which a change between hardening mechanisms take place [9-12]. According to Jin et al. [13] the base material used in the tests has a SFE of $\Gamma = 33$ mJ/m^2 and is therefore a full austenitic TWIP steel. Taking into account the research results of Song et al. [14] the limit between the formation of mechanical twins and strain induced martensite is $\Gamma = 20$ mJ/m^2. Accordingly, following the mechanism maps of Song et al. [14] a fusion rate of $\Phi = 13$ % would result in a weld metal which is still in range of the TWIP effect. At the higher fusion rate of $\Phi = 50$ % the SFE is below $\Gamma = 20$ mJ/m^2. The resulting hardening mechanism in the weld metal would therefore be the TRIP effect.

The Al concentration of the base material is about 1.5 wt.% (table 1). At $\Phi = 13$ % the increasing ferrite content in part 5 shows that the Al content in the weld metal (WM) decreases to values between 0.7 - 1.1 wt.%. At a fusion rate of $\Phi = 50$ % Al is further reduced to values between 0.3 - 0.9 wt.% as shown in field 6. The investigations of Senk et al. [15] show that the element is subject to a certain solidification kinetics. As a result of this, Al is deposited anticyclically in the microstructure during phase formation. In areas of Mn depletion, therefore, Al enrichment takes place and vice versa. In combination with the concentration differences of Mn the behavior shown schematically in figure 8 occurs. As a result, an increased Al content in the dendrite can primarily lead to a formation of ferrite or martensite.

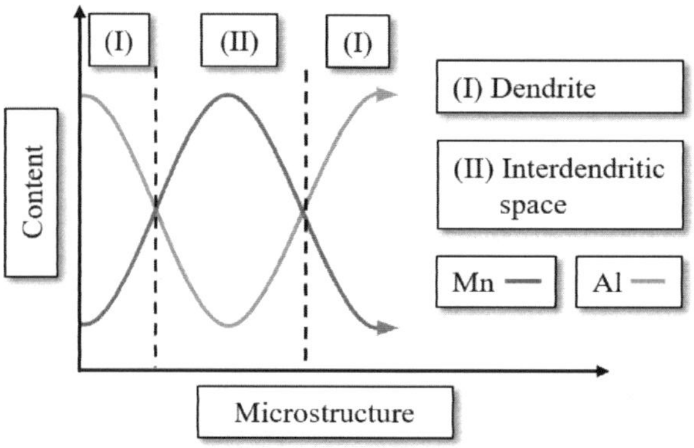

Figure 8. Anticyclically solidification behavior of Mn and Al in the weld metal.

4.2 Micro hardness

The previous observation has shown that depending on the fusion rate an increased ferrite content in the weld metal consequently leads to a chemical composition which differs from the one of the base materials. According to this result, the hardening behavior of the weld metal and the surrounding areas is to be further investigated. For this purpose, conventional shear tensile test specimens according to [5] are subjected to different external loads after welding so that discrete strain states are created. The samples are prestrained in the quasi-static range to certain force levels. Table 3 shows the unstrained initial state (F_0) and three further values at 4 kN (F_1), 8 kN (F_2) and F_3. The state F_3 is just below the maximum force of the welded specimen. It was determined in preliminary tests and specified as $0.9 \cdot F_{max}$. It can be seen that different forces are achieved at this state. While the rupture of the specimens with a large quantity of ferrite ($\Phi = 50$ %) already starts at $F_{max} = 9.8$ kN, the material separation at a lower fusion rate ($\Phi = 13$ %) begins at about $F_{max} = 12.6$ kN. This behavior according to the mechanical properties is discussed later in section 5. Subsequently, an UCI hardness measurement is performed on both unstrained and prestrained welded joints as illustrated in figure 9.

Table 3. Prestrained conditions of UCI-specimens for hardness measurements.

State	Fusion rate $\Phi = 13$ %	Fusion rate $\Phi = 50$ %
F_0	0 kN	0 kN
F_1	4 kN	4 kN
F_2	8 kN	8 kN
$F_3 = 0.9 \cdot F_{max}$	11.3 kN	8.8 kN

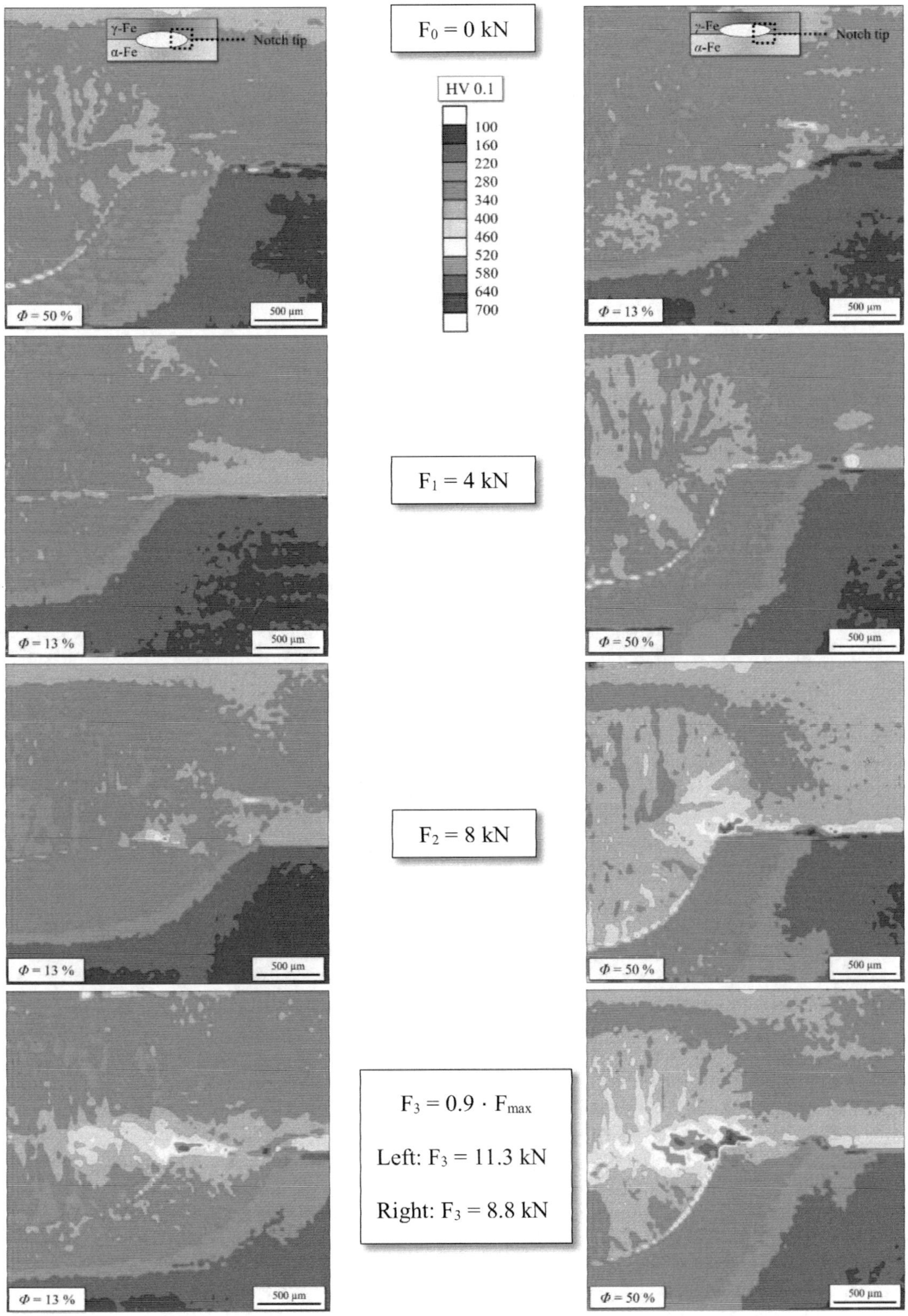

Figure 9. Micro hardness (UCI) at the notch tip of the joint (left: $\Phi = 13$ %; right: $\Phi = 50$ %).

The comparison of the UCI measurements of both fusion rates show clear differences in the hardness distribution in the weld metal with increasing prestraining states. While the hardness of the weld metal with a low fusion rate of ɋ = 13 % for the state F_0 is approximately at the base material level of the FeMn steel, a brittle, seam-like area forms along the austenitic-ferritic fusion line at ɋ = 50 %. In addition, hardness differences can be seen directly in the weld metal along the solidification structure.

In the prestrained state F_1, a homogeneous hardness level is still evident at ɋ = 13 %. No hardness peaks can be detected at the solidification line, nor are pronounced gradients formed along the dendritic solidification structure. For this condition it can be stated that no deformation induced phase transformation in the microstructure of the weld metal has occurred, which can be verified by means of hardness measurement. This is due to the fact that the high proportion of austenite stabilizing elements in the weld metal, in addition to the usual gliding of dislocations, primarily supports the formation of mechanical twins. However, before the TWIP effect is initiated in the material, a critical dislocation density must be exceeded. According to Fonstein [16] and Remy [17], this value depends on the SFE and other factors such as temperature and should therefore be considered for each alloying concept. In the state F_1 the first signs of hardening can already be seen in the weld metal at ɋ = 50 %. No change can be documented in the seam along the fusion line which is present from the beginning. On the other hand, an increase in hardness is observed in the weld metal in the stressed areas around the notch tip of the joining plane.

The state F_2 for ɋ = 13 % differs only slightly from case F_1. A slight increase in hardness can be measured in the notch tip of the joining plane. A hardness peak can be detected in one measuring point. In contrast, the measurement with a high fusion rate of ɋ = 50 % shows a significant increase in hardness in the weld metal. Large areas are between 340 HV and values of 460 HV. Directly in the notch tip even values between 520 HV and 700 HV are achieved. The evidence of such brittle microstructure suggests at this point that, in addition to the increase in dislocation density, deformation induced martensite has occurred due to the high stress. According to the high quantity of ferrite in the weld metal, the SFE decreases and the originally full austenitic phase is no longer sufficiently stabilized. Metastable austenite is formed which transforms under stress. To what extent this transformation is ε or α' martensite cannot be further determined by the UCI measurement. If the results of Schumann [18] are taken into account, the area in which the indirect phase transformation of γ (fcc) $\rightarrow$ ε (hcp) $\rightarrow$ α' (bcc) takes place is in a concentration range between 10 and 15 wt.% Mn.

In the last state considered (F_3) the hardness measurement is carried out after a load immediately before failure of the specimens. At ɋ = 50 % a hardness between 500 and 700 HV can be determined in the area in front of the notch tip at a shear tensile force of F_3 = 8.8 kN. This area shows clear signs of embrittlement. Also at $0.9 \cdot F_{max}$, an increase in hardness at a similar point in the weld metal can be demonstrated at ɋ = 13 %. In absolute terms this condition is reached at F_3 = 11.3 kN. This delayed hardening can be explained by the dynamic Hall-Petch effect. In which the twin density in the austenite grains increases with increasing elongation, thus reducing the free path between the twins. The distances available for dislocation glide become shorter until finally a blockage occurs [16], [17].

4.3 Phase analysis

In the investigations carried out so far, differences in the concentrations of the main alloying elements as well as in the mechanical properties on a microstructural level in the highly stressed areas of the nugget could be demonstrated. The question of whether hardening is due to an increase in dislocation or twin density or martensite formation will be investigated by EBSD analysis. A decisive influencing factor is the chemical composition of the weld metal. Based on this, the two fusion rates of the states F_0 = 0 kN and $F_3 = 0.9 \cdot F_{max}$ shall be considered in accordance with table 3. These four specimens are used to investigate how the weld metal behaves during mechanical loading. Furthermore, the question is to be answered, which microstructure modifications support an early failure and whether a change of the hardening mechanisms occurs.

EBSD analysis can be used to determine the existing phases and their orientation. In addition to the thermally induced phase formation during welding, alloys with a high Mn content can also undergo a

IOP Conf. Series: Materials Science and Engineering **882** (2020) 012008 doi:10.1088/1757-899X/882/1/012008

phase transformation induced by deformation. Thus, lattice shear during martensite formation can be facilitated by an externally applied load. As a consequence it is difficult to characterize the resulting microstructure and its mechanical properties, since aspects such as dislocation and twin density as well as SFE play a role. These mechanisms can influence each other during the loading process.

Figure 10 shows the phase analysis for the state F_0 of the two fusion rates $\Phi = 13$ % and $\Phi = 50$ %. In this case, the scan is carried out on a part of the fusion line near the notch tip. For $\Phi = 13$ %, the left part of the figure shows a clear separation of both phases from each other along the solidification line. From the ferritic HAZ only a small amount of ferrite has grown into the weld metal. In the right part of figure 10, however, it is clearly visible that the proportion of ferrite increases at a higher fusion rate. In this area the distribution is island-like within the austenite phase.

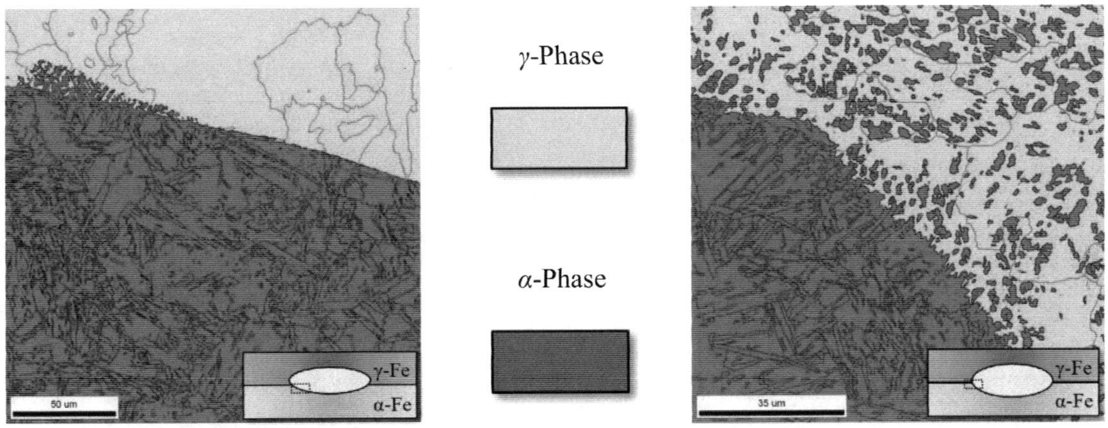

Figure 10. Phase analysis at the notch tip of the joining plane at $F_0 = 0$ kN (left: $\Phi = 13$ %; right: $\Phi = 50$ %).

Figure 11 shows the scan field in the center of the nugget. The phase analysis of the two fusion rates in reference condition F_0 show clear differences in the proportion of ferrite in the weld metal. The left part of the figure shows a full austenitic microstructure for $\Phi = 13$ %. No modification of the α-phase can be detected at this point. In contrast, in the right part of the figure at $\Phi = 50$ %, it can be seen that the proportion of ferrite increases significantly. The formation of ferrite follows the solidification structure and takes place for the most part in the dendrites.

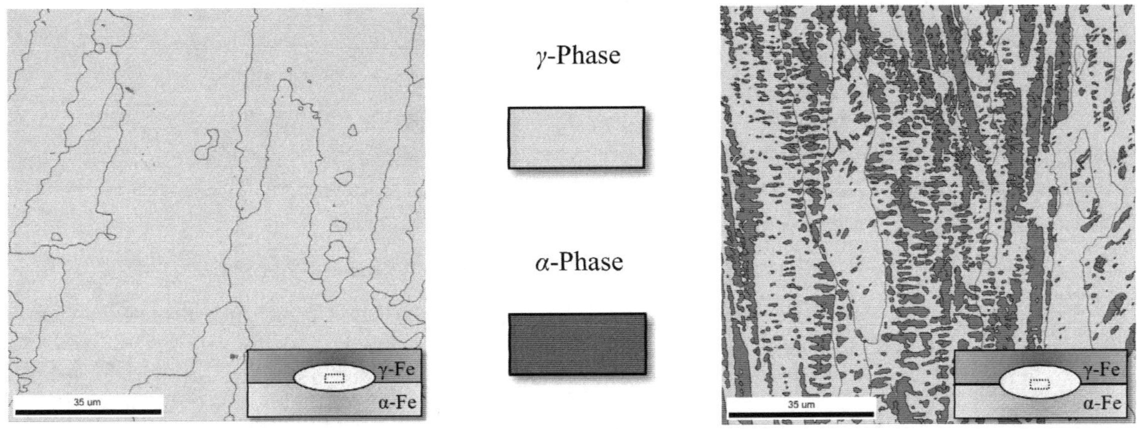

Figure 11. Phase analysis in the center of the weld metal at $F_0 = 0$ kN (left: $\Phi = 13$ %; right: $\Phi = 50$ %).

Due to the high crystallographic similarity of ferrite and martensite, the detection and difference of both phases is not trivial, even with electron diffraction based methods. As in the present case, the conclusion that martensite is necessarily involved cannot be drawn directly from the detection of α-Fe in the weld metal. Furthermore, the question about the origin (deformation or thermally induced) and the type (α'- or ε-martensite) is remains open.

The misorientation relationships between the austenitic and martensitic modifications are subject to discrete laws that describe a parallelism between planes and directions of the face-centered cubic austenitic (fcc) and tetragonal body-centered martensitic (bct) cell as shown in table 4. On the basis of the misorientation distributions between the different phases measured by EBSD analysis, a pronounced maximum between 42° and 46° can be seen in figure 12, which essentially corresponds to the range described by the misorientation relationships listed in table 4. This supports the assumption that the occuring phase is martensite.

Table 4. Misorientation relationships between austenite and martensite [19].

Misorientation relationship	Misorientation angle	Reference
Kurdjumow-Sachs	42,85°	[20]
Greninger-Troiano	44,23°	[21]
Nishiyama-Wassermann	45,98°	[22]

Furthermore, the statement can be derived that the transformation into a cubic cell has taken place. The detected phase is therefore α'-martensite, because ε-martensite has a hexagonal cell and could not be detected in the weld metal.

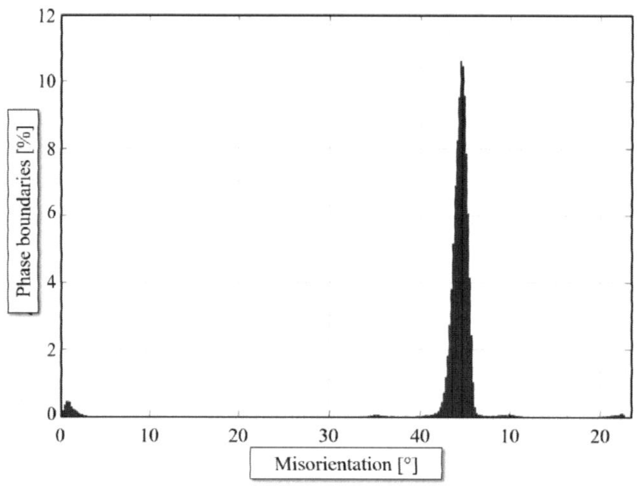

Figure 12. Misorientation distribution between the fcc and bct phases.

In the following investigation two metallurgical states are subjected to an external load. Before the phase analysis, the samples are prestrained to $F_3 = 0.9 \cdot F_{max}$ as described in the previous section. For $\Phi = 13$ % the EBSD analysis is shown in figure 13. The image quality (IQ) map shows that deformation twins in the austenitic phase have occurred due to stress. Only in a few spots martensitic formation can be detected along the twin structures. The weld metal consists almost entirely of austenite.

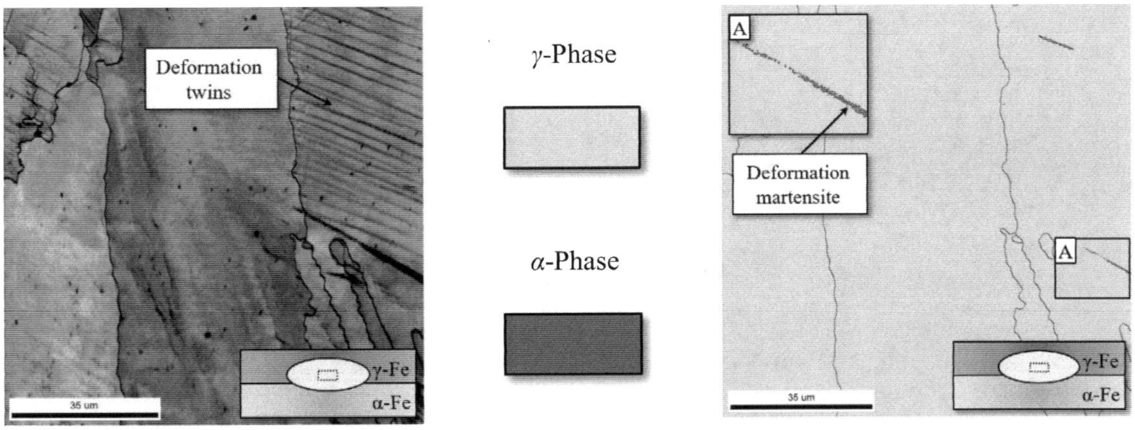

Figure 13. IQ map (left) and phase analysis (right) in the weld metal of a prestrained specimen (Φ = 13 %; F = 0.9 · F_{max}).

A similar observation can be made in figure 14 along the joining plane between ferrite and austenite. The highly stressed area in the weld metal shows also a significant increase in twin density. This is an indicator that the SFE of the microstructure in the fusion zone is still in range of the TWIP effect as the primary hardening mechanism.

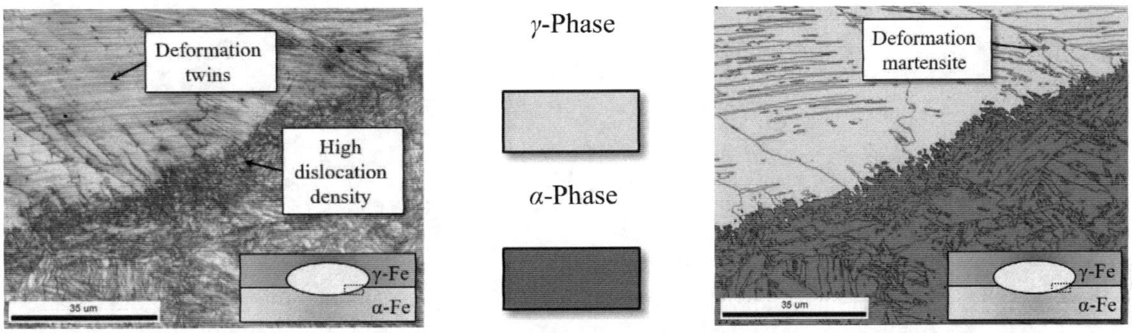

Figure 14. IQ map (left) and phase analysis (right) at the solidification line of a prestrained specimen (Φ = 13 %; F = 0.9 · F_{max}).

For the higher fusion rate figure 15 shows that several effects are superimposed during loading. In addition to the thermally formed α'-martensite, there is also deformation induced martensite. Due to the energetically better conditions, this deformation induced martensite is deposited primarily along the deformation twins within the grains [23]. According to figure 16, both types of this microstructure can be detected at the crack tip of the prestrained sample. From this it can be deduced that this effect supports the failure of the joint. The thermal martensite resulting from the welding process, in combination with the deformation induced martensite and twins resulting from the loading, can cause a ductility loss of the structure that crack formation occurs along the microstructural interface between areas of different mechanical properties.

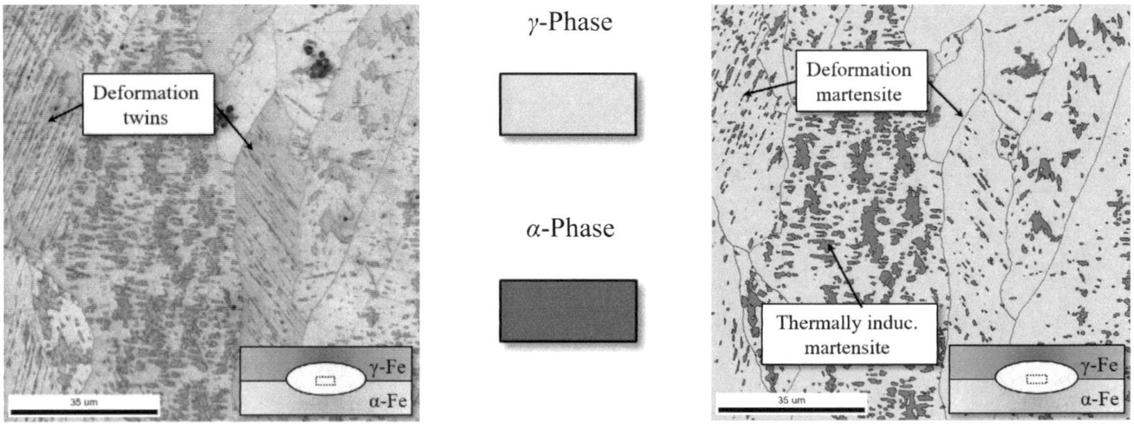

Figure 15. IQ map (left) and phase analysis (right) in the weld metal of a prestrained specimen (Φ = 50 %; F = 0.9 · F_{max}).

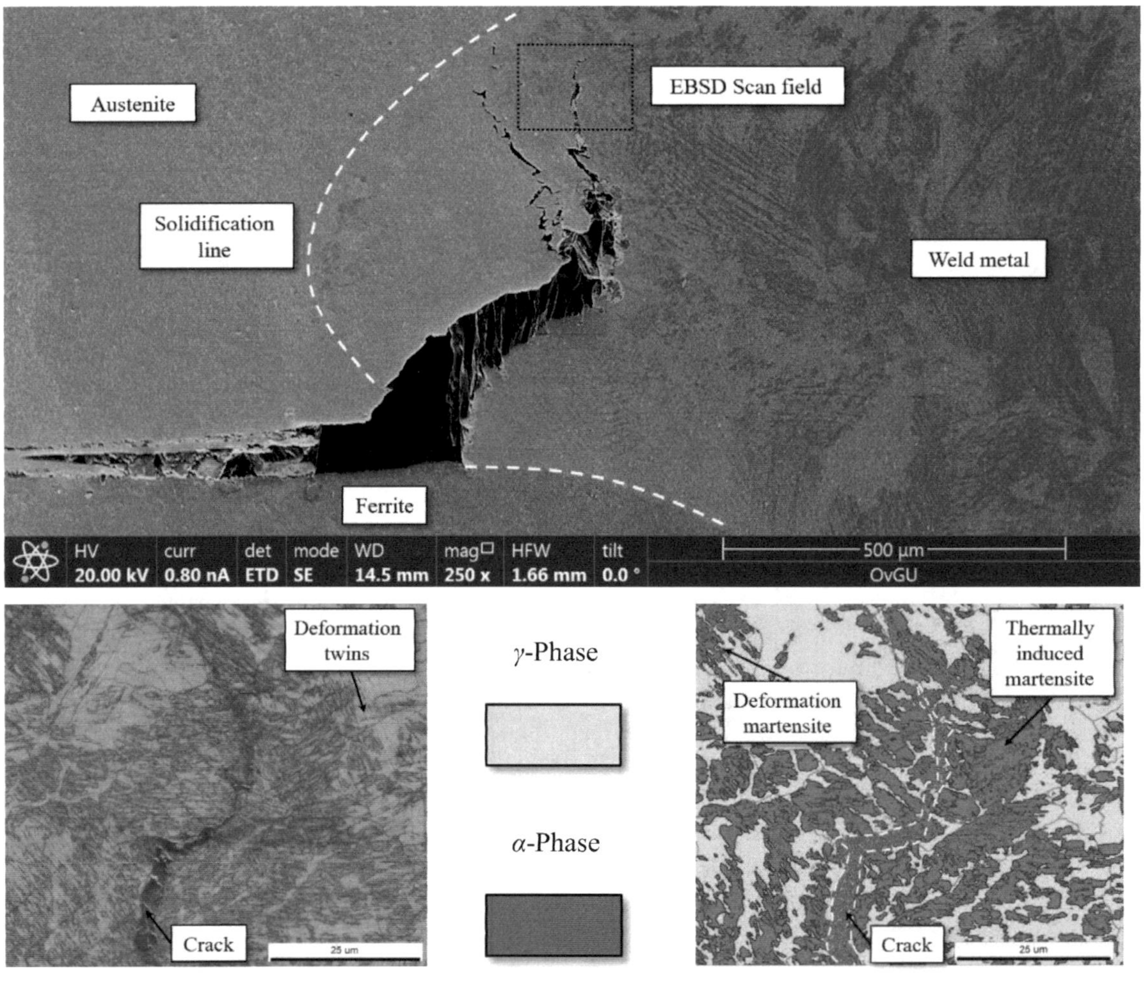

Figure 16. IQ map (left) as well as SEM illustration (above) and phase analysis (right) at the notch tip of a crack of a prestrained specimen (Φ = 50 %; F = 0.9 · F_{max}).

5 Mechanical properties

In the previous section, metallurgical aspects were examined with regard to different fusion rates. It was demonstrated what resulting weld metal can be expected with an increasing quantity of ferrite. In this part, the effects of material degradation on the load capacity are examined. In order to determine the mechanical properties of the resistance spot welded joints shear and cross tensile tests under quasi-static load are carried out.

5.1 Shear and cross tensile properties

According to the joint properties, figure 17 represents a comparison of different fusion rates under shear and cross tensile load. For both cases the results show a reduced load capacity with an increasing fusion rate. Although the shear tensile force decreases from 12.6 kN at $\triangleleft$ = 13 % to 9.8 kN at $\triangleleft$ = 50 % it only provides limited information about the load capacity. Compared to the determined energy under shear tensile stress, the force is only influenced on a lower level. In contrast, the difference of the energy absorption between both fusion rates is about 50 % and thus represents a more significant degradation of the investigated spot welds according to the field of application.

The cross tensile test shows a similar behavior. With an increasing fusion rate the cross tensile force decreases from 4.6 kN at $\triangleleft$ = 13 % to 1 kN at $\triangleleft$ = 50 %. As mentioned above the energy absorption capacity represents more reliable parameter to evaluate the mechanical properties. In the present case this means a decrease of 95 % from an average energy of 44.4 J at $\triangleleft$ = 13 % to 2.1 J at $\triangleleft$ = 50 %.

Furthermore, this investigation shows that the ductility of a spot weld is subjected to degradation. Figure 18 illustrates cross sections of failed spot welds after mechanical testing. Comparing the two fusion rates it can be concluded that the failure behavior of the specimen with a fusion rate of $\triangleleft$ = 13 % is much more ductile due to higher plastic deformation of the weld metal and the surrounding areas. The other specimen with a fusion rate of $\triangleleft$ = 50 % shows in contrast a brittle failure through the weld metal along the joining plane.

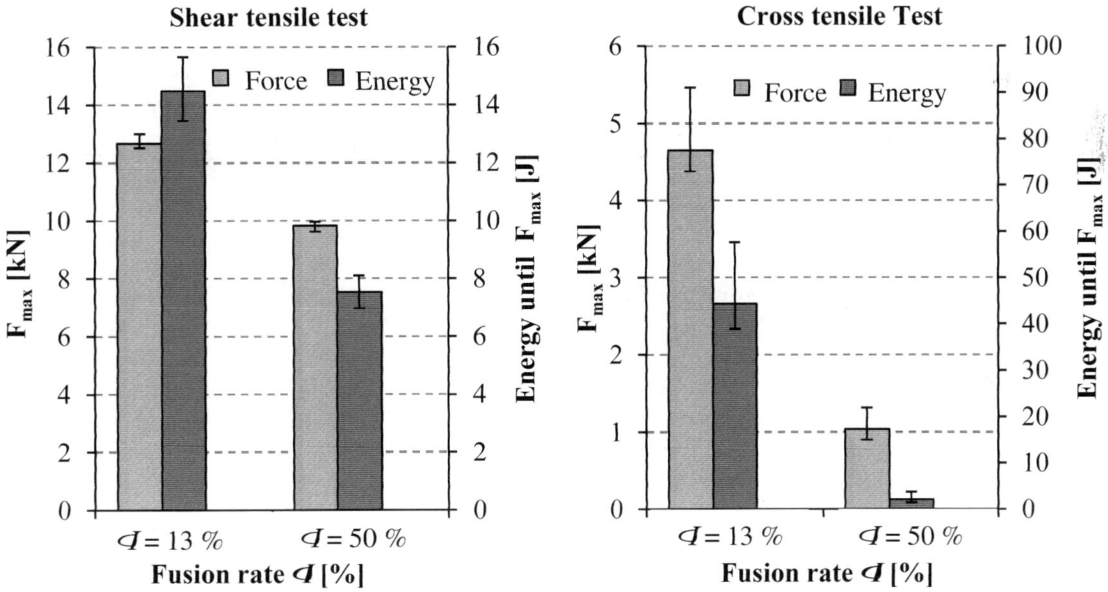

Figure 17. Shear tensile test (left) and cross tensile test (right) in dependence of $\triangleleft$.

IOP Conf. Series: Materials Science and Engineering **882** (2020) 012008 doi:10.1088/1757-899X/882/1/012008

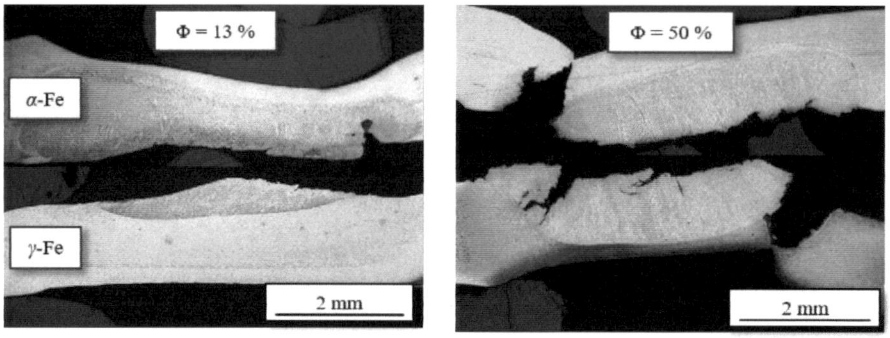

Figure 18. Cross sections of failed spot welds after shear tensile testing.

A similar behavior can be observed in the cross sections of figure 19. For $\Phi = 13$ % the failed spot weld shows a pull-out failure with a high degree of deformation due to the detected necking. For the higher fusion rate of $\Phi = 50$ % a brittle separation of the two joining partner through the weld metal takes place. Neither in the nugget nor in the HAZ or the BM an essential plastic deformation can be determined.

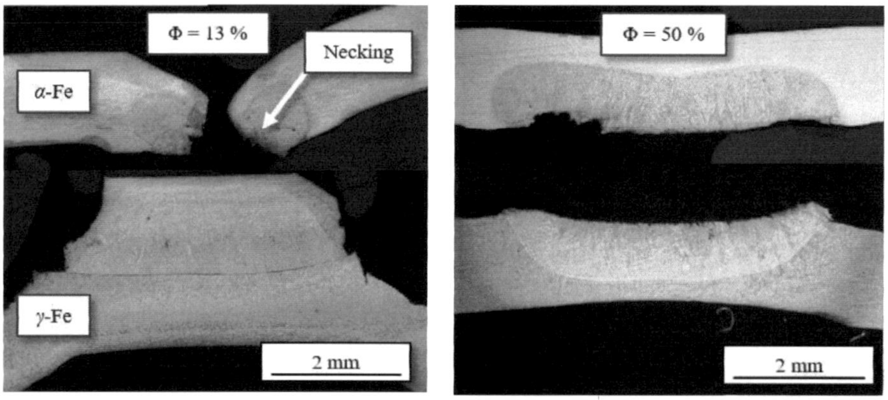

Figure 19. Cross sections of failed spot welds after cross tensile testing.

5.2 Multiple spot geometry

In order to reduce the degree of abstraction between a standardized shear test specimen and a resistance spot welded assembly, a multiple spot geometry as presented in figure 2 was used in the following experiment. The focus of the analysis is primarily on the influence of the fusion rate on the load transfer behavior between the individual spot welds. The results of the shear tensile tests are shown in figure 20. Each curve is a representative case for $\Phi = 13$ % and $\Phi = 50$ % out of n = 5 experiments.

IOP Conf. Series: Materials Science and Engineering **882** (2020) 012008 doi:10.1088/1757-899X/882/1/012008

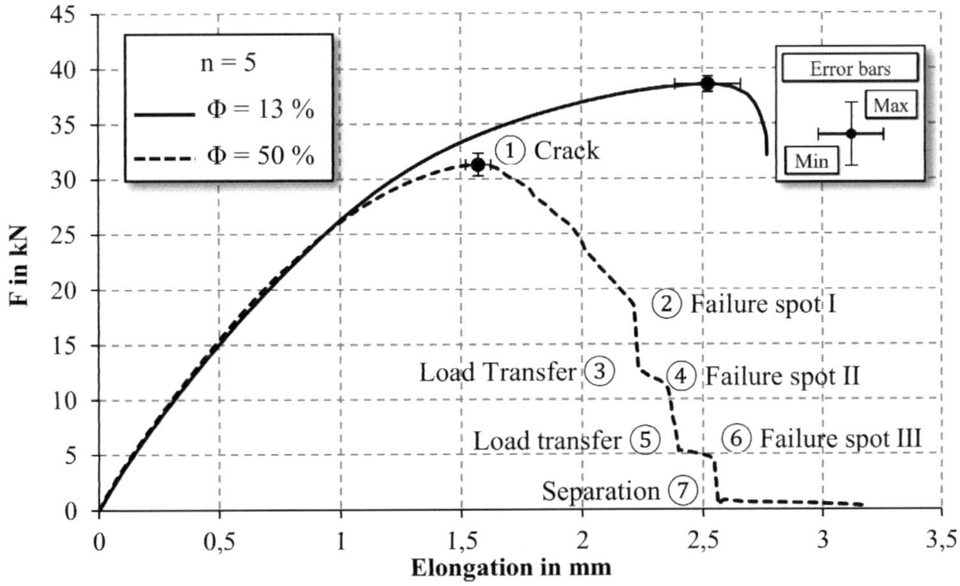

Figure 20. Shear tensile test of multiple spot geometry in dependence of Φ.

As shown in the previous section, due to a low fusion rate a high ductility of the weld metal enables a sufficiently large relative movement between the parts to be joined under shear tensile stress. This aspect leads to an even loading of all weld spots. For $\Phi = 13$ % assembly failure does not occur in the form of individual, successive failure, but occurs collectively after the tensile strength is exceeded.
In contrast, the increased ferrite content in the weld metal of $\Phi = 50$ % shows that the failure passes through seven discrete stages as follows:

① reaching F_{max} and crack initiation in spot I,
② failure of spot I,
③ load transfer and crack initiation in spot II,
④ failure of spot II,
⑤ load transfer and crack initiation in spot III,
⑥ failure of spot III and
⑦ complete failure.

This results in the so-called "zipper effect" at higher fusion rates. Although a load transfer to adjacent spot welds takes place, but only to a small extent into the base material, the first spot weld fails early. As a consequence of that the subsequent spot weld is also subjected to maximum stress until a brittle fracture occurs. This behavior continues until the components are finally separated. As illustrated in the graph of figure 20 the energy absorption capacity is.

6 Conclusion

The experimental results confirmed that the ferrite amount in the weld nugget influences the mechanical properties. Furthermore, the here defined fusion rate Φ offers an appropriate possibility to characterize the joint integrity of dissimilar spot welds of FeMn and low alloyed steels. With an increasing fusion rate respectively quantity of ferrite in the weld metal the SFE shifts to values which allow other hardening mechanism such as TRIP-Effect to take place. As a result brittle martensitic phases occur besides dislocation glide and the formation of mechanical twins under external load. In addition to the deformation induced martensite after prestraining, the specimens martensite formed during welding can be detected. The consequence is a degradation of the mechanical properties and a brittle failure of the joint with no essential plastic deformation. This behavior was demonstrated on standardized as well as modified multiple spot specimens and must be further investigated by SEM of fracture surface topography. From these results it can be concluded that the fusion rate ϑ must be adequately taken into account when processing these materials.

Acknowledgments
The authors would like to thank the Deutsche Forschungsgemeinschaft (DFG) for the funding of the used DualBeam FEI Scios within the framework of the funding of large-scale facilities according to GG§91.

References

[1] Cornette D, Cugy P, Hildenbrand A, Bouzekri M and Lovato G 2005 *Revue de Métallurgie* **102** p 905
[2] Cooman B C, Chen L, Kim H S, Estrin Y, Kim S K and Voswinckel H 2009 *Microstructure and Texture in Steels* (Springer) p 165
[3] Saeed-Akbari A, Imlau J, Prahl U and Bleck W 2009 *Metallurgical and Materials Transactions A* **40** 3076
[4] Behm V, Höfemann M, Hatscher A, Springer A, Kaierle S, Hein D, Otto M and Overmeyer L 2014 *Physics Procedia* **56** p 610
[5] SEP 1220 Testing and Documentation Guideline for the Joinability of thin sheet of steel – Part 2: Resistance Spot Welding (Düsseldorf 2011)
[6] Adib H, Jeong J and Pluvinage G 2004 *Strength of Materials* **36** p 353
[7] Song C, Yu H, Li L and Lu J 2016 *Materials Letters* **174** p 75
[8] Rajinikanth V, Mukherjee K, Ghosh Chowdhury S, Schiebahn A, Harms A and Bleck W 2013 *Science and Technology of Welding and Joining* **18** p 485
[9] Frommeyer G 2007 *Stahl und Eisen* **127** p 67
[10] Kuntz M, Lempenauer K, Staufner W and Ross E. 2007 *Stahl und Eisen* **127** p 83
[11] Dumay A, Chateau J P, Allain S, Migot S and Bouaziz O 2008 *Materials Science and Engineering A* **484** p 184
[12] Allain S, Chateau J P, Dahmounb D and Bouaziz O 2004 *Materials Science and Engineering A* **387** p 272
[13] Jin J E and Lee Y K 2009 *Materials Science and Engineering A* **527** p 157
[14] Song W, Ingendahl T and Bleck W 2014 *Acta Metallurgica Sinica* **27** p 546
[15] Senk D, Emmerich H, Rezende J and Siquieri R 2007 *Advanced Engineering Materials* **9** p. 695
[16] Fonstein N 2015 *Advanced High Strength Sheet Steels - Physical Metallurgy, Design, Processing, and Properties* (Springer) p 369
[17] Remy L 1978 *Acta Metallurgica* **26** p 443
[18] Schumann H 1974 *Kristall und Technik* **9** p 1141
[19] Zilnyk K, Almeida Junior D, Sandim H, Rios P and Raabe D 2018 *Acta Materialia* **143** p 227
[20] Kurdjumow G and Sachs G 1930 *Zeitschrift für Physik* **64** p 325
[21] Greninger B and Troiano A 1940 *Trans. AIME* **140** p 307
[22] Wassermann G 1933 *Archiv für das Eisenhüttenwesen* **6** p 347
[23] Spencer K, Véron M, Yu-Zhang K and Embury J D 2009 *Materials Science and Technology* **25** p 7

IOP Conf. Series: Materials Science and Engineering **882** (2020) 012009 doi:10.1088/1757-899X/882/1/012009

Quantitative evaluation of global and local chromium contents with the EPR test on ferritic and martensitic stainless steels

A Heyn and P Rosemann

Otto-von-Guericke-University Magdeburg, Faculty of Mechanical Engineering, Institute of Materials and Joining Technology, Universitätsplatz 2, 39106 Magdeburg, Germany

andreas.heyn@ovgu.de

Abstract. The cause of localised corrosion phenomena in stainless steels is often related to the element chromium, which is essential for the formation and resistance of the passive layer. A lower than nominally stated content of chromium occurs when the element is highly concentrated and locally bound in chromium-rich phases or precipitates, mostly carbides, and thus locally or globally absent for passivation. The method of electrochemical potentiodynamic reactivation (EPR) is uniquely suited to demonstrate the effect of chromium depletion on the passivation. Here, the steel surface is initially active in sulfuric acid and passivation is achieved by external dynamic polarization. After passivation, the polarization direction is reversed and in chromium-depleted areas a reactivation takes place, which is detected by an increase in current. The detection limit for a reduced chromium content depends largely on the strength of the sulfuric acid used in the EPR test and the polarization speed. The relationship between the sulfuric acid concentration and the electrochemical parameters from the EPR test is shown here using the example of ferritic Fe-Cr alloys with graded Cr contents in the range of approx. 6 to 18 %. From the passivation and reactivation current densities of the EPR data, models for the detection of global and local chromium contents (detection limits) are derived. The developed models are applied to the martensitic stainless steel X20Cr13 in different heat treatment conditions and the influence on the distribution of chromium in the microstructure, which is responsible for the passivation and thus the corrosion resistance, is shown.

1 Introduction

In ferritic and martensitic stainless steels, the element chromium is the decisive alloying element for corrosion resistance, as it enables the surface to be passivated. The so-called resistance limit is a Cr content of at least 10.5 %, above which a passive layer forms under mild atmospheric conditions, which protects the iron-based material from corrosion and the formation of rust in a humid atmosphere. Ferritic stainless steels have chromium contents of more than 12 % (more frequently between 16 and 18 %) and are low in carbon (< 0.1 % C). Typical representatives of martensitic stainless steels are alloyed with 13 to 15 % (more rarely 17 %) chromium and, in contrast to the ferritic stainless steels, have increased carbon contents of more than 0.1 % so that the steel can be hardened by martensitic phase transformation. Both types of steel are almost free of nickel and therefore form a ferritic or martensitic matrix. As chromium is decisive for corrosion resistance, special attention must be paid to establish an evenly distribution and sufficient concentration in the solid solution. Phases rich in chromium (e.g. chromium carbides) can locally concentrate and bind larger amounts of chromium, so that chromium is no longer supporting the passivation of the matrix. Chromium is removed from the solid solution during carbide formation and the chromium content is often reduced in the immediate vicinity of chromium-rich precipitates.

For the experimental proof of chromium depletion and the resulting decrease in corrosion resistance, several corrosion tests in specific media exist, such as the various tests according to ASTM A 262, Practice A-E [1]. These tests are carried out in strong acids and at elevated temperatures and have the disadvantage that they are time-consuming, destructive and do not allow an accurate quantification of chromium depletion. In industrial practice, they are nevertheless frequently used when only simple yes/no statements on the corrosion resistance are required.

Significantly faster, more accurate and more meaningful are electrochemical methods, such as the EPR test (electrochemical potentiodynamic reactivation), which is standardized in ISO 12732 as a double-loop method [2] and is based on the work of Cihal and numerous other colleagues, which are summarized in [3]. The same procedure is described in ASTM G108-94 [4]. The parameters recommended in ISO 12732 are mainly suitable for high alloyed austenitic and duplex stainless steels. Test solutions with H_2SO_4 concentration in the range of 0.5 to 2.5 mol/L and addition of potassium thiocyanate (KSCN) between 1 and 50 mmol/L as well as a polarization rate between 2 and 15 V/h are listed. For the most frequently used austenitic stainless steel 1.4301 (AISI 304) with approx. 18 % chromium, a test electrolyte with 0.5 mol/L H_2SO_4 and 10 mmol/L KSCN and a polarization rate of 6 V/h (1.67 mV/s) is recommended. This approach has also proved successful in detecting sensitization in duplex stainless steel [5]. The electrolytes or experiment process are often further varied to study alloys with higher or lower alloy contents or to cause selective dissolution at special phases or precipitates.

For ferritic stainless steels of type 1.4016 (AISI 430) with 17 % Cr, Sidhom et al. [6] have varied the test parameters very extensively to find suitable conditions for the detection of sensitization. The optimum test condition they determined was 3 mol/L H_2SO_4 without further addition of an activator and 2.5 mV/s as polarization rate. However, the distance between the experiments parameters was chosen very broadly in this study (0.1 to 6 mol/L H_2SO_4, 0.5 to 5 mV/s polarization rate) and it is therefore difficult to understand how an optimum could be determined from the results. Kim et al. used a concentration of 0.14 mol/L H_2SO_4 with 2.5 mmol/L KSCN and 1.67 mV/s as polarization rate for EPR experiments on a ferritic stainless steel X2CrTi12 (1.4512, AISI 409L), but did not specify why these parameters were chosen in this way [7]. Kumar et al. also vary the test parameters of the EPR test to investigate intergranular corrosion on heat-treated martensitic stainless steels [8]. The gradations of the H_2SO_4 concentration between 0.03 and 0.25 mol/L are quite reasonable. The optimum is a concentration of 0.03 mol/L H_2SO_4 without activator and a polarization rate of 1.67 mV/s. EPR investigations on martensitic stainless steels have been conducted and published at the Otto-von-Guericke University since 2013 [9-11]. The use of a 0.1 mol/L H_2SO_4 solution without further additives and a polarization rate of 2 mV/s for investigations at laboratory temperature (22 °C) with the use of top-mounted measuring cells has proved to be successful.

Martensitic stainless steels, with 12-15 % chromium, are closer to the resistance limit than most other stainless steels and have a significantly higher carbon content. Chromium-rich carbides can bind significant amounts of chromium. It is therefore important to know how much chromium is available in the solid solution for passive layer formation, which is essential for the corrosion protection ability. This "available" chromium content for corrosion protection cannot be determined by atomic emission spectroscopy, because chromium carbides are also ionized, and this proportion is therefore determined analytically in an integral way. Analytical EDS measurements on the other hand do only characterize the small areas of the microstructure with a broad interaction volume of several microns. It can therefore be assumed that significantly less effective chromium is present in chromium steels containing carbon, especially in the soft-annealed state with a high fraction of carbides within the microstructure. Only through austenitisation at typically 1050 °C are the carbides dissolved in the austenite due to the increased carbon solubility. The chromium of the dissolved carbides is distributed homogeneously in the solid solution by diffusion and this homogeneous element distribution is "frozen" by subsequent quenching [12]. Subsequent tempering at excessively high temperatures and for long periods can lead to the formation of new chromium rich carbides and reduce the content of "available" chromium in the solid solution. Therefore, in the case of martensitic stainless steels, not only the effect of chromium depletion in the immediate vicinity of carbides is significant, but also the "available" content of chromium for passive layer formation and its resistance.

The aim of this study was to determine whether the EPR test could be used to quantify the "available" chromium content in the matrix and at the same time the locally reduced chromium content in the immediate vicinity of the carbides. The metrological background of the so-called "degree of sensitization" (DOS) in the standardized EPR test is based on the ratio of the reactivation and passivation current densities. It is assumed that the reactivation current results only from active metal dissolution of the sensitized portion of the surface and that the dissolution kinetic is always the same. The rest of the surface in the experiment remains passive and its share in the total current is neglected. The reactivation of the sensitized areas, however, depends on the acid concentration and the exposure time of the acid (and thus the polarization rate) and, of course, the local chromium content. Therefore, the basic idea of this work is to use the chromium and acid-dependent passivation and reactivation for the determination of the "available" chromium content as well as to develop a detection limit for the degree of local chromium depletion by a specific variation of chromium content and acid concentration in the EPR test. The model generated by the investigations was tested on the martensitic stainless steel X20Cr13 in different heat treatment states.

2 Materials and Experimental procedures

2.1 Materials

To produce Fe-Cr alloys with chromium contents of 6 to 18 %, samples were prepared in an arc melting furnace (Compact Arc Melter MAM-1, Edmund Buehler GmbH). The different chromium contents were produced by mixing flakes of pure (Armco) iron, 1.4003 (X2CrNi12) and 1.4509 (X2CrTiNb18). This results in slightly varying contents of Ni, Mn and Si. The melting chamber was previously evacuated three times and flushed with argon to minimize oxidation during melting. Approximately 20 grams of each reference alloy were melted twice in the arc, which then solidified on a water-cooled copper plate. The chemical composition of all reference alloys was determined by atomic emission spectrometry. The main alloying and accompanying elements are listed in Table 1.

Table 1. Chemical composition of the ferritic reference alloys in wt.-% [13].

Element	Cr	Ni	Mo	Mn	Si	N	C	P	S
5.8 %	5.26	0.20	0.01	0.59	0.20	0.008	0.016	0.009	0.004
7.3 %	7.32	0.28	0.01	0.78	0.28	0.005	0.019	0.011	0.003
9.3 %	9.34	0.36	0.01	0.99	0.37	0.009	0.025	0.013	0.002
11.2 %	11.19	0.43	0.01	1.17	0.44	0.007	0.029	0.015	0.002
13.6 %	13.56	0.33	0.02	0.90	0.46	0.019	0.025	0.016	0.002
15.8 %	15.80	0.25	0.03	0.68	0.49	0.015	0.029	0.020	0.002
17.4 %	17.36	0.18	0.03	0.48	0.49	0.024	0.022	0.018	0.002
18.0 %	17.96	0.16	0.03	0.42	0.51	0.023	0.025	0.019	0.002

The reference alloys were then homogenized at 1050 °C for 30 minutes and quenched in water. All reference alloys were then annealed at 800 °C to convert the martensitic structure into ferrite at low chromium contents (< 10 %). The samples were then cold rolled to 50 % of their initial height and annealed again at 800 °C for 1 hour to achieve a crystallised ferritic microstructure. The steel X20Cr13, with the chemical composition listed in Table 2, was used for further investigations.

Table 2. Chemical composition of the investigated material X20Cr13 [12].

Fe	C	Cr	Mo	N	V	Ni	Si	Mn	Cu	P	S
bal.	0.23	13.3	0.02	0.01	0.02	0.1	0.29	0.33	0.02	0.025	0.001

The steel X20Cr13 was examined in the soft-annealed delivery condition and two hardened and differently tempered conditions. For hardening, sheet metal samples (35 x 35 x 2 mm) were austenitized for 15 min at 1050 °C and quenched in water. One sample was tempered for 2 h at 200 °C and the other sample for 2 h at 500 °C. A clearly different corrosion behaviour is to be expected in the three conditions due to the content of free chromium in the solid solution and the effect of chromium depletion by carbides precipitation.

2.2 EPR-testing

To carry out the EPR tests, the samples were freshly ground in advance (wet grinding with abrasive paper of appropriate grain sizes), rinsed with water and ethanol, dried and then covered with a perforated (Ø 14 mm) foil so that a defined and always equally large area (1.54 cm^2) of the sample is exposed to the test electrolyte. An acrylic glass tube of slightly larger diameter with a silicon sealing ring was placed on the perforated foil and clamped so that the test electrolyte cannot leak out after the measuring cell has been filled. Immediately after filling the cell with approx. 4 mL test solution, the measuring electrodes were positioned in the solution directly above the sample surface and the measurement was started. A saturated (3 M KCl) Ag/AgCl electrode was used as reference electrode (reference potential +197 mV$_{SHE}$). As counter electrode an electrode of Ti/Ti mixed oxide grid with a surface that is approx. 5 times larger than the working electrode was used so that the currents can flow unhindered and without necessary compensation of a IR drop.

A series of electrochemical parameters can be obtained from an EPR test (double loop), which can be used to understand the influence of various experimental factors on the passivation and the resistance of the passive layer. Figure 1 shows schematically which electrochemical parameters were determined from an EPR test.

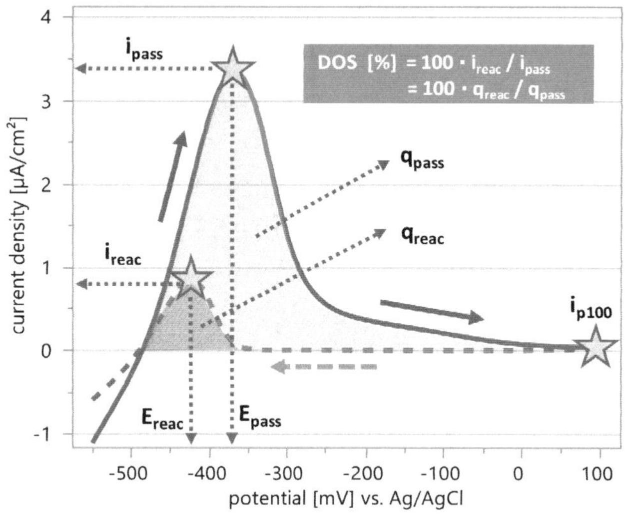

Figure 1. Course of an EPR curve with marking of the determined electrochemical characteristic values.

Usually the passivation and reactivation current densities (i_{pass}, i_{reac}) or the corresponding charge values (q_{pass}, q_{reac}) are used to calculate the degree of sensitization (DOS [%] = 100 * i_{reac} / i_{pass}). Due to the ratio of the two current densities, this value is area-independent and therefore less sensitive if the area is unknown or cannot be precisely maintained. Nevertheless, important findings can be derived from the values of current densities alone, which is why a constant and precise measuring area is indispensable for experimental implementation.

2.3 Designed experiments and evaluation

In order to clarify how great a possible influence of the sample pre-treatment and the polarization speed is on the results of the EPR test, the steel X20Cr13 was examined in the soft-annealed state and the preparation was carried out with different grinding papers, which lead to differently rough surfaces. Grinding papers with grits of 180, 320 and 500 were used to ground the samples under running water.

This resulted in average R_a values of 0.18 µm, 0.12 µm and 0.06 µm and R_z values of 1.7 µm, 1 µm and 0.54 µm. The polarisation rate was varied with 1, 2 and 4 mV/s. The sulfuric acid concentration was kept constant at 0.1 mol/L for these experiments. The full-factorial and thus orthogonal experiment plan comprised 3 x 3 = 9 experiments and was conducted randomly with two repetitions.

To study the interaction of acid concentration and chromium content, the ferritic Fe-Cr alloys, described under 2.1, with eight graded chromium contents from 5.3 to 18 % were used. The sulphuric acid concentration was varied in gradations of 0.01, 0.0316 and 0.1 mol/L. The samples were uniformly wet ground with 500 grit paper, then dried and measured immediately. The polarization rate was 2 mV/s. The full-factorial and thus orthogonal experimental design comprised 8 x 3 = 24 experiments and was conducted randomly with two repetitions. Test points at 9.3 and 11.2 % chromium were repeated 3 to 4 times, as this is the content where reactivation changes most.

Furthermore, tests were carried out on the steel X20Cr13 in different heat treatment conditions (see 2.1). The sulphuric acid concentration was varied in gradations of 0.01, 0.0316 and 0.1 mol/L. The samples were uniformly wet ground with paper of grain size 500, then dried and measured immediately. The polarization rate was 2 mV/s. The full-factorial and thus orthogonal experimental design thus comprised 3 x 3 = 9 experiments and was conducted randomly with three repetitions.

The statistical evaluation of the influencing factors from all designed experiments was performed with simple regression models with a "standard least squares" personality.

3 Results

3.1 Influence of roughness and polarization rate

The influence of the polarization rate on the course of the EPR curves is exemplarily shown in Figure 2. The height of the passivation current density increases slightly, but especially the reactivation is different in its height and position on the potential axis.

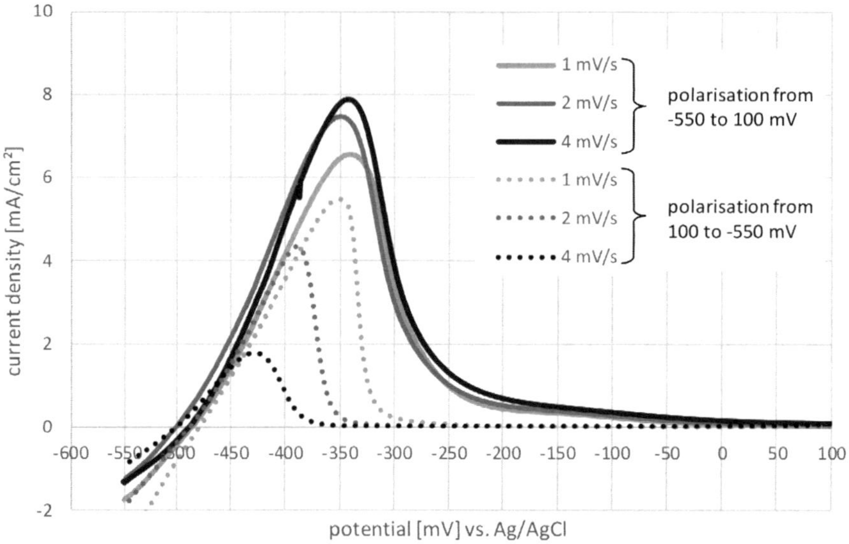

Figure 2. Examples of cyclic polarization curves from EPR tests on the steel X20Cr13 (soft-annealed state, grit 320) in 0.1 mol/L H_2SO_4 solution, variation of polarization rate: 1, 2 and 4 mV/s.

The evaluation of the EPR curves allows conclusions about the influence of grit size and polarization rate on the characteristic values of the EPR test shown in Figure 3.

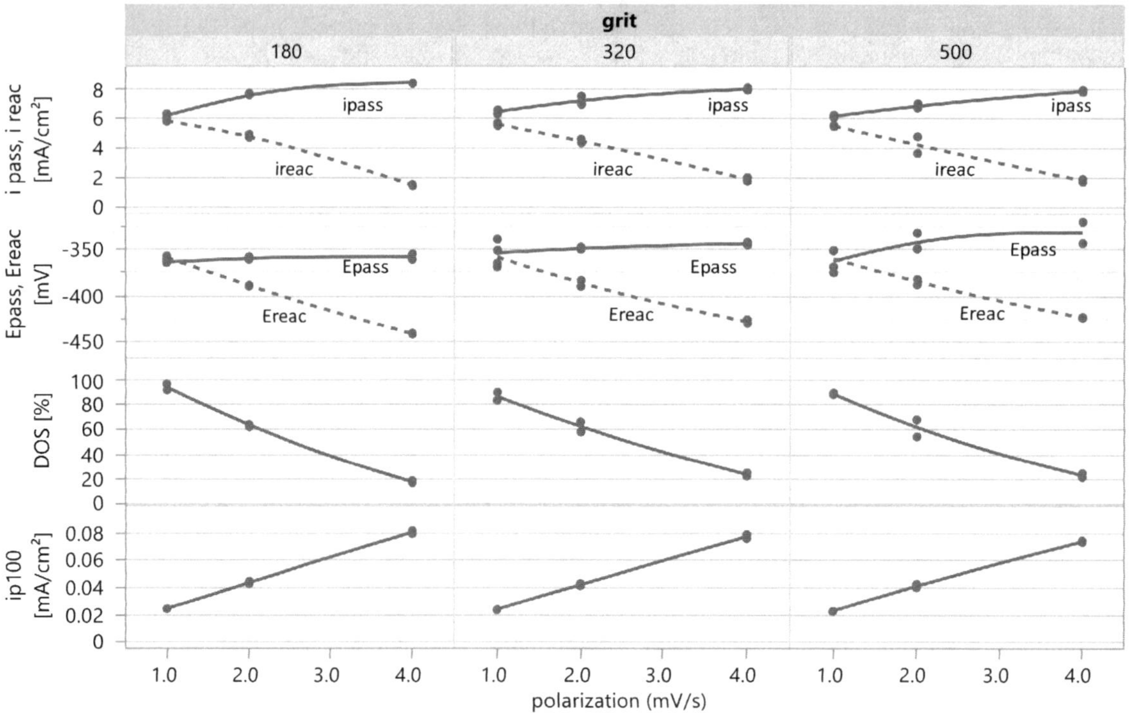

Figure 3. Electrochemical characteristics of the tests carried out on X20Cr13 (soft-annealed state) in 0.1 mol/L H_2SO_4, variation of polarization rate (1, 2, 4 mV/s) and surface preparation.

IOP Conf. Series: Materials Science and Engineering **882** (2020) 012009 doi:10.1088/1757-899X/882/1/012009

The dominant influence of the polarization rate on all determined characteristic values is clearly visible. The effects on the DOS value are mainly caused by the different reactivation current densities, since the influence of time is stronger on reactivation compared to passivation. The reactivation potentials (E_{reac}) are shifted to more negative values, which can also be evaluated as a time effect. The exposure time of the sulphuric acid, which causes the freshly formed passive layer to dissolve and thus reactivate the surface, plays a decisive role. The passive current density at the reversal point at 100 $mV_{Ag/AgCl}$ is decreases using slower polarization rates as the passive layer is further strengthened by the longer time in the passive range and thus further inhibits the charge transfer of metal ions. The influence of roughness is not clearly visible in the diagram and was investigated by statistical evaluation. It is only significantly detectable during passivation (i_{pass}) and is generally low. This is understandable, since the surface is actively corroded for a short time during passivation and a levelling effect is to be expected, whereby no significant influence of the surface roughness can be recognized afterwards.

The polarization rate is therefore a decisive factor in the EPR experiment. It will also be related to the acid concentration because the strength of the acid and its exposure time influence the passive layer stability and thus the occurrence of reactivation in the EPR experiment. For the following investigations with variation of chromium content and sulfuric acid, the polarization rate was set to 2 mV/s in order not to further increase the scope of the test.

3.2 Variation of chromium content and sulphuric acid concentration

The effects of the chromium content of the Fe-Cr alloys and the acid concentration on the course of the EPR curves are presented in the following figures. Figure 4 shows that the passivation current density decreases with increasing chromium content while the acid concentration remains constant, i.e. the passivation process is easier, i.e. with lower charge conversion. A current density maximum due to reactivation can be seen at 9.3 and 11.2 % chromium. At 13.6 % chromium, on the other hand, the passive layer is resistant to reactivation under the selected test conditions and no increase in current density occurs during reverse polarization. The influence of the acid concentration is shown in Figure 5 using the Fe-Cr alloy with 11.2 % Cr. With a constant chromium content, the passivation current density increases with the acid concentration, i.e. a higher charge conversion is necessary for passivation. At 11.2 % Cr reactivation only appear at the highest acid concentration of 0.1 mol/L. With the two weaker acids, however, reactivation does not occur.

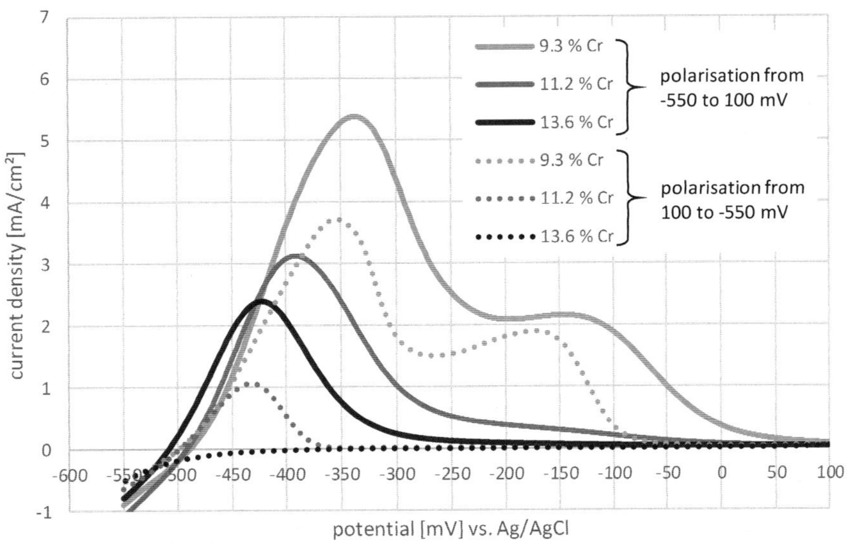

Figure 4. Examples of cyclic polarization curves from EPR tests of Fe-Cr-alloys in 0.1 mol/L H_2SO_4 solution, polarization rate 2 mV/s, variation of Cr-content.

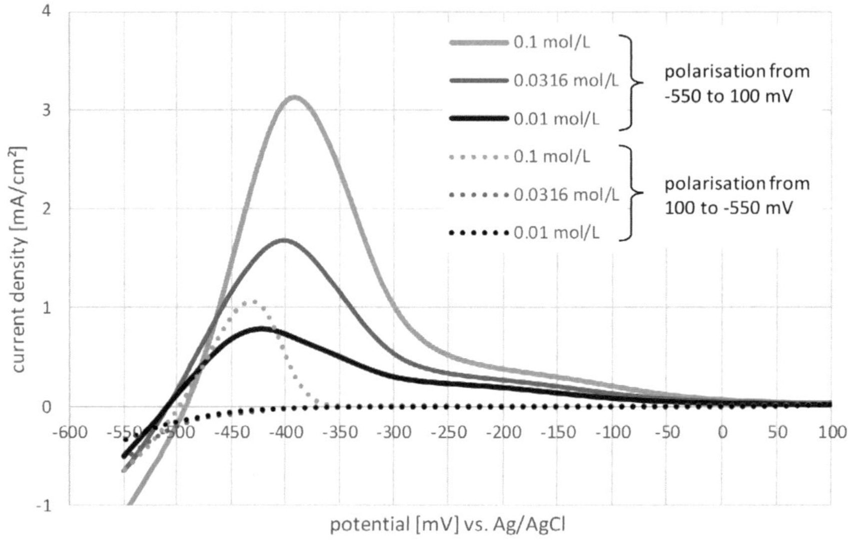

Figure 5. Examples of cyclic polarization curves from EPR tests of the Fe-Cr-alloy with 11.2 % Cr, polarization rate 2 mV/s, variation of molarity of H_2SO_4.

The passivation and reactivation current densities (i_{pass}, i_{reac}) were determined from the measurement curves and compared as a function of the eight different chromium contents in Figure 6 in relation to the applied H_2SO_4 concentrations.

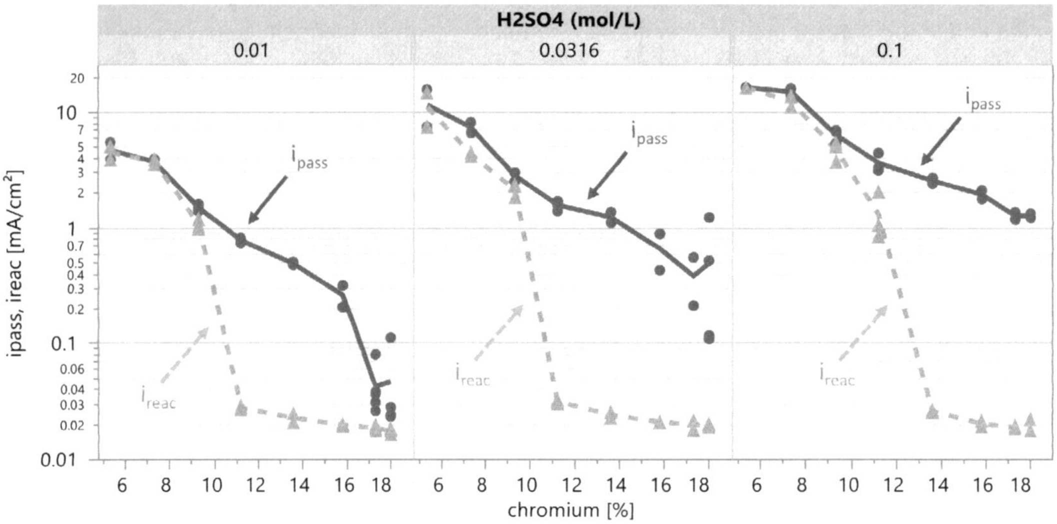

Figure 6. Passivation and reactivation current densities as a function of the test parameters with variation of the chromium content of the Fe-Cr alloy and the acid concentration.

An exponentially decreasing course of the passivation current density with increasing chromium content can be seen. At the lowest chromium content of 5.3 %, the curve flattens out somewhat, which probably indicates a problem with an IR drop due to the relatively high currents. A stronger scatter of the individual values occurs for the medium and lowest acid concentration at high chromium contents of 16 to 18 %. This can be explained by the fact that the steel strives to passivate themselves during the test in the weaker acids and are therefore less active during passivation. This is particularly noticeable with 0.01 mol/L H_2SO_4.

Up to a certain chromium content, the reactivation current densities also show an exponential decrease in current density with increasing chromium content. However, there is a transition range where the current density is abruptly very low. No reactivation occurs and the value reflects the passive current density (note: in the absence of reactivation, the average passive current density was used as a value to avoid the value "0").

The influence of the acid concentration can be seen in the level of the current density values. The average values of the passivation current densities over all chromium contents are 4.86 mA/cm² for 0.1 mol/L, 2.64 mA/cm² for 0.0316 mol/L and 1.13 mA/cm² for 0.01 mol/L and are statistically significantly different. The reactivation current densities have a similar gradation with respect to the acid concentration. However, they are no longer comparable with the passivation current densities due to the abrupt drop from a certain chromium content. With the low chromium contents, they are still at a comparable level, which shows that there is an almost uniform surface reactivation with approximately the same dissolution kinetics. This is understandable, since chromium is evenly distributed in a defined quantity, which also explains the sudden transition between reactivated and non-reactivated.

3.3 Modelling for the determination of the chromium content from EPR tests

The passivation current densities shown in Figure 6 correlate very well with the chromium content and acid concentration. Applying the ordinary least squares method to the test series showed the significance of both effects, but also a significant interaction between chromium content and acid concentration. From the model created, the graphical representation is shown in Figure 7 and formulas for predicting the total chromium content "available" for passivation from the passivation current density were derived. Since the formulas were derived from experimental results, they are only valid for the investigated ranges and selected metrological settings.

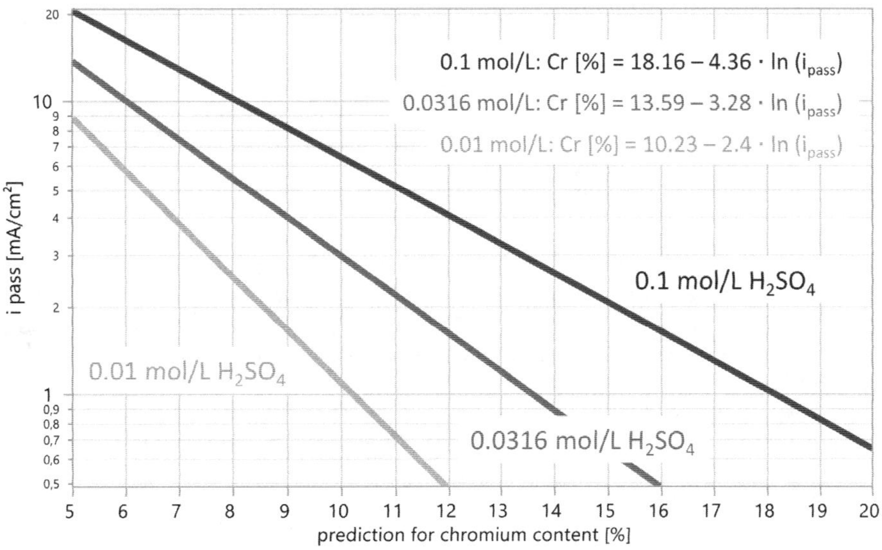

Figure 7. Graphical representation of the model from the test series with variation of Cr content and acid concentration with formulas for predicting the Cr content from passivation current densities.

The reactivation current densities shown in Figure 6, on the other hand, show an unsteady course and cannot be satisfactorily adjusted by the ordinary least squares method. Therefore, a logistic adjustment was used, i.e. it was modelled according to the probability of exceeding (or not exceeding) a defined limit value. The limit value used was 0.06 mA/cm². Figure 8 shows a graphical representation of the logistic model derived from this.

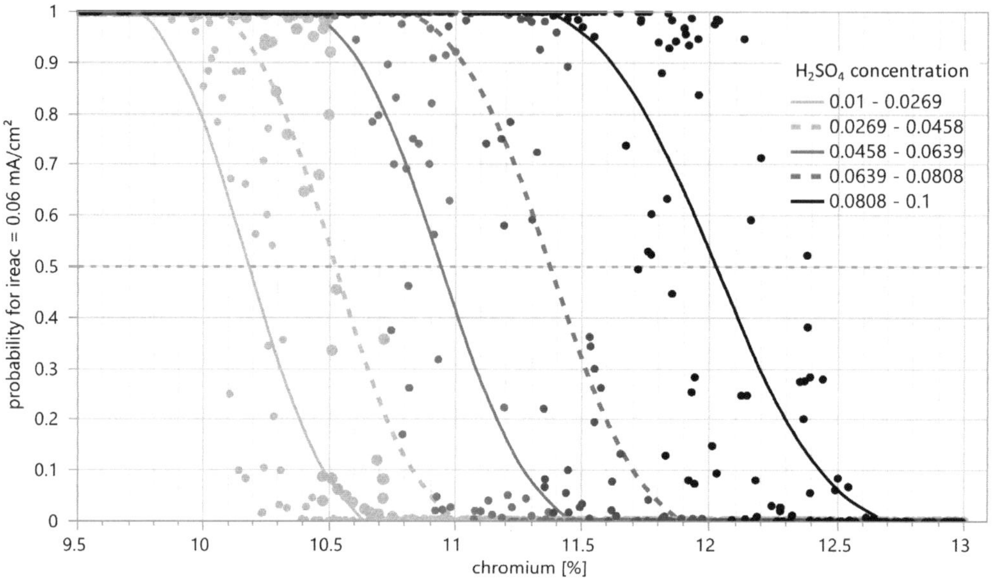

Figure 8. Graphical representation of the model from the test series with variation of chromium content and acid concentration for predicting the chromium content from reactivation current densities.

From the model for the reactivation current density it can be seen at what chromium content and at what acid concentration reactivation is likely. The model can be used, for example, to set a specific acid concentration to detect any chromium depletion and thus achieve a specific detection limit. It is also possible to determine the degree of depletion of a sample with unknown chromium depletion more precisely by experiments with graded acid concentration.

3.4 Application of the models on the X20Cr13 in different heat treatment conditions

In order to apply the models, investigations were carried out on the steel X20Cr13 in the soft-annealed state and in two hardened and tempered conditions. The corresponding microstructures and Vickers hardness values are presented in Figure 9.

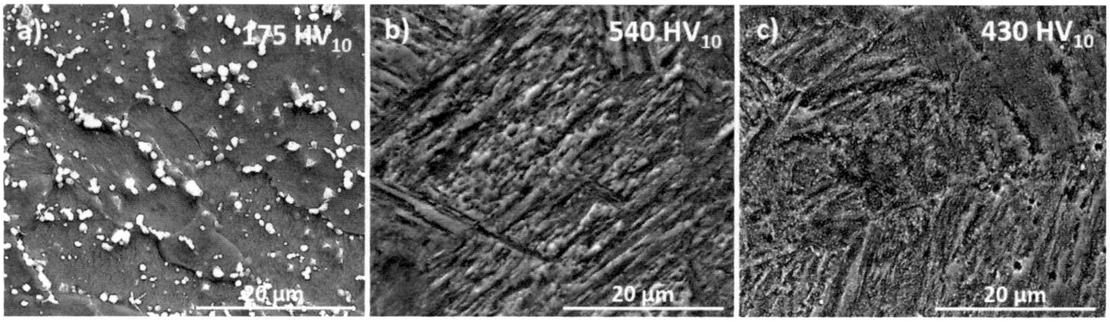

Figure 9. SEM images of the different microstructures of the studied stainless steel X20Cr13: **a)** soft annealed, **b)** hardened and tempered for 2 h at 200 °C, **c)** hardened and tempered for 2 h at 500 °C.

The soft annealed condition has many large carbides rich in chromium in a ferritic matrix which has a lower chromium content than the nominal chemical composition of the steel. This soft annealed condition was austenitized at 1050 °C for 15 min and then quenched in water. During this process, all carbides are dissolved in the austenite, chromium is homogeneously distributed in the matrix and a homogeneous chromium content is assured in the martensitic matrix formed by quenching. Tempering at 200 °C for 2 h merely relaxes the distorted martensite and does not cause significant carbide formation and sensitization. Tempering at 500 °C, on the other hand, produces numerous microcarbides throughout the matrix, which lead to strong sensitization.

Figure 10 shows the corresponding values for the passivation and reactivation current densities of the stainless steel X20Cr13 in dependence on tested microstructure and applied H_2SO_4 concentration.

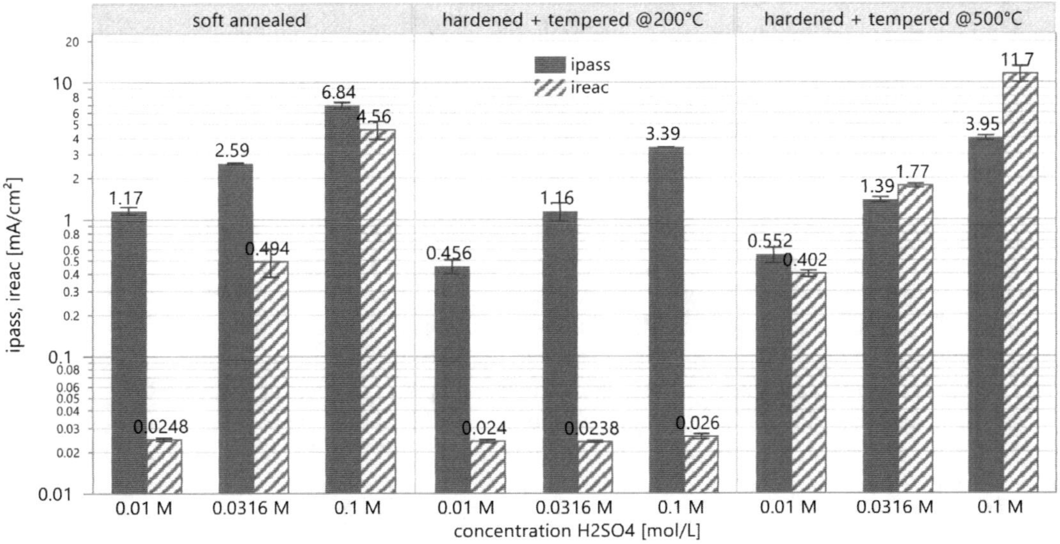

Figure 10. Passivation and reactivation current densities from the EPR tests on the three heat-treated states of the martensitic stainless steel X20Cr13.

The dependence of the current densities on the acid concentration is clearly visible in three states. The passivation current densities of the soft-annealed condition are significantly higher than those of the other two hardened states, which differ only slightly from each other, but are nevertheless statistically significant. In contrast, the reactivation current densities are quite different. The soft-annealed state shows reactivation at 0.1 mol/L and 0.0316 mol/L, but not at 0.01 mol/L. In contrast, the hardened and at 200 °C tempered state shows no reactivation at all three acid concentrations since chromium is homogeneously distributed in the martensitic matrix. The pronounced carbide formation during tempering at 500 °C causes a noticeably clear reactivation at all three acid concentrations. If the relationships and generated formulas of Figure 8 are applied to the measured passivation current densities, the calculated contents of "available" chromium involved in the passivation reactions can be estimated and is presented in Table 3.

Table 3. Predicted chromium content from the formulae in Figure 7.

sample	H_2SO_4 [mol/L]	i_{pass} [mA/cm²]	predicted Cr [%]	Cr [%] (average)
Soft annealed	0.1	6.84	9.87	10.03
	0.0316	2.59	10.46	
	0.01	1.17	9.85	
Hardened + tempered at 200 °C	0.1	3.39	11.66	12.29
	0.0316	1.16	13.10	
	0.01	0.456	12.11	
Hardened + tempered at 500 °C	0.1	3.95	12.17	12.11
	0.0316	1.39	12.51	
	0.01	0.551	11.66	

Using the reactivation current densities, the limit value of 0.06 mA/cm² can be used to decide whether reactivation has occurred or not. The soft-annealed sample shows reactivation at 0.0316 mol/L, but no more at 0.01 mol/L. In conjunction with Figure 8 it can be deduced from this (probability = 1, second curve from the left) that reactivated areas have a high probability of a chromium content of about 10 %. In comparison with the mean value for the "available" chromium content in Table 3, which is also approx. 10 %, it can be concluded that there is no local chromium depletion, but that the ferritic matrix of the soft annealed condition is uniformly depleted in chromium and the derived chromium content is clearly below the chromium content of the chemical analysis in Table 2. The missing chromium is thus obviously bound in carbides and the solid solution is uniformly depleted.

A completely different picture emerges with the hardened sample, which was tempered at 200 °C. Even at an acid concentration of 0.1 mol/L, no reactivation is observed. This means that, considering the relationships shown in Figure 8 (probability = 0.5, first curve from the right), the chromium content is higher than 12 % everywhere in the martensitic matrix, which also corresponds to the average "available" chromium content of 12.29 % in Table 3.

The hardened and tempered at 500 °C condition shows a clear reactivation even at 0.01 mol/L. If the correlations in Figure 8 are applied (probability = 1, first curve from the left), the result is a chromium value of 9.8 % or even less, which is clearly below the value of the determined "available" chromium content of 12.11 %. This means that there is a strong and localised chromium depletion at this heat treatment condition.

4 Conclusions

The EPR test has several advantages in comparison to exposure tests in specific test media. In addition to a significantly faster test procedure, which can also be used in mobile applications, a further advantage is that the degree of sensitization (DOS) can be quantified. However, due to the integral measuring method (current related to the entire measuring surface), it is not yet possible to distinguish whether the reactivation is due to a larger reactivated area or a more strongly reduced chromium content. Both effects are mixed in the DOS characteristic value and complicate the scientific clarification of the mechanisms of chromium depletion and their effects on corrosion behaviour.

With the investigation strategy shown in this study for ferritic Fe-Cr alloys, the first step was to determine the chromium content that is effectively available for passivation. A mathematical model was derived from the results obtained and tested on a martensitic stainless steel. The passivation current density correlates with the acid concentration and the chromium content which is effectively available for passivation. On the steel X20Cr13 it was shown that this does not necessarily correspond to the nominal Cr content of the spectrometric analysis. In the soft annealed condition, more than 3 % chromium are bound in carbides and therefore are not available for passivation. With an "available" chromium content of only about 10 %, the corrosion resistance is severely diminished, as this value is even below the resistance limit. After heat treatment, significant amounts of chromium diffuse back into the matrix and more than 12 % chromium are available for passivation.

In the case of unfavourable tempering treatment, renewed chromium depletion is possible, but in this case, it is mainly localised in the vicinity of microscopic carbides. The primary aim of the EPR test is to detect chromium depletion as a result of this local chromium depletion. The selected test conditions are decisive for the success of this detection, which is why there are always only recommendations for certain groups of stainless steels. Local chromium depletion is indicated when the chromium content in the depletion zone is sufficiently reduced to cause reactivation, or the acid is strong enough to trigger reactivation. This was investigated and modelled in the investigations with the Fe-Cr alloys under variation of the sulphuric acid concentration. With the help of the established logistic model, a detection limit for the desired detection of chromium can now be set and suitable test conditions selected. In addition, it is possible to estimate by how much the chromium content in these depletion zones has been reduced on unknown samples by varying the acid concentration. In conjunction with the "available" chromium content determined from the passivation current density, a much better explanation of the underlying mechanisms is possible.

Acknowledgements
The funding of parts of this work by the German Research Foundation (DFG) under grant no. 330472124 is gratefully acknowledged!

Thanks go to Mrs. B.Sc. Marie Becker for conducting a series of electrochemical experiments.

5 References

[1] *ASTM A262 – 15* Standard Practices for Detecting Susceptibility to Intergranular Attack in Austenitic Stainless Steels
[2] *DIN EN ISO 12732:2008-07* Corrosion of metals and alloys - Electrochemical potentiokinetic reactivation measurement using the double loop method (based on Cihal's method) (ISO 12732:2006); German version EN ISO 12732:2008
[3] Chihal V, Lasek S, Blahetova M, Kalabisova E and Krhutova Z 2007 *Chem. Biochem Eng.* Q. 21 (1) 47-54
[4] *ASTM G108-94* 2015 Standard Test Method for Electrochemical Reactivation (EPR) for Detecting Sensitization of AISI Type 304 and 304L Stainless Steels
[5] Kauss N, Heyn A, Halle T and Rosemann P 2019 *Electrochimica Acta* Vol 317 17-24
[6] Sidhom H, Amadou T and Braham C 2010 *Metallurgical and Materials Transactions A* Vol 41A 3136-3150
[7] Kim JK, Kim YH, Uhm SH, Lee JS and Kim KY 2009 *Corrosion Science* 51 2716-2723
[8] Kumar BS, Kain V and B Vishwanadh 2017 *Corrosion* Vol 72 No 4 362-378
[9] Rosemann P, Müller T, Babutzka M and Heyn A 2015 *Materials and Corrosion* Vol 66 No. 1 45-53
[10] Müller T, Heyn A, Babutzka M and Rosemann P 2015 *Materials and Corrosion* Vol 66 No. 7 656-662
[11] Babutzka M, Heyn A and Rosemann P 2018 *Materials and Corrosion* Vol 69 No. 10 1338-1349
[12] Rosemann P 2017 *Dissertation Shaker Verlag* ISBN: 978-3-8440-5404-0
[13] Rosemann P and Kauss N 2018 *WOMag - Competence in materials and functional surfaces* **7** 1

IOP Conf. Series: Materials Science and Engineering **882** (2020) 012010 doi:10.1088/1757-899X/882/1/012010

Comparison of liquid and gel electrolytes for the investigation of pitting corrosion on stainless steels

A Heyn

Otto-von-Guericke-University Magdeburg, Faculty of Mechanical Engineering, Institute of Materials and Joining Technology, Universitätsplatz 2, 39106 Magdeburg, Germany

andreas.heyn@ovgu.de

Abstract. Electrochemical methods have become established for the investigation and testing of stainless steels for their resistance to pitting corrosion. In contrast to liquid electrolytes, which are normally used for this purpose, gel-like electrolytes have clear advantages regarding their usability. However, it must be clarified whether both types of electrolyte lead to the same or different results. In the investigations presented here, various agar-based gel electrolytes were characterized regarding their corrosion-relevant properties, such as chemical composition, rheological properties and the electrolyte film forming on the surface. To compare the two electrolyte types, electrochemical investigations were carried out with graduated agar and NaCl concentrations on the material 1.4301 (AISI 304). Pitting and repassivation potentials were determined by dynamic polarization experiments and the varied factors were statistically evaluated regarding their influence. The results show that liquid and gel electrolytes provide comparable results. However, the pitting corrosion phenomena under external polarization are shallower with gel electrolytes and therefore their repassivation is supported. Also, the use of gel electrolytes with indicators for iron ions, which at the same time act as oxidizing agents and allow polarization into critical areas, gives comparable results. Gel electrolytes are therefore a very good alternative to liquid electrolytes if the detection of corrosion-sensitive surfaces is in focus.

1 Introduction

Stainless steels are among the most commonly used materials when corrosion resistant surfaces and related functionalities are required. However, already during production, but especially during further processing, the materials can be unintentionally modified in such a way that their very good corrosion resistance is no longer guaranteed. A number of testing and examination methods are available to check this. Most of these are, however, laboratory applications, which are costly, time-consuming and destructive. On the part of the processors and users of these steels, however, there is a desire to have a testing method for use on site (point-of-care) to determine the corrosion resistance of a semi-finished or end product quickly and reliably. Electrochemistry is virtually predestined for such questions. However, it is difficult to transfer laboratory methods to practical use because of the necessary equipment and electrodes and, not least, because of the handling with liquid electrolytes. For this reason, research efforts have been made over several years to solve these problems and provide users with a simple and functional testing tool.

The use of gel electrolytes instead of liquid test electrolytes is an important part of a point-of-care strategy in corrosion testing. The liquid electrolyte is immobilized by a network of polymer fibers.

IOP Conf. Series: Materials Science and Engineering **882** (2020) 012010 doi:10.1088/1757-899X/882/1/012010

Agar/Agarose is used as a gel-forming polymer because it forms stable gels even in very low concentrations. The gel also has the property of forming a thin electrolyte film on the gel surface by syneresis. This enables electrochemical reactions and measurements on metallic surfaces. However, the question is, are the measurements as well feasible as with liquid electrolytes and are they comparable at all. For this reason, investigations of the pitting corrosion resistance of stainless steel 1.4301 were carried out by means of dynamic polarization and various gel and liquid electrolytes were compared with each other. Furthermore, the "KorroPad" method was tested as possible point of care method and is compared with the polarization results.

2 State of knowledge

Agar offers itself as the basic structure for a gel electrolyte since it leads to stable gels even in very low concentrations in water. Agar is a macromolecular polysaccharide which is obtained from the cell walls of various red algae by extraction with hot water, freeze-thaw processes or spray drying. It consists of about 70 % agarose (gelling) and about 30 % agaropectin (sulphated, non-gelling). A separation to pure agarose is carried out for use in gel electrophoresis. Agar can be mixed with water but is insoluble at room temperature. It melts between 80 and 100 °C and crosslinks physically by cooling. The structure of the solidified macromolecules is described as double helices, which cross-link spatially and, depending on the concentration of the agarose, form cavities in the range of several hundred nanometers, which are filled with the solvent.

Although the specific research and use of agar gels for corrosion investigations is still comparatively limited, their application for so-called electrographic processes has been known for a long time. As early as 1955, Laque et al. used gel electrolytes mixed with indicators to visualize partial reactions of corrosion processes for teaching material science [1]. In a similar way, many years later Isaacs et al. added chloride and pH indicators to agar gel pads and used them to study aluminum and aluminum alloys [2]. They were able to visualize the location and course of anodic and cathodic partial reactions in localized corrosion on aluminum by means of indicators in the gel. The visualization of corrosion by an indicator for iron ions is also the basic principle of the electrographic Ferroxyl test. This test is still used today to determine the porosity of metallic coatings [3] or to test the porosity of phosphating's on steel, whereby the use of gelatin instead of filter paper as electrolyte carrier is often recommended. As early as 1959, the Ferroxyl test with gels was used as a test method for implant materials made of stainless steels and other high alloyed materials [4]. A few years ago, the method was further developed for the detection of local corrosion on stainless steels by the Federal Institute for Materials Research and Testing (BAM) [5], patented [6] and can be purchased as "KorroPad" test equipment. It is based on a gel pad with an agar content of 3 % in water and contents of 0.1 mol/L NaCl and 0.001 mol/L potassium ferricyanide. The latter leads to the formation of a stable redox potential which, as a critical test potential if the material is sensitive to corrosion. The chloride ions activate weak points in the passive layer and a local metal dissolution takes place. The iron ions that are dissolved in the solution are complexed by the potassium ferricyanide to form Prussian Blue, a poorly soluble pigment that indicates the local corrosion sites by color and makes them assessable. In recent years, the KorroPad has been increasingly used as a testing agent in different industrial applications. It has been applied to questions regarding the corrosion resistance of stainless steels in industry and research [7], e.g. to optimize their heat treatment [8], to show the influence of surface treatments [9] or for proof of sensitization [10].

An application of gel-like electrolytes to simulate special corrosive conditions has been demonstrated e.g. by Newton and Sykes [11]. They used an agar gel mixed with NaOH and NaCl to examine steel in mortar-like electrolytes and were able to document significant differences to measurements in liquid electrolyte solutions. Spark et al. [12] describe in their studies how corrosion processes in clay soils with the help of agar gels can be simulated under the complex physical-chemical conditions prevailing there. The investigations were carried out on an unalloyed steel and free corrosion potentials and current densities were determined from polarization curves. In further publications the authors describe investigations with agar gels to which peptides have been added [13] in order to simulate microbiologically induced corrosion of pipelines in soils [14]. Vanbrabant et al. also used agar-based gel electrolytes as a corrosion medium to simulate special media conditions, with simultaneous electrochemical instrumentation to perform electrochemical measurements [15]. They see the advantage of using gels as being closer to media conditions relevant to practice, especially where the transport of corrosion products is limited. Another advantage they see is that the individual stages of corrosion can be visualized in the course of the tests. Applications were presented that better simulate conditions of steel fibers in concrete, galvanized steel (zinc and zinc-aluminum alloys) in agricultural atmospheres and in animal farming.

Agar/agarose gels are also used in corrosion research for tailor-made, often miniaturized reference electrodes in which the bridge electrolytes are immobilized by the gel to minimize the leak rate [16]. In recent years, numerous authors have reported on the use of the gels as electrolyte with electrochemical instrumentation. Monrabal et al., for example, have added agar gels with a high proportion of glycerin (20 to 70 %) in order to realize more flexible top-mounted measuring cells [17] and to better couple them to shaped components. In this way, they are trying to identify weak points in welds of stainless steels [18] and have carried out numerous other electrochemical investigations [19, 20]. However, it is uncertain whether comparable results to liquid, purely water-based electrolytes are possible due to the high glycerin content. Cano et al. also developed a measuring cell based on agar gel as a connecting electrolyte, which makes it possible to carry out electrochemical impedance measurements on monuments and sculptures without causing any significant damage to them [21]. Di Turo et al. pursued the same goal by using gel-based electrolytes for the electrochemical characterization of bronze patina on historical coins [22]. The first own electrochemical investigations with gel electrolytes were started in 2014 and led, among other things, to a methodical approach to determine electrochemical parameters of galvanized steels under atmospheric exposure in order to determine the current state of the protective coatings [23].

The use of gels for corrosion tests can be summarized as follows:

- Triggering and visualization of corrosion processes by indicators in the gel electrolyte
- Electrochemical instrumentation of gel electrolytes to facilitate the characterization of corrosion protection systems and to determine characteristic values
- Simulation of specific corrosion conditions (building materials, soils, food, biological tissue) with solid/liquid phases and reduced mass transfer

3 Corrosion relevant properties of agar gels

For the application of gels as electrolyte for corrosion investigations it is important to determine the corrosion relevant properties of the gels in advance. In the case of the gel former, the chemical composition of the raw products is important, as a certain percentage is added to the aqueous electrolyte and can thus change its corrosivity. The mechanical properties of the gel also play an important role. When used as a test electrolyte, the gel must have a sufficient contact with the surface and the other electrodes but must not change or be damaged in an uncontrolled manner. Another important factor is the electrolyte film that forms on the gel and between the gel and the test surface since this can influence corrosion reactions and electrochemical measurements. It is important to find out how the wetness on the surface is related to the gel former and the electrolyte components. In general, the agar concentration determines the pore sizes and thus most properties, including mass transfer in the gel. This means that the transport of ions and corrosion products within the gel and between the gel and the surface can be influenced in different ways.

To produce an agar gel, a certain amount of agar powder is added to the desired electrolyte, mixed and heated with stirring to a temperature at which the agar dissolves and melts in the aqueous solution. The melting temperature varies according to the type of agar, e.g. 45 °C for low melting agarose, approx. 85 °C for standard agarose or up to 98 °C for agar. After a certain holding time, the melted agar-water mixture is poured off, where the gel-formation starts when the gel transition temperature is reached. The agar gel can be poured into almost any shape and cut into a specific form (e.g. circular pads) after reaching the gel state.

3.1 Properties of agar raw products

The chemical composition of the gel electrolyte consisting of water and agar/agarose is specifically adjusted to the respective metrological requirements. Usually ionic constituents are added in a certain concentration, which have an influence on the physical-chemical properties and thus also on the electrochemical reactions at the phase boundary to the surface. The agar substance, even with only a few percent agar in water, brings its own ions of different types and quantities and adds them to the gel electrolyte. The type and quantity can vary depending on the origin and quality of the agar and must be determined analytically in advance. The corrosion relevant effects can be measured on the finished gel, e.g. as a change in conductivity or in the redox potential. Table 1 shows the contents of corrosion-relevant ions for three commercially available agar substances, the conductivity of a 3 % mixture with distilled water and the redox potentials of the mixture and the gels produced from it.

Table 1. Ion contents of agar products and effects on conductivity and redox potentials of agar-water solution and prepared 3% agar gels.

type of agar	Cl^- [ppm]	SO_4^{2-} [ppm]	PO_4^{3-} [ppm]	NO_3^- [ppm]	γ 3% in H2O [μS/cm]	pH	ERedox 3% in H2O [mV, NHE]	ERedox 3% Gel, fresh [mV, NHE]	ERedox 3% gel, after 3 h [mV, NHE]
Agar Extra Pure (Merck, 2015)	176	25.3	73.5	3.7	406	7.7	483	382	462
Agar Gallery Extra Pure (Merck, 2017)	50.3	9.7	0.7	0.7	233	7.6	539	414	472
Agarose Basic (AppliChem, 2017)	12.3	3.3	1.0	0.7	24.5	8.0	551	440	489

The difference between the two agar substances, which were purchased as "Extra Pure" from the same manufacturer but at different times, is quite striking. The older batch contains significantly more chlorides as well as sulfates/phosphates, which is also reflected in the conductivity of the solutions produced. The agarose, on the other hand, has a very high purity, which is due to the more complex production (additional separation of the agaropectin). Regarding pH-value the substances differ only slightly and hardly change the neutral pH-value of the distilled water used. Interesting are also the redox potentials, especially after the single steps of gel production. By heating to 88 or 95 °C oxygen is expelled from the mixture, which leads to approx. 100 mV lower potentials in freshly produced gels. Already after 3 h the redox potentials are more positive again due to the oxygen uptake. A check after 72 h showed no significant increase or decrease. As a conclusion of the investigations on the corrosiveness of the agar base substances it can be stated that they behave relatively neutral. Mixtures with 3 % in distilled water, even with 176 ppm chloride in the agar powder, in the end only result in approx. 5.3 ppm chloride in the finished gel, which is at a very low level. For particularly chloride-sensitive applications, however, the introduction via the gel former would have to be considered.

3.2 Mechanical properties

Agar gels are regarded as viscos-elastic, i.e. they combine the properties of solids and liquids. The proportion and type of polymer as well as the electrolytic components can influence these properties. Rheological measurements were carried out to investigate the viscos-elastic properties. A rheometer type MCR301 from Anton Paar was used in plate-plate mode. Using oscillation rheometry, so-called amplitude sweep curves were recorded at a displacement frequency of 10 rad/s (corresponds to 1.6 Hz) and a steady increase in deformation. The parameters of the investigated agar gels were varied according to polymer type, concentration and electrolyte composition, which are of interest for corrosion investigations. The gel samples had a diameter of 20 mm and a thickness of 4 mm. Figure 1 shows as an example the course of storage and loss modulus versus deformation for a neutral 2 % and 4 % agarose gel with addition of 0.1 mol/L NaCl.

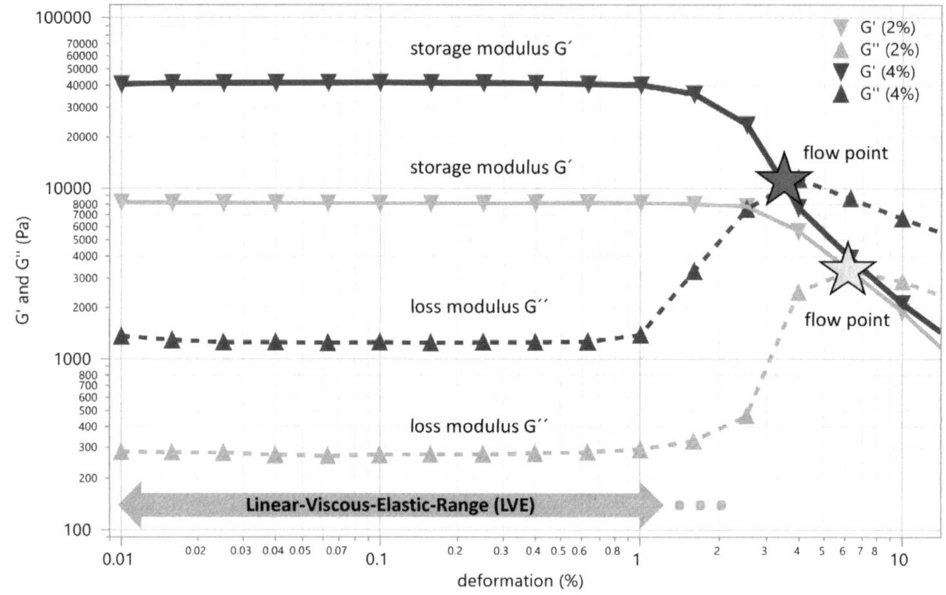

Figure 1. Loss (G") and storage modulus (G') curves in an amplitude sweep test for agar gels containing 2 % or 4 % agar and 0.1 mol/L NaCl.

It can be clearly seen that the height of the modules correlates with the agar concentration, as does the position of the intersection of the storage and loss module, which corresponds to the so-called crossover point or yield stress at which the gel structure is destroyed. Figure 2 shows the effect of various parameters on the yield stress. The orthogonal experimental designs were evaluated using the ordinary least squares method.

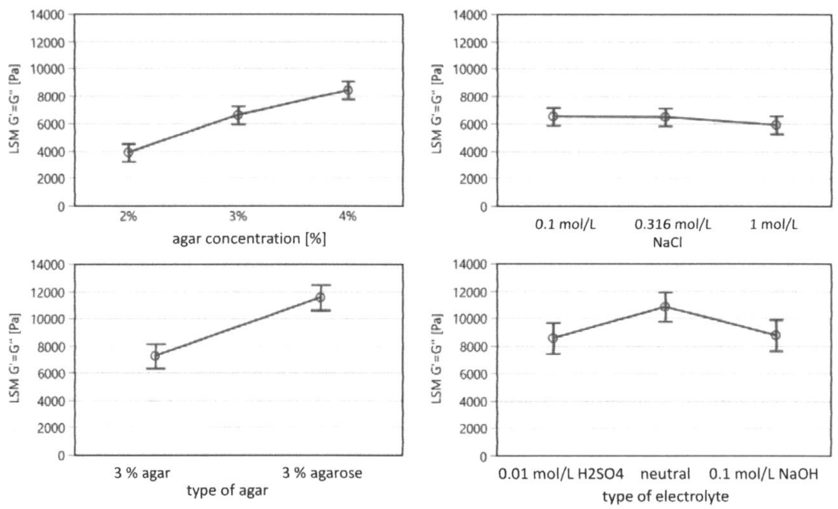

Figure 2. Influences of agar type, agar concentration and electrolytic additives on the flow voltage (G'=G", least squares mean - LSM) at which gel destruction begins.

A dominant dependence of the yield stress on the agar concentration is shown. Compared to the standard agar, agarose forms stiffer gels. Neutral salts such as NaCl have no significant influence in the mechanical properties up to 1 mol/L. On the other hand, acidic and alkaline solutions have been shown to weaken the gel network, which can be seen in the decrease in yield stress. The highest concentration of H_2SO_4 in the gel, which still allowed stable gels, was 0.1 mol/L.

3.3 Syneresis-based electrolyte film

In order to determine the supply of electrolytes at the surface of the gels, a methodology for extracting the electrolyte film from the surface was developed, which is relatively simple and functional. For this purpose, circular gels with a diameter of 15 mm were extracted from a gel plate. These are placed in a suitable plastic container and weighed with a precision balance. The container with the gel is then placed in a device in which a Whatman Grade 1 filter paper (11 μm pore size) is pressed down with 45 cN for 3 seconds (corresponds to a contact pressure of approx. 0.25 N/cm² in relation to the surface). This contact pressure is below the yield stress of the gels determined by rheological measurements and is sufficient to absorb the electrolyte film on the surface of the filter paper without damaging the gel. The gel in the container is then weighed again. The difference in mass is proportional to the volume of the removed electrolyte film and can be converted to an equivalent of the electrolyte film thickness, considering the geometry and electrolyte density.

In this way, a large number of different gels with variations in agar concentration and electrolyte components were investigated. In addition, the top and bottom sides of the gels were also examined separately. On the bottom side the gels solidify on a solid (polished acrylic glass) and on the top side in air. The factors with the greatest influence were the gel concentration and the solidification side or the surface that is formed. Based on the dependence of the electrolyte film thickness on the concentration of the gel former, a strong correlation with the gel network and the pore size is also likely. Figure 3 shows the equivalents for the electrolyte film thickness as a function of the agar concentration and the side on which it was measured in a diagram.

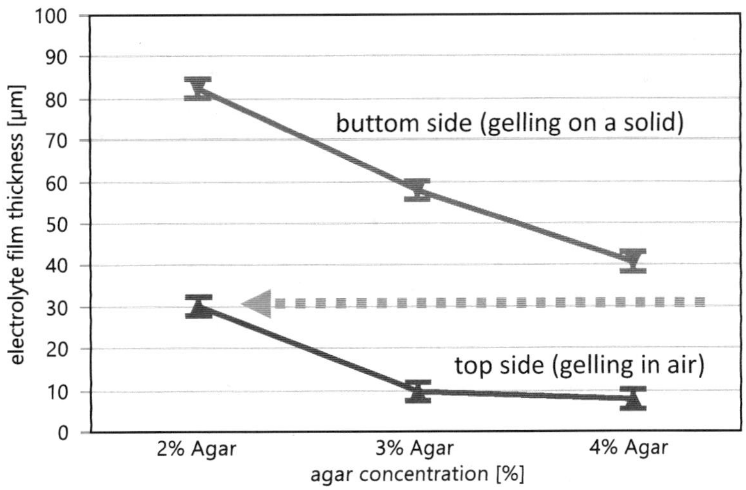

Figure 3. Determined equivalents for electrolyte film thicknesses of agar gels as a function of agar concentration and side of the gels.

A possible explanation why the top and bottom sides are so different from each other can first be deduced from the roughness of the surfaces. The gels were poured onto polished acrylic glass in a liquid state and solidified there. They thus adopt the smooth surface profile on the underside. On the upper side, on the other hand, they can dry out and shrink slightly during solidification, as there is virtually no air resistance to these processes. This makes the surface somewhat rougher and wavier, which means that a smaller proportion of the electrolyte film may be extracted.

If one excludes an influence of the surface, there is another, equally plausible and possible reason for the observed behavior, which can be found in the inner gel structure just below the surface. Drying during solidification can cause the polymer fibers to concentrate and lead to a zone with a denser gel network. This was already observed in 2007 by Labille et al. [24]. Diffusion experiments have shown that a zone approx. 120 µm thick exists at the edge of the gel, which has a significantly denser network. According to the authors, an agarose gel with a nominal 1.5 % polymer content at the edge has a diffusion behavior that corresponds to a gel with 5 % agarose and thus a denser network. This finding can be transferred to the context in Figure 3 (dotted arrow). If, for example, one assumes that a gel with 2 % agar on the top side actually corresponds to a gel with approx. 4.5 % agar, the determined courses fit together very well. It can be seen that with increasing agarose concentration, the equivalent for the electrolyte film thickness decreases more and more, i.e. the amount of liquid at the surface decreases. It should therefore be noted that the side solidified in air has a zone with a denser gel network and is therefore somewhat drier.

The influence of different ions on the electrolyte film thickness is, on the other hand, significantly less and in some cases not statistically relevant, e.g. for NaCl contents between 0.1 and 1 mol/L NaCl. Compared to distilled water as electrolyte base, acidic and alkaline electrolytes have slightly higher equivalents for the electrolyte film thickness. As in the rheological investigations, the relationship to the gel network or pore size is shown here. If the gel network is slightly weakened by certain ions in the electrolyte, the syneresis-induced leakage of electrolytes from the gel increases.

4 Application of Agar-based Gels for corrosion experiments on stainless steels

Corrosion tests on stainless steels are often carried out with the aim of determining the resistance to a particular type of corrosion. In most cases this is pitting corrosion. Resistance to pitting corrosion depends largely on the stability of the passive layer against the adjacent electrolyte and its ability to reform after mechanical or chemical degradation. This, in turn, depends on a number of alloy and material-related, surface-related and media conditions. The complex interaction of these factors in a corrosion system makes corrosion testing of stainless steels a recurring task. In industrial practice, the aim is to be able to carry out these corrosion tests on semi-finished and end products for quality assurance, as quickly, easily and virtually non-destructively as possible. The use of gel electrolytes instead of liquid electrolytes is one way to achieve this. Whether the use of liquid and gel electrolytes leads to the same or different results is shown by the following investigations.

4.1 Materials and experimental procedures

The stainless steel type X5CrNi18-10 (1.4301 or AISI 304) was used for the comparative investigations on the pitting corrosion resistance. The chemical composition of the steel was determined by spark spectrometry and is listed in Table 2. The element concentrations are within the range of the standard. It is noticeable that the sulfur content is at the upper permissible limit.

Table 2. Chemical composition of the investigated material X5CrNi18-10 (1.4301).

Fe	C	Cr	Ni	Mo	N	P	S	Mn	Si	Cu	V
bal.	0.04	18.0	9.51	0.47	0.03	0.02	0.01	1.49	0.35	0.13	0.05

Samples measuring 65x24 mm were cut from 2 mm thick metal sheets and prepared for the tests. The surface was wet ground with 180 grit sandpaper and finally with 500 grit. Before the tests, the samples were stored for 24 h in an air atmosphere with more than 95 % relative humidity at 22 °C.

NaCl concentrations of 1, 0.316 and 0.1 mol/L were selected as the electrolyte component of the liquid and gel electrolytes that causes pitting corrosion. The electrolytes were prepared with distilled water and pure NaCl (99.95 purity). For the gel electrolytes an agar concentration of 2, 3 and 4 % was chosen. Agar of the brand ExtraPure Empore (Merck, see table 1, product of 2017) was used for this purpose, which was weighed and then mixed with the NaCl solution. The mixture was then heated to 95 °C with stirring until the agar was melted and a transparent solution was obtained. The mixture was poured into acrylic glass bowls, resulting in a gel thickness of about 6 mm. Circular gels with a diameter of 14 mm were pierced out of the gel plates for the experiments. The bottom side of the gels was always coupled to the sample surface.

For the investigations with liquid electrolytes, a perforated foil was first glued to the prepared surface to limit the working electrode area (1.54 cm²), which proved to be insensitive to crevice corrosion. A measuring cell for an electrolyte volume of 4 mL was clamped onto the foil. The counter and reference electrode were immersed into the cell from above and positioned approx. 5 mm from the surface. A platinum sheet was used as counter electrode and a saturated Ag/AgCl electrode (reference potential +197 mV vs. NHE) as reference electrode. For the investigations with the gel electrolytes, the sample surface was also covered with a perforated foil. The gel electrolyte was positioned on the separated working electrode surface (1.54 cm²). A counter electrode made of a titanium mixed oxide grid was applied to the top of the gel and the Ag/AgCl reference electrode was coupled to its ceramic diaphragm. A constant contact pressure of approx. 0.25 N/cm² was set using a balance located under the sample. Figure 4 shows the two measurement setups.

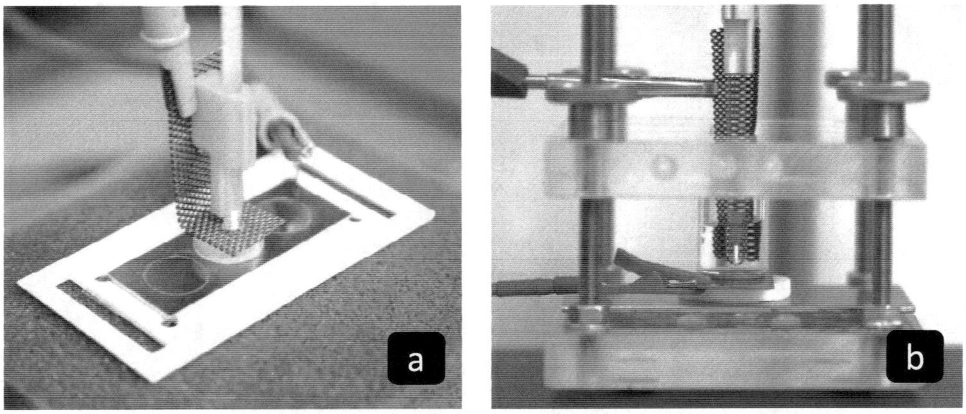

Figure 4. Experimental setup for electrochemical measurements with gel electrolyte (a) and liquid electrolyte (b).

Cyclic polarization experiments were carried out to determine the pitting corrosion and repassivation potentials. First the free corrosion potential was measured for 5 min and then the specimen was polarized from -150 $mV_{Ag/AgCl}$ vs. OCP with 1 mV/s in anodic direction. If a current density of 100 µA/cm² was exceeded, polarization was performed back in the cathodic direction at the same speed. The critical potential was determined when exceeding 10 µA/cm². The repassivation potential was determined when the current density dropped below 10 µA/cm² after reversing the polarization direction.

Furthermore, experiments were carried out with gel electrolytes of the same NaCl content, to which potassium ferricyanide ($K_3[Fe(CN)_6]$) was additionally added (KorroPad method). The addition causes on the one hand an increase of the free corrosion potential and on the other hand released iron ions are complexed to Prussian Blue, which leads to circular blue indications at the site of pitting corrosion. The gradations here were 1 mmol/L, 3.16 mmol/L and 10 mmol/L. The variant with 1 mmol/L $K_3[Fe(CN)_6$ and 0.1 mol/L NaCl corresponds to the contents of the "normal" Korropad [5], all other variations have a higher corrosiveness. Gels with 3 % agar and a thickness of about 2 mm were prepared from the brand ExtraPure (Merck, see table 1, product of 2017). The maximum achieved corrosion potential was measured on the applied gel pads in order to identify the level of polarization. After the 20-minute run, the gel pads were scanned and subjected to a semi-automatic image analysis to determine the percentage of indications from the total area. For this purpose, a self-developed software was used, which is based on color space separation and automatic recognition of circular structures.

After the experiments, the surfaces of the investigated samples were examined by scanning electron microscopy in order to compare the morphology of the pitting corrosion defects that occurred. The investigations were performed with the scanning electron microscope Evo 15 (Zeiss).

4.2 Determination of the pitting corrosion behaviour of stainless steel 1.4301

To investigate the pitting corrosion behavior of a stainless steel, the determination of critical potentials and repassivation potentials by means of dynamic polarization is an established procedure. The results are subject to a certain range of fluctuation depending on the material and test. In order to find out whether there is a statistically significant difference between gel and liquid electrolytes and whether the gel concentration has an influence, a full-factorial experimental design was set up and carried out three times randomly. The evaluation was carried out by analysis of variance based on the method of least squares. Figure 5 shows an example of the cyclic polarization curves for liquid and 3 % agar electrolyte with the graded NaCl contents.

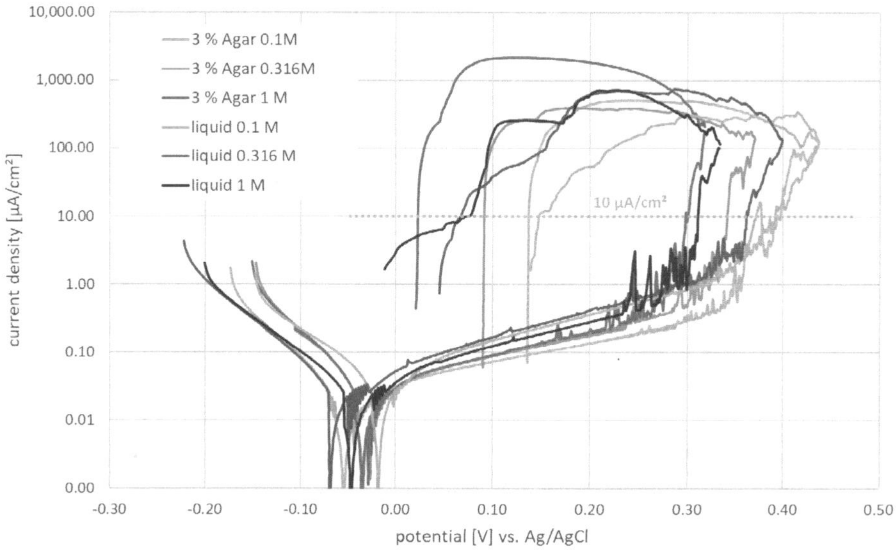

Figure 5. Cyclic polarization curves on the material 1.4301 when using gel and liquid electrolytes with different NaCl concentrations.

As expected, the NaCl content has the greatest influence, but regardless of whether a gel or liquid electrolyte is used. It can also be seen that the current densities in the passive area tend to be lower when using the gel electrolyte. Whether the critical potentials differ significantly depending on the type of electrolyte can only be shown by statistical evaluation of all the experiments, even if in Figure 5 it initially appears as if the current density increase in the case of the gel electrolytes occurs at slightly lower potentials. In the case of the repassivation potentials, likewise only the strong dependence on the NaCl content can be seen from individual test data.

From the experimental data the values for E_{crit} and E_{rep} were determined and subjected to statistical analysis. Figure 6 first shows the dependencies of the pitting corrosion potential on the NaCl content and the medium used, calculated as least squares mean values with corresponding error differences. As expected, a large influence and a high significance for the NaCl content can be seen. The distance between the individual NaCl variations is approx. 38 mV, between 0.1 and 1 M NaCl it is 77 mV. With regard to the type of medium or the content of agar, however, only a minor influence can be observed. The liquid electrolyte does not differ significantly from the agar variants with 3 and 4 %. In contrast, the agar gel with 2 % differs slightly, but statistically significant, from all other variants. The explanation for this was found when carrying out the experiment. In the gel variant with 2 % agar, a small amount of electrolyte escaped between the edge of the gel and the sample surface when pressed against the surface. It is likely that some of the water evaporated during the experiment, causing the NaCl concentration at the edge to rise slightly. As a result, pitting corrosion will occur at these points even at slightly more negative potentials. The effect was stronger at low NaCl contents and, averaged over all variants, leads to approx. 10 mV lower pitting corrosion potentials.

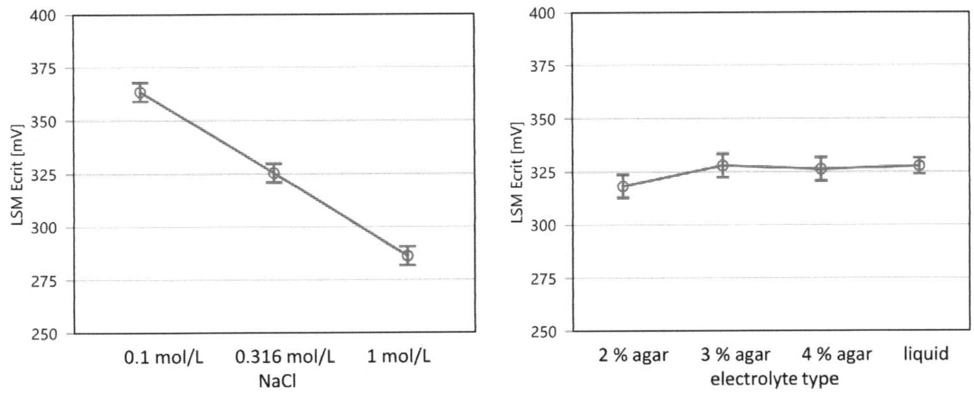

Figure 6. Least squares mean values for E_{crit} as a function of NaCl concentration and medium (agar gel of different concentration and liquid electrolyte).

Figure 7 shows the dependence of the repassivation potential on the NaCl content and the medium used, calculated as least-squares mean values with corresponding error differences. As expected, a large influence and a high significance for the NaCl content can also be seen here. The distance between the individual NaCl variations is approx. 54 mV, between 0.1 and 1 M NaCl it is 108 mV. With regard to the type of medium or the content of agar, however, only a minor influence can be observed. The repassivation potential for the liquid electrolyte is 20 to 24 mV lower compared to agar gels of different concentrations. The reason for this lies in a different morphology of the resulting holes, see chapter 4.4.

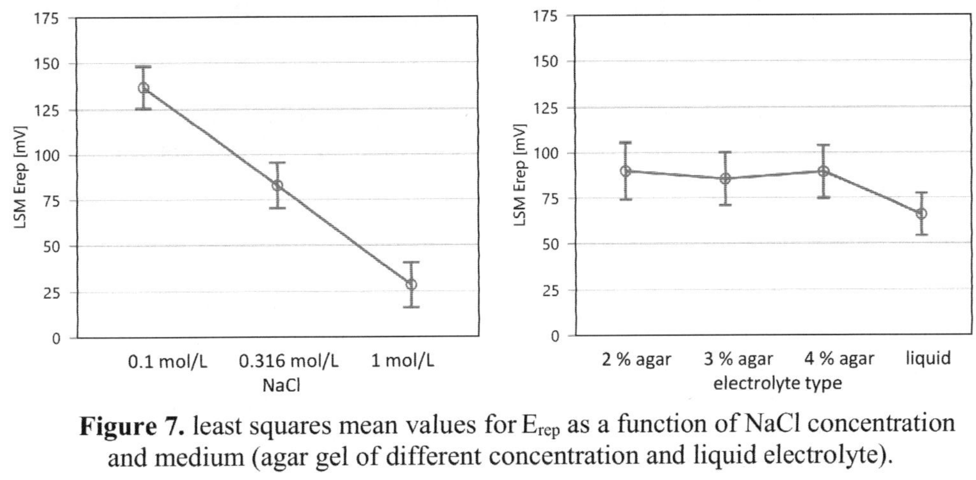

Figure 7. least squares mean values for E_{rep} as a function of NaCl concentration and medium (agar gel of different concentration and liquid electrolyte).

4.3 Use of indicators for the detection and localization of corrosion

To investigate the pitting corrosion resistance of stainless steels, gel-based electrolytes can also be used without external polarization. In contrast to external polarization, an appropriately concentrated oxidizing agent is required to set critical pitting potentials. The indication of pitting corrosion, characterized by a current increase under external polarization, is done here by an indicator for iron ions. This makes point-of-care application much easier, as no further measurement technology is required. Polarization effect and indication of iron ions is realized by $K_3[Fe(CN)_6]$. Starting from the standard concentration of the so-called Korropad [5] with 0.1 mol/L NaCl and 1 mmol/L $K_3[Fe(CN)_6]$, a test matrix with graded higher contents of NaCl and $K_3[Fe(CN)_6]$ was investigated on the material 1.4301.

Since this is a test method without external polarization, it is important to know the location of the open corrosion potential (OCP) that is established on the samples. The potential position determines whether the respective critical potential is reached or whether the sample is in the passive state or in the area of metastable pitting corrosion. For this reason, the maximum OCP values achieved by all gel variants were measured as well as the redox potentials on platinum. Figure 8 shows a summary of the results.

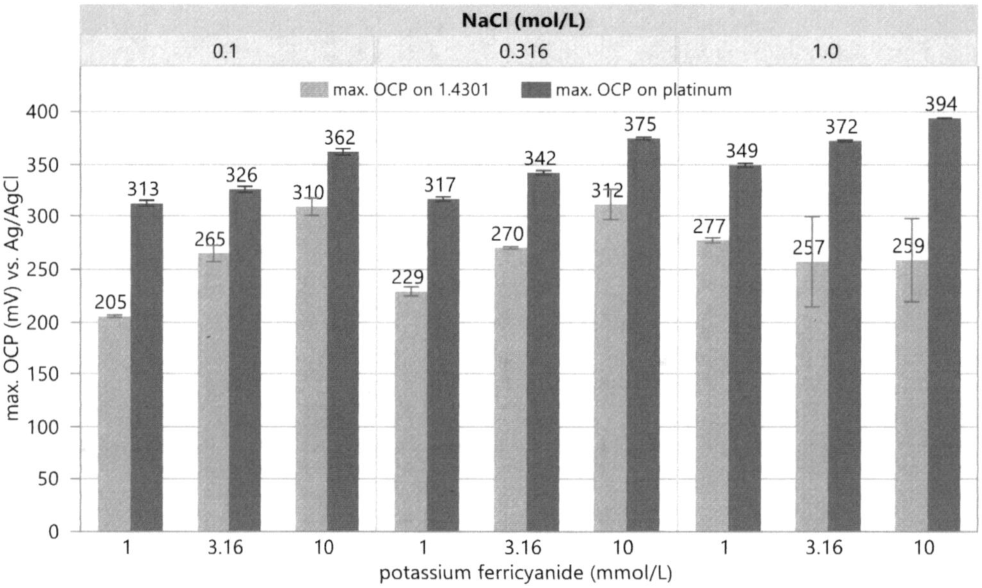

Figure 8. Maximum OCP values measured immediately after applying the gel pads to the passive surface of the samples of material 1.4301 and on platinum.

It can be seen that the content of $K_3[Fe(CN)_6]$ significantly and NaCl slightly influences the location of the OCP on stainless steel as well as on a platinum electrode. At high contents of NaCl and $K_3[Fe(CN)_6]$ the values scatter considerably more on stainless steel and do not reach the expected high potential values. This is due to the fact that stable pitting corrosion already occurs there. As a result, the OCP is shifted in a negative direction, since a high proportion of active corrosion sites leads to a strong electron surplus in the metal.

The gel pads were scanned after the experiments and evaluated by means of image analysis. Figure 9 shows exemplary images of the corrosion indications for all investigated gel variants with the mean area fraction and compares the critical pitting potentials from the potentiodynamically determined values. Stable pitting corrosion is indicated on some gel pads, recognizable by isolated large displays and a comparatively high area fraction. However, only those variants with simultaneously high NaCl and high $K_3[Fe(CN)_6]$ concentrations are affected. At 10 mmol/L $K_3[Fe(CN)_6]$ and 0.1 mol/L NaCl, for example, only metastable pitting corrosion occurs, since the E_{crit} of approx. 360 mV, determined from the potentiodynamic tests, is not reached. With 1 M NaCl and 1 mmol/L $K_3[Fe(CN)_6]$ a transition from metastable to stable pitting corrosion seems to have been detected, since the OCP and E_{crit} are close together. Some larger indications are already visible, but also many smaller ones, whose area is small.

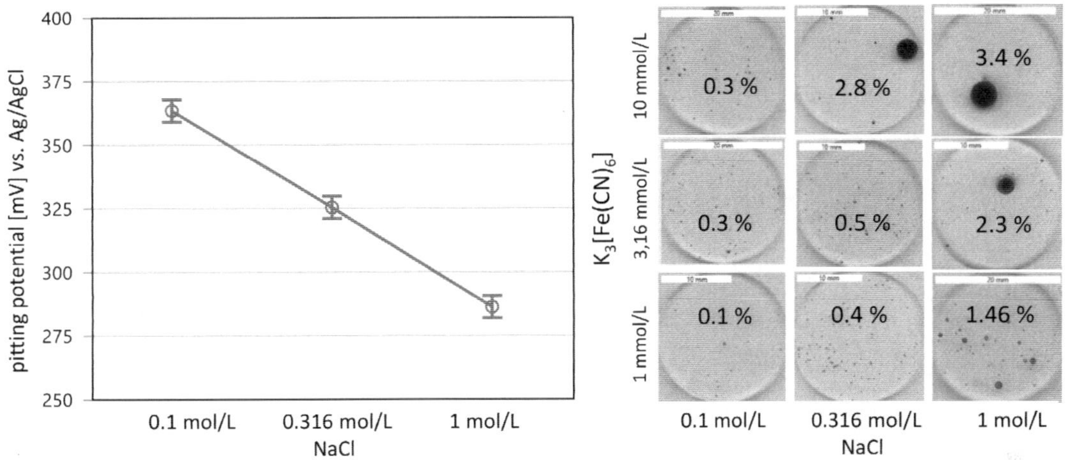

Figure 9. Comparison of results on the critical potentials of the material 1.4301 determined by potentiodynamic tests (left) and results from Korropad tests with graded NaCl and $K_3[Fe(CN)_6]$contents.

4.4 Comparison of pitting corrosion appearance in gel and liquid electrolytes

The results of the polarization experiments with gel and liquid electrolytes proved to be comparable, with only slight difference in the repassivation. Even in the gel pads used without external polarization, the $K_3[Fe(CN)_6]$ was able to achieve approximately the same critical potentials and thus pitting corrosion occurs. To clarify whether the pitting corrosion appearance differ when gel and liquid electrolytes are used, the samples were microscopically examined after the experiments, see Figure 10.

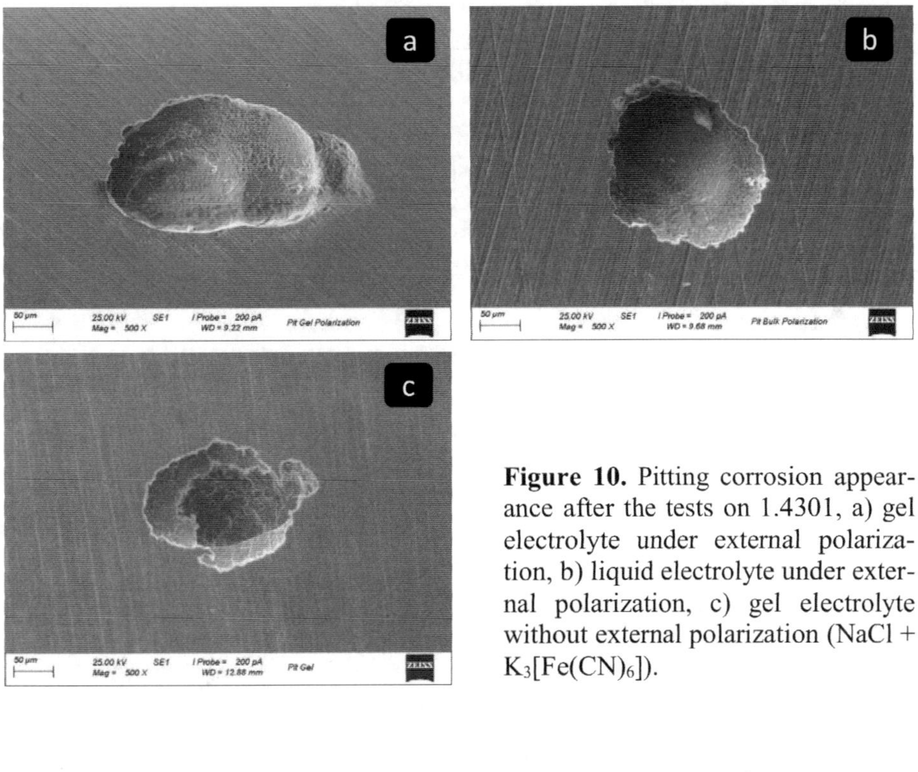

Figure 10. Pitting corrosion appearance after the tests on 1.4301, a) gel electrolyte under external polarization, b) liquid electrolyte under external polarization, c) gel electrolyte without external polarization (NaCl + $K_3[Fe(CN)_6]$).

If one compares the appearance of the resulting pits, it is noticeable that in gel electrolytes under external polarization (Figure 10 a) the metal dissolution sites tends to be more shallow, open and extended. This makes it more likely that the electrolyte is less concentrated with chlorides and that a pH drop through hydrolysis is less pronounced. The repassivation can proceed faster, which is expressed by more positive repassivation potentials, see figure 7.

In contrast, liquid electrolytes and external polarization (Figure 10 b) tend to produce hemispherical and thus deeper holes, in which remnants of a perforated hole cover can still be seen at the edge, indicating initially hidden pitting corrosion. This allows the occluded electrolyte to decouple from the bulk and mix less well. In addition, a voltage drop due to geometrical reasons delays the repassivation at the bottom of the hole.

However, such flat corrosion appearance has only occurred with the gel electrolytes in the case of external polarization. If pitting corrosion is initiated under free conditions, e.g. by the presence of a sufficient amount of $K_3[Fe(CN)_6]$and NaCl in the electrolyte, then the corrosion appearance in the gel electrolytes are comparable to those in the liquid electrolytes (Figure 10 c). At an earlier stage, the pits still have a perforated cover, which dissolves or detaches as pitting corrosion progresses.

5 Conclusions

Agar-based gel electrolytes are a viable electrolyte alternative for the investigation and testing of stainless steels with regard to pitting corrosion. They have the advantage that the liquid electrolyte is immobilized in the gel and a thin electrolyte film is formed on the surface, which enables electrochemical measurements. NaCl contents of different concentrations in the gel can be used without significantly changing the gel matrix.

Potentiodynamic experiments to determine pitting and repassivation potentials on gel and liquid electrolytes show a very good agreement between the electrolyte types. At low contents of 2 % agar in the gel, however, there is a risk that electrolyte will leak out at the edge, evaporate and thus cause pitting corrosion too early. When using gels under external polarization conditions, stable pitting corrosion leads to shallower pits, which in turn leads to more positive repassivation potentials, as repassivation takes place more easily. In contrast, these shallow pitting corrosion appearances do not occur in gel electrolytes without external polarization. Here the pit morphology is comparable to that of liquid electrolytes.

Agar gels can be used very well as test electrolyte for testing the pitting corrosion resistance of stainless steels even without external polarization. By adding $K_3[Fe(CN)_6]$to the chloride-containing solution in the gel, specific potential ranges can be set which are in the passive range (with metastable events) or already in the stable pitting corrosion range depending on the material and surface condition. Using the example of the used material 1.4301 with graded $K_3[Fe(CN)_6]$-NaCl contents, the position of the potentiodynamically determined critical potential ranges could be confirmed. It should be noted that the material used here has a relatively poor pitting corrosion resistance compared to other batches of the same type, which is due to the slightly increased sulfur content.

Acknowledgement

The author would like to thank Mrs. M.Sc. Christine Becker (OvGU Magdeburg) for performing the electrochemical measurements and electrolyte film thickness, Mr. M.Sc. Ludwig Gropler (OvGU Magdeburg) for manufacturing of setups for measuring electrolyte film thickness, iLF Magdeburg GmbH for performing the rheological measurements and BAM Berlin (FB 7.6) for the chemical analysis of the agar raw materials. Special thanks also go to Mr. M.Sc. Sebastian Hütter (OvGU) for programming the image analysis.

Funding

Funded by the German Research Foundation (DFG) - Project number 330472124

IOP Conf. Series: Materials Science and Engineering **882** (2020) 012010 doi:10.1088/1757-899X/882/1/012010

References

[1] Laque F L, May T P and H H Uhlig, "Corrosion in Action", *International Nickel Company of Canada*, Ltd, 1955

[2] Isaacs H S, Adzic G and C S Jeffcoate, *Corrosion* 56 (2000) Issue 10, S. 971-978

[3] *DIN EN ISO 10309:2016-08*: Metallic coatings - Test methods for determination of porosity - Ferroxyl test (ISO 10309:1994); German version EN ISO 10309:2016

[4] Petersen P and H. Emnéus, *Acta Orthopaedica Scandinavica*, 1959, 29:1-4, 331-340

[5] Lehmann J, Burkert A, Müller T, Bohlmann T and A Burkert: *Final report on the IGF project 17136 N/1* "Detection of corrosion-sensitive surfaces of stainless steels by processors

[6] *Patent specification DE 10 2010 037 775 B4*, patent granted on 8.5.2014

[7] Rosemann P, Kauss N and A Heyn: KorroPad testing - applications from industry and research, *Proceedings: 3-Länder-Korrosionstagung - Korrosion ist kein Zufall - Neue Messmethoden, Analytik und Simulation*, May 2019, Frankfurt a. Main

[8] Rosemann P, Müller T, Babutzka M and A Heyn, *Materials and Corrosion*, 2015, 66, 45–53

[9] Reinemann S, Babutzka M, Rosemann P, Lehmann J and A Burkert, *Materials and Corrosion*, 2019; 70: 1776– 1787

[10] Kauss N, Heyn A, Halle T and P Rosemann, *Electrochimica Acta* 317 (2019) 17-24

[11] Newton C J and J M Sykes, *Corrosion Science*, 28 (1988) Nr. 11, S. 1051-1074

[12] Spark A J, Cole I, Law D and L Ward, *Corrosion Science* 110 (2016), S. 174-181

[13] Spark A J, Cole I, Law D, Marney D and L Ward, *Materials and Corrosion* 67 (2016), S 7–12

[14] Spark A J, Law D W, Ward L P, Cole I S and A S Best, *Environmental Science & Technology* 2017 51 (15), 8501-8509

[15] Vanbrabant J and N van de Velde: Industrial application of an electrochemical corrosion test using a gel matrix as simulation for atmospheric and solid media. *Proceedings: European General Galvanizers Association Intergalva Berlin*, Vol. 19 pp 29/1 to 29/13, June 7th, 2000

[16] Shao X M and J L Feldman, *Journal of Neuroscience Methods* 159 (2007) 108–115

[17] Monrrabal G, Guzmán S, Hamilton I E, Bautista A F and F Velasco, *Electrochimica Acta*, Volume 220, December 2016, Pages 20-28

[18] Monrrabal G, Ramírez-Barat B, Bautista A, Velasco F and E Cano, *Metals* 2018, 8, 500

[19] Monrrabal G, Huet F and A Bautista, *Corrosion Science*, Volume 148, March 2019, Pages 48-56

[20] Monrrabal G, Bautista A and F Valesco, *CORROSION*, December 2019, Vol. 75, No. 12, pp. 1502-1512

[21] Cano E, Crespo A, Lafuente D and B Ramirez Barat, *Electrochemistry Communications*, 41 (2014), S. 16-19

[22] Di Turo F, De Vito C, Coletti F, Mazzei F, Antiochia R and G Favero, *Microchemical Journal*, Volume 134, 2017, Pages 154-163

[23] Babutzka M, Burkert A and A Heyn: Corrosion investigations with gel-like electrolytes to de scribe the corrosion protection effect of zinc coatings. *Proceedings: 16th Summer Course Materials and Joining*, Magdeburg, September 08 and 09, 2017, pages 119-128

[24] Labille J, Fatin-Rouge, N and J Buffle, *Langmuir* 2007, 23, 2083-2090

IOP Conf. Series: Materials Science and Engineering **882** (2020) 012011 doi:10.1088/1757-899X/882/1/012011

Recent advances in additive manufacturing of Mo-Si-B alloys – A status report on the cooperative project LextrA -

M Krüger[1], J Schmelzer[1], D Fichtner[2], C Heinze[3], Y Küsters[2], S K Rittinghaus[4], A Weisheit[4], R Heinz[5], L Gerking[5], K Gruber[6], S Schmigalla[7] and S Schultze[7]

[1]Institute of Materials and Joining Technology, Faculty of Mechanical Engineering, Otto-von-Guericke University Magdeburg, Universitätsplatz 2, 39106 Magdeburg, Germany
[2] Siemens AG, Corporate Technology Research in Energy and Electronics Coatings & Additive Manufacturing, Siemensdamm 50, 13629 Berlin, Germany
[3] Siemens AG, Energy Generation Large Rotation Equipment, Additive Manufacturing Materials and Technology, Huttenstraße 12, 10553 Berlin, Germany
[4] Fraunhofer - Institute for Laser Technology, Steinbachstraße 15, 52074 Aachen, Germany
[5] NANOVAL GmbH & Co.KG, Kienhorststraße 61-65, 13403 Berlin, Germany
[6] Dr. Kochanek Entwicklungsgesellschaft, Freiheitstr. 57, 67434 Neustadt a. d. Weinstraße, Germany
[7] Institut für Korrosions- und Schadensanalyse, Schwiesaustr. 11, 39124 Magdeburg, Germany

manja.krueger@ovgu.de

Abstract. Mo-Si-B alloys are potential materials for ultra-high temperature applications, e.g. as turbine blades. Due to their excellent mechanical properties they are subject to basic research for about 20 years. The next step towards industrial application of this type of alloys is to use the current knowledge on microstructure-properties relationships and combine this with innovative laser additive manufacturing (AM). This way of processing is very challenging since the melting point of Mo-Si-B materials is > 2000 °C and the brittle-to-ductile transition temperature is typically about 900 °C. The authors demonstrate that gas atomized Mo-Si-B powders of three different compositions could be successfully processed by different AM processes, namely Direct Energy Deposition (DED) and Laser Powder Bed Fusion (LPBF). It is verified that the mechanical and oxidation properties provided by the AM materials are comparable and competitive to similar alloy compositions from conventional processes.

1 Introduction

The demand for novel materials for structural high temperature applications results in extensive research on high strength metallic and intermetallic materials. Aiming in resource-saving, sustainable and cost-efficient industrial processes the development of new materials must be combined with innovative processing methods. For improving the efficiency of gas or airplane turbines, new materials beyond the capability of Ni-based superalloys are under consideration [1–3]. Besides others, multi-phase Mo-Si-B alloys are seen as potential substitutes, as they provide improved creep resistance, good fracture toughness and high thermal resistance. Simulations and comparative experimental assessments against Ni-base superalloys already demonstrated the outstanding performance of Mo-Si-B alloys [4–6].

In the past two decades basic research was dedicated to basic understanding of microstructural evolution, phase equilibria, different options of processing and application-related properties as fracture toughness, creep response and oxidation behavior [7–10]. In terms of bulk processing two general approaches were followed in the past, i.e. processes that are based on phase transformations from a molten phase (classical ingot metallurgy, directional solidification) or processes based on diffusion and/or phase transformation mechanisms in the solid state (powder metallurgical processes). The classical melting processes are very energy-intensive and costly due to ultra-high melting points of Mo-based materials up to 2600°C, and does not allow a high degree of freedom in terms of tailoring properties due to inhomogeneous microstructural evolution of the individual phases. An exception are eutectic or near-eutectic alloys having a melting point < 2100 °C, which show very fine network-like microstructures. Such eutectic microstructures can be grown directionally to provide anisotropic materials properties [11–13]. The more established processes are powder metallurgical (PM) processes that avoid phase formation from the melt and therefore, allows the formation of very fine-grained and homogenous microstructures with typical isotropic properties. Furthermore, by PM techniques different additional elements or compounds can be incorporated in the alloys to modify their properties in a wide range [7]. However, using the above-mentioned processing methods extensive mechanical finishing will be necessary to produce complex parts like turbine blades with cooling structures. This is challenging due to the extremely hard silicide phases in Mo-Si-B alloys.

In our present approach we use the basic knowledge on the microstructural evolution of Mo-Si-B alloys and combine this with new processing methods aiming in manufacturing of net-shaped materials (Fig. 1). In the framework of the project LextrA (full title in German: Laserbasierte additive Fertigung von Bauteilen für extreme Anforderungen aus innovativen intermetallischen Werkstoffen) a consortium of industrial partners (Siemens AG, Nanoval, Kochanek Entwicklungsgesellschaft, IfKorr) and research institutes (Fraunhofer ILT Aachen, Otto-von-Guericke University Magdeburg) focuses on the industrial application of additive manufacturing (AM) of new materials, i.e. Mo-Si-B, V-Si-B and Fe-Al alloys. These groups of intermetallic materials were identified as potential candidates for turbine blade materials in different stages of gas turbines.

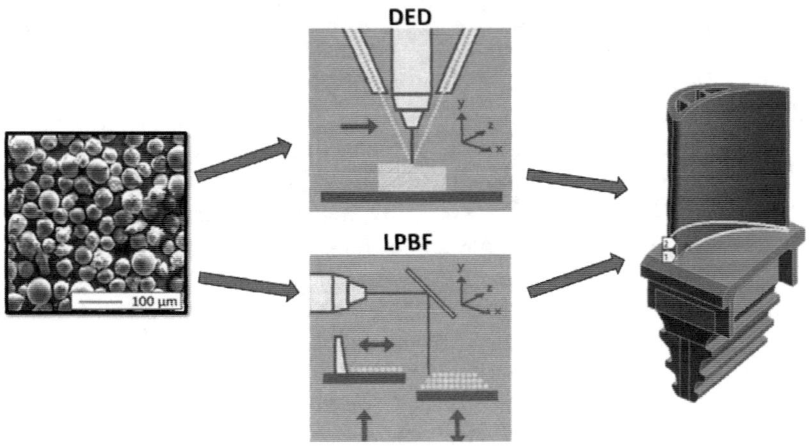

Figure 1. Schematic process chain showing the gas atomized Mo-Si-B powders, both laser-based processes used for this study and a potential turbine blade geometry (FEM model from [14]).

To date there is just a small number of publications on additive manufacturing of Mo-based alloys worldwide, because this field of research is very challenging from the technical and metallurgical point of view [15–17]. The present manuscript will concentrate on the advances that were achieved within the project LextrA with respect to DED (Direct Energy Deposition) and LPBF (Laser Powder Bed Fusion) processing of Mo-Si-B materials. Experimental findings on AM of other high temperature materials, namely multi-phase V-Si-B alloys, are published recently elsewhere [18].

IOP Conf. Series: Materials Science and Engineering **882** (2020) 012011 doi:10.1088/1757-899X/882/1/012011

2 Materials and Methods

The alloys Mo-9Si-8B, Mo-13.5Si-7.5B and Mo-16.5Si-7.5B (in at.%) were chosen for this study. Alloy Mo-9Si-8B is a reference alloy for the Mo-rich Mo-Si-B materials, for which a large database is available in literature [6,7,10,19]. Both alloys with higher silicon and boron concentrations are close to the eutectic point and provide therefore, a lower melting point as compared to the reference alloy, which could make manufacturing more energy-efficient and at least provides a very defined microstructure [11,20]. The powders for the AM processes were processed by gas atomization (GA). The solid raw materials (Mo > 99.95%, Si > 99.6% and B > 99.4% purity) were molten and superheated in a crucible (inductive heating system) and subsequently atomized into alloyed powders. More details on this process are given in [21]. The powders were investigated by SEM in combination with EDS to analyze the morphology, the chemical homogeneity and potential agglomeration or satellite formation. Microscopic images of the three Mo-Si-B powders are presented in Figure 2, which show mainly spherical morphology that is mandatory for a good flowability during the AM process.

Figure 2. Micrographs of gas atomized powders of alloys (a) Mo-13.5Si-7.5B for processing via DED as well as (b) Mo-9Si-8B and (c) Mo-16.5Si-7.5B for AM via LPBF.

After GA the powders were sized to fractions of +45/−90 µm for the DED and +15/–45 µm for the LPBF process. It is known from refractory powders that they absorb impurities from the environment over time depending on storage [22], especially embrittling oxygen. Due to these reasons, the powders were heat treated before LPBF processing to "clean" the surfaces and therefore, avoid crack formation in the AM builds. A specific technical requirement for successful processing of crack-free Mo-Si-B samples is the need of high pre-heating temperatures, i.e. approx. 700 °C for the DED process and up to 1200 °C in the substrate for LPBF process. More details on the LPBF are described more specifically in [23,24], the description of the DED process is provided in [18,21,25].

The microstructures of the AM builds were investigated in terms of their microstructural evolution by SEM and EBSD methods. Phase fractions were determined and compared to literature data of existing phase diagrams and other Mo-Si-B alloys. For analyzing the porosity, image analysis was performed by using both the software Stream Motion (Olympus) and ImageJ, working on the principle of separating different grey values.

The high temperature mechanical properties were determined by means of compressive creep tests. Parallelepiped-shaped specimen were electro-discharge machined from the AM builds with dimensions of 2 mm x 2 mm x 3.5 mm. All creep tests were carried out under constant true stress or constant strain using a Zwick electromechanical testing device equipped with a Maytec furnace at temperatures at 1093 °C, which involved continuous monitoring of creep strain and stress throughout the test.

Cyclic oxidation tests were performed at 800 °C, 1100 °C and 1300 °C to get an idea on the oxidation response in the so-called pesting regime [9] and at potential application temperatures, respectively. Therefore, samples were put into a hot furnace under atmospheric conditions, exposed for a certain period of time and then removed. Afterwards, the surface of the materials was recorded by optical microscopy and the weight change was monitored by weighing the samples after each removal from the furnace treatment.

3 Results and Discussion

This paragraph provides an overview on the microstructures of AM Mo-Si-B builds evolving from the DED and LPBF processes as well as selected results on the mechanical behavior of AM materials in comparison to data from differently processed materials. Additional materials properties determined at room temperature and elevated temperatures can be found in [21,24,25]. Furthermore, first insights into the oxidation response of AM Mo-Si-B materials will be given.

3.1 Additively manufactured compacts and microstructures

Typical macroscopic cross sections of AM alloys, i.e. LPBF Mo-16.5Si-7.5B, DED Mo-13.5Si-7.5B, and DED Mo-9Si-8B as well as a hollow AM structure are presented in Figure 3. Samples with different geometries show the processability of this type of material. In general, typical defects in such high melting multi-phase intermetallic materials are cracks or ablation in consequence of internal stresses as well as porosity. In another study internal microcracks were observed which are assumed to reduce the hardness as compared to as-cast alloys and might also be responsible for premature failure of AM alloys [17]. Due to pre-heating of substrates in LPBF process [24] and using a movable induction coil for DED processing [25] cracks could be avoided in our experiments (Fig. 3). However, there is residual porosity, which in case of LPBF specimen is typically <1 %, while DED samples have a porosity of about 2 %.

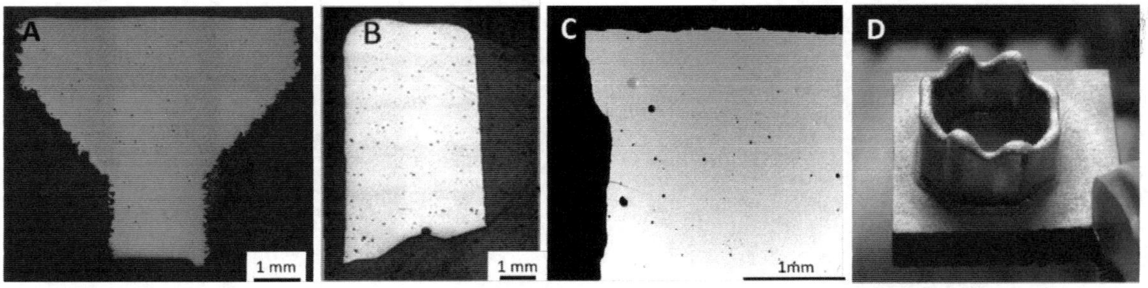

Figure 3. Macroscopic cross sections of (A) LPBF build of alloy Mo-16.5Si-7.5B and DED builds of (B) alloy Mo-13.5Si-7.5B and (C) alloy Mo-9Si-8B as well as (D) hollow structure of Mo-9Si-8B produced by DED.

The microstructure of the builds was investigated by SEM in combination with EBSD, as exemplarily shown in Figure 4 for three different AM alloys. Distinct zones resulting from the layer-wise laser processing were also observed and described in [21]. However, the overall microstructure evolution in AM processes seems to be quite familiar, since it has a similar morphology as compared to microstructures as solidified from other melting processes. In case of alloy Mo-9Si-8B, which is located in the primary solidification area of Mo solid solution (Mo_{ss}), large dendrites of this phase (red in Fig. 4) are formed, before the residual melt solidifies in eutectic structures, finally reaching the ternary eutectic Mo_{ss}-Mo_3Si-Mo_5SiB_2. Interestingly, the observed microstructural evolution is in good agreement with the sequence of solidification analyzed by arc-melting experiments as well as the predictions of phase equilibria from the Mo-rich corner of the ternary Mo-Si-B phase diagram [26,27]. However, our results are in contrast to Makineni et al. [16], who found the constituents Mo_{ss}, Mo_5SiB_2 and Mo_5Si_3 in ball-milled and laser-melted alloy Mo-8.5Si-5.6B, of which the composition is very close to alloy Mo-9Si-8B. This difference is assumed to be due to rapid quenching and accompanying non-equilibrium conditions in the process of Makineni [16].

Alloys with higher Si and B concentrations are located closer to the binary eutectic valleys and/or to the ternary eutectic point as described recently by Hasemann et al. [26]. Consequently, the content of the Mo_{ss} primary phase in Mo-13.5Si-7.5B is less than in Mo-9Si-8B (see Fig. 4). Furthermore, the alloy Mo-16.5Si-7.5B entirely provides a ternary eutectic microstructure consisting of the phases Mo_{ss} (red), Mo_3Si (blue) and Mo_5SiB_2 (yellow), which again is in very good agreement with observations from other melting processes [26,28]. This gives evidence that the additive manufacturing processes used in this study result in (near-) equilibrium microstructures. It is well-known that phase transformations are not existing within the three-phase region of Mo_{ss}-Mo_3Si-Mo_5SiB_2 up to temperatures of at least 1800 °C [28]. Therefore, it can be assumed that significant and sudden modifications of the properties due to phase transformations will not occur during static or cyclic thermal loadings at potential application temperatures of turbine blades.

The present results show that Mo-rich multi-phase Mo-Si-B alloys with ultra-high melting points between ~ 2050 °C < T < 2400 °C can be processed not only with near-equilibrium microstructures but also free of cracks by using pre-alloyed gas atomized powders and the tailored additive manufacturing techniques DED and LPBF.

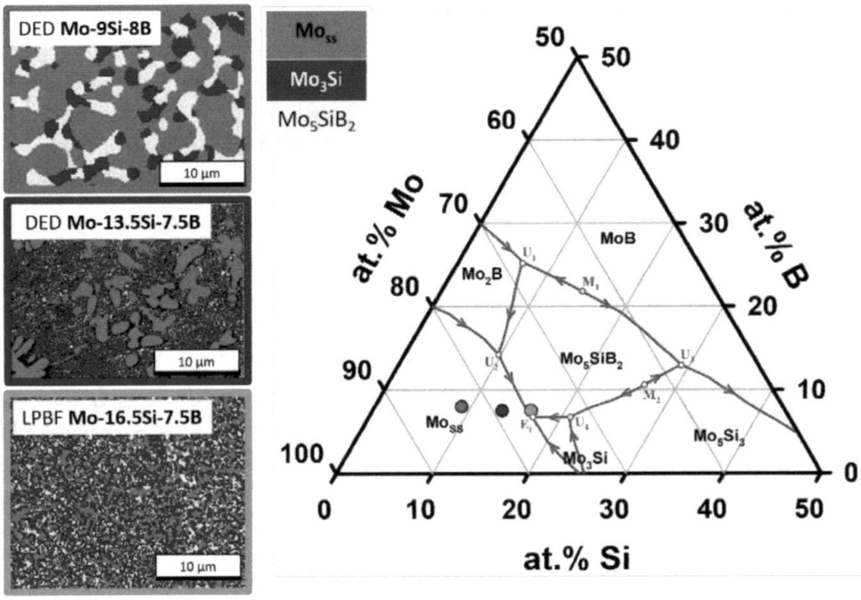

Figure 4. Microstructures of AM Mo-Si-B alloys with primary Mo solidification (Mo-9Si-8B and Mo-13.5Si-7.5B) as well as near eutectic (Mo-16.5Si-7.5B) in context of a recently published liquidus projection (reproduction from [26] provided by G. Hasemann).

3.2 Mechanical properties in comparison to literature data

The mechanical properties of the alloys were analyzed from room temperature up to high potential application temperatures. In previous studies on PM Mo-Si-B materials it was found that alloys with higher Mo_{ss} concentrations tend to provide improved mechanical characteristics at ambient temperatures, e.g. a higher fracture toughness and reduced brittle-to-ductile-transition temperatures (BDTT), while alloys with higher silicide concentrations have an improved creep resistance and reduced oxidation rates [7].

Hardness measurements of DED materials as presented in [21] as well as for LPBF samples [24] show a very good agreement with the values for Mo-Si-B alloys achieved by other processes [7,29]. It is important to note, that the macrohardness of the alloy is typically in between the hardness of the Mo_{ss} phase (2…10 GPa, depending on the concentration of dissolved elements and the grain size) and the hardness of the eutectic regions (16…20 GPa). These results indicate that the strength of AM alloys at room temperature could also be competitive compared with conventionally processed Mo-Si-B materials, but was not proven by tensile tests, yet. However, first bending tests provide new insights into the temperature-dependent flexural stress-strain data and, therefore, the BDTT. These results, which will be presented in detail in [24], show that there is even an improvement of the ductility of the AM material at 950 °C in comparison with PM Mo-Si-B alloys.

A very important issue in terms of application in a gas turbine is the creep resistance of Mo-Si-B alloys. In our earlier work we found out that the creep resistance is improved against the state-of-the-art turbine blade materials nickel-based superalloys at their typical application temperatures around 1100 °C. This advantage becomes even more pronounced at higher temperatures of 1200 °C, which could be a potential application temperature of turbines using new high-temperature resistant Mo-Si-B materials [6,12].

For the first time, the time-dependent deformability of AM alloys at constant applied stresses was investigated within the project LextrA by the same experimental procedure as used for other Mo-Si-B alloys in previous studies. This enables a direct comparison of AM Mo-Si-B alloys with powder metallurgically (PM) processed as well as arc-melted (ARC) and zone melted (ZM) alloys [12,30]. The results are presented as a double-logarithmic representation of the minimum creep rate vs. applied stress (so-called Norton plot) in Figure 5. All Mo-Si-B alloys show a significantly improved creep resistance compared to pure Mo, which is due to the much more creep resistant silicide phases Mo_3Si and Mo_5SiB_2 vs. bcc Mo or the Mo_{ss} phase. Furthermore, there is a strong divergence between the creep strength of alloys Mo-9Si-8B (with ~ 50 % Mo_{ss}) and the near-eutectic alloys with ~ 20 – 30 % Mo_{ss}. The reasons are seen mainly in the Mo_{ss} volume fraction. Additionally, the distribution of the phases (network-like or interpenetrating structure of PM alloys vs. typical cast microstructures of ARC and AM alloys) and the size-scale of the constituents play a role to understand the creep response of the alloys. Due to the ultra-fine grained structure of eutectic AM alloys, the lower creep resistance of AM alloys against ARC alloy could be rationalized. TEM analyzes should help to understand the creep mechanism of AM alloys and are subject of ongoing research. However, from other research on fine-grained Mo-Si-B materials it is already known that heat treatments targeted in grain growth will improve the creep resistance. This must be proven for the novel AM materials.

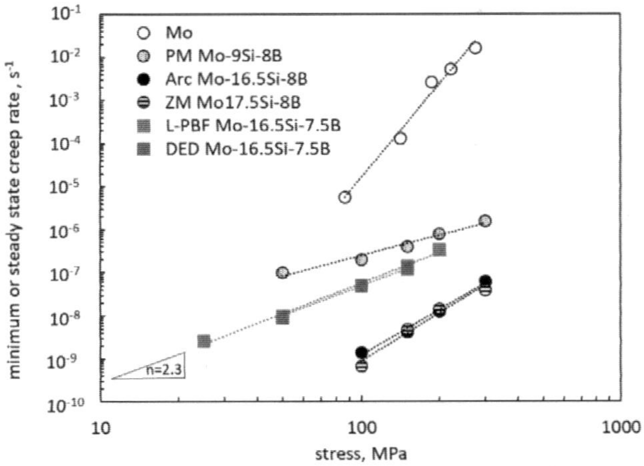

Figure 5. Norton plot of differently processed Mo-Si-B alloys at a potential application temperature of 1093 °C.

Figure 6. DED Mo-9Si-8B after oxidation for 100 h at 1300 °C.

3.3 Oxidation response

The interaction of Mo-Si-B alloys with the environment at different temperatures plays a key role for potential applications as turbine blade materials. It is well-known that Mo-Si-B alloys suffer from the so-called pesting phenomenon, which is accompanied with catastrophic mass loss at intermediate temperatures due to MoO_3 formation and evaporation [9]. Depending on the microstructures, especially on the fraction, size scale and distribution of the Mo_{ss} phase, the quantity of mass loss will be affected. At higher temperatures, e.g. 1100 °C, the formation of a well-sticking borosilicate glass layer may protect the surface from further oxidation if the Mo_{ss} concentration in the alloy is not higher than 50% and the phases size is in the lower µm-range [7,9].

In this work the AM materials were exposed to air for certain periods and the mass change was determined in correlation to the surface area of samples. The AM materials show a similar behavior as compared to other Mo-Si-B alloys in dependence on the microstructural features as addressed before. Therefore, it is not surprising that the highest mass loss of the AM Mo-Si-B samples is observed after treatment at 800 °C as it is a known mechanism [9]. In contrast, there is just a small mass loss observed at 1100 °C and 1300 °C and the shape of samples was preserved after oxidation treatment due to the self-protection mechanism by protective glass formation at the surface. This behavior is, of course, dependent on the alloy composition and other parameters. An example of the macroscopic morphology of the DED Mo-9Si-8B alloy after oxidation treatment at 1300 °C for 100 h is provided in Figure 6. More details on the oxidation response at different temperatures and data on specific mass change of AM Mo-16.5Si-7.5B material will be presented elsewhere [31].

However, there are some potential approaches to protect Mo-Si-B alloys from environmental degradation by different coating strategies, like sputtering techniques, pack cementation and dip-coating methods [32–34]. These strategies typically follow the target to protect Mo-based alloys with layers of oxidation-resistant silicides, which have a similar thermal expansion coefficient as the substrate materials. Therefore, it is expected that coatings will solve the problem of oxidation sensitivity of Mo_{ss}-rich Mo-Si-B alloys in the future.

4 Summary and Conclusions

This work shows the successful additive manufacturing of Mo-Si-B bulk materials by Laser Powder Bed Fusion (LPBF) and Direct Energy Deposition (DED). This way of processing is very challenging since the melting point of Mo-Si-B materials is > 2000 °C and the brittle-to-ductile transition temperature is typically higher than 900 °C. Three different alloys taken from the Mo-rich corner of the Mo-Si-B system were investigated: Mo-9Si-8B, Mo-13.5Si-7.5B and Mo-16.5Si-7.5B. For both AM processes size-fractioned gas atomized powders were used. An important outcome is that the bulk materials of the three alloys were processed free of cracks, but they have a residual porosity of < 1 % in case of LPBF and < 2 % in case of DED. It could be shown that the microstructure of all alloys correspond to the equilibrium phase fractions of Mo_{ss}, Mo_3Si and Mo_5SiB_2 as expected by the phase diagram. Therefore, no additional heat treatment was necessary to achieve equilibrium microstructures.

In this study the creep resistance of LPBF and DED Mo-Si-B materials were comparatively assessed against Mo-Si-B materials from powder metallurgical, arc-melting and zone melting processes. This data shows that the high temperature mechanical properties provided by the AM materials are comparable and competitive to similar alloy compositions from conventional processes. There is an additional improvement of the creep resistance possible by grain growth annealing of the ultra fine-grained eutectic alloys. Furthermore, the AM materials were exposed to laboratory air for certain periods of time and at different application-related temperatures. In general, the oxidation response of the AM alloys is similar to other well-known Mo-Si-B alloys. The highest mass loss was observed at 800 °C, which is well-known as the so-called pesting phenomenon. At 1100 °C as well as at 1300 °C a protective borosilicate glass layer forms at the surface of the Mo-Si-B materials, which prevents the materials from further oxidation and, therefore, avoids further mass losses.

The results of this project provide new insights into the possibilities given by additive manufacturing processes in connection with the challenging class of ultra-high melting multi-phase Mo-Si-B alloys. First evaluation of mechanical and oxidation properties show the great potential of AM Mo-Si-B materials as they have shown to be competitive to conventional processes. More detailed analyses of the mechanical behavior and the oxidation behavior of AM Mo-Si-B will be published soon [24,31].

Acknowledgements

This work was supported by the German Ministry of Education and Research BMBF (funding number: 03XP0094) within the funding program "ProMat_3D".

References

[1] Soboyejo W O and Srivatsan T S 2006 *Advanced Structural Materials Properties, Design Optimization, and Applications* (CRC Press)
[2] Dimiduk D M and Perepezko J H 2003 Mo-Si-B Alloys : Developing a Revolutionary Turbine-Engine Material *MRS Bull.* **28** 639–45
[3] Perepezko J H 2009 The hotter the engine, the better *Science (80-.).* **326** 1068–9
[4] Brindley K A, Priddy M W and Neu R W 2019 Integrative Materials Design of Three-Phase Mo-Si-B Alloys *Integr. Mater. Manuf. Innov.* **8** 1–16
[5] Morris D G, Lopez-Delgado A, Padilla I and Munoz-Morris M A 2015 Selection of high temperature materials for concentrated solar power systems: Property maps and experiments *Sol. Energy* **112** 246–58
[6] Kauss O, Tsybenko H, Naumenko K, Hütter S and Krüger M 2019 Structural analysis of gas turbine blades made of Mo-Si-B under transient thermo-mechanical loadings *Comput. Mater. Sci.* **165** 129–36
[7] Krüger M, Jain P, Kumar K S and Heilmaier M 2014 Correlation between microstructure and properties of fine grained Mo-Mo3Si-Mo5SiB2 alloys *Intermetallics* **48** 10–8
[8] Yoshimi K, Nakatani S, Nomura N and Hanada S 2003 Thermal expansion, strength and oxidation resistance of Mo/Mo5SiB2 in-situ composites at elevated temperatures *Intermetallics* **11** 787–94
[9] Rioult F A, Imhoff S D, Sakidja R and Perepezko J H 2009 Transient oxidation of Mo-Si-B alloys: Effect of the microstructure size scale *Acta Mater.* **57** 4600–13
[10] Berczik D M 1997 US Patents No. 5595616 and 5693156
[11] Kazemi O, Hasemann G, Krüger M and Halle T 2016 Phase field simulation of a directional solidification of a ternary eutectic Mo-Si-B alloy *IOP Conf. Ser. Mater. Sci. Eng.* **118** 012028
[12] Hasemann G, Bogomol I, Schliephake D, Loboda P I and Krüger M 2014 Microstructure and creep properties of a near-eutectic directionally solidified multiphase Mo-Si-B alloy *Intermetallics* **48** 28–33
[13] Gombola C, Sprenger I, Kauffmann A, Bolbut V, Hasemann G, Oehing M, Blankenburg M, Schell N, Staron P, Pyczak F, Krüger M and Heilmaier M 2019 FlexiDS: In-situ investigation of microstructural evolution during directional solidification *DESY User Meeting, Hamburg*
[14] Tsybenko H 2018 *Numerische Analyse des Verformungsverhaltens von Gasturbinenschaufeln bei instationären thermo-mechanischen Beanspruchungen* (Otto-von-Guericke University Magdeburg)
[15] Kaserer L, Braun J, Stajkovic J, Leitz K H, Tabernig B, Singer P, Letofsky-Papst I, Kestler H and Leichtfried G 2019 Fully dense and crack free molybdenum manufactured by Selective Laser Melting through alloying with carbon *Int. J. Refract. Met. Hard Mater.* **84** 105000
[16] Makineni S K, Kini A R, Jägle E A, Springer H, Raabe D and Gault B 2018 Synthesis and stabilization of a new phase regime in a Mo-Si-B based alloy by laser-based additive manufacturing *Acta Mater.* **151** 31–40
[17] Zhou W, Sun X, Tsunoda K, Kikuchi K, Nomura N, Yoshimi K and Kawasaki A 2019 Powder fabrication and laser additive manufacturing of MoSiBTiC alloy *Intermetallics* **104** 33–42
[18] Schmelzer J, Rittinghaus S-K, Gruber K, Veit P, Weisheit A and Krüger M 2020 Printability and microstructural evolution of a near-eutectic three-phase V-based alloy *Addit. Manuf.* **34** 101208
[19] Krüger M, Franz S, Saage H, Heilmaier M, Schneibel J H, Jéhanno P, Böning M and Kestler H 2008 Mechanically alloyed Mo-Si-B alloys with a continuous alpha-Mo matrix and improved mechanical properties *Intermetallics* **16** 933–41
[20] Hasemann G, Krüger M, Palm M and Stein F 2018 Microstructures of ternary eutectic refractory Me-Si-B (Me=Mo, V) alloy systems *Mater. Sci. Forum* **941** 827–32

IOP Conf. Series: Materials Science and Engineering **882** (2020) 012011 doi:10.1088/1757-899X/882/1/012011

[21] Schmelzer J, Rittinghaus S-K, Weisheit A, Stobik M, Paulus J, Gruber K, Wessel E, Heinze C and Krüger M 2019 Printability of gas atomized Mo-Si-B powders by laser metal deposition *Int. J. Refract. Met. Hard Mater.* **78** 123–6

[22] Braun J, Kaserer L, Stajkovic J, Leitz K H, Tabernig B, Singer P, Leibenguth P, Gspan C, Kestler H and Leichtfried G 2019 Molybdenum and tungsten manufactured by selective laser melting: Analysis of defect structure and solidification mechanisms *Int. J. Refract. Met. Hard Mater.* **84** 104999

[23] Fichtner D, Schmigalla S, Schmelzer J, Heinze C and Krüger M 2019 *Intermetallics 2019 (30.09.-04.10.2019): Laser-based additive manufacturing of Mo-Si-B alloys* (Bad Staffelstein)

[24] Fichtner D, Schmelzer J, Wessel E, Yang W, Heinze C and Krüger M L-PBF of a near-eutectic Mo-Si-B alloy: Processing and resulting properties *Intermetallics.* paper in preparation

[25] Rittinghaus S-K, Schmelzer J, Wilms M B and Krüger M 2019 Laser additive manufacturing of intermetallic alloys for high-temperature applications *Metal Additive Manufacturing Conference* p contribution ID 15

[26] Hasemann G, Ida S, Zhu L, Iizawa T, Yoshimi K and Krüger M 2020 Experimental assessment of the microstructure evolution and liquidus projection in the Mo-rich Mo-Si-B system *Mater. Des.* **185** 108233

[27] Nowotny E, Dimakopoulou and Kudielka H 1957 Untersuchungen in den Dreistoffsystemen : Molybdän-Silizium-Bor, Wolfram-Silizium-Bor und in dem System : VSi2-TaSi2 *Monatshefte für Chemie und verwandte Teile anderer Wissenschaften* **88** 180–92

[28] Ha S H, Yoshimi K, Nakamura J, Kaneko T, Maruyama K, Tu R and Goto T 2014 Experimental study of Moss-T2, Moss- Mo3Si-T2, and Mo3Si-T2 eutectic reactions in Mo-rich Mo-Si-B alloys *J. Alloys Compd.* **594** 52–9

[29] Jain P and Kumar K S 2010 Dissolved Si in Mo and its effects on the properties of Mo-Si-B alloys *Scr. Mater.* **62** 1–4

[30] Krüger M, Schliephake D, Jain P, Kumar K S, Schumacher G and Heilmaier M 2013 Effects of Zr additions on the microstructure and the mechanical behavior of PM Mo-Si-B alloys *JOM* **65** 301–6

[31] Becker J, Fichtner D, Schmigalla S, Schultze S, Heinze C, Küsters Y, Hasemann G, Schmelzer J and Krüger M 2020 Oxidation response of additively manufactured eutectic Mo-Si-B alloys *IOP Conf. Ser. Mater. Sci. Eng.* submitted

[32] Lange A, Braun R and Heilmaier M 2014 Oxidation behavior of magnetron sputtered double layer coatings containing molybdenum, silicon and boron *Intermetallics* **48** 19–27

[33] Perepezko J H, Bero J M, Sakidja R, Talmy I G and Zaykoski J 2012 Oxidation resistant coatings for refractory metal cermets *Surf. Coatings Technol.* **206** 3816–22

[34] Smokovych I, Hasemann G, Krüger M and Scheffler M 2017 Polymer derived oxidation barrier coatings for Mo-Si-B alloys *J. Eur. Ceram. Soc.* **37** 4559–65

IOP Conf. Series: Materials Science and Engineering **882** (2020) 012012 doi:10.1088/1757-899X/882/1/012012

Influence of Nb, Ti and Mo on Microstructure and Mechanical Properties of Vanadium Solid Solutions

C Müller, R S Touzani, U Betke and M Krüger

Otto-von-Guericke-University Magdeburg, Faculty of Mechanical Engineering, Institute of Materials and Joining Technology, Universitätsplatz 2, 39106 Magdeburg, Germany

christopher.mueller@ovgu.de

Abstract. The present study addresses the microstructural evolution and mechanical properties of three different cast V-5X (X = Nb, Ti and Mo) alloys. All alloy compositions are located in the vanadium solid solution (V_{SS}) phase field in their respective phase diagrams. While the alloys V-5Ti and V-5Nb solidified as single-phase V_{SS} after arc-melting, the alloy V-5Mo additionally shows some amount of molybdenum solid solution (Mo_{SS}) in the resulting as-cast microstructure. To improve the mechanical database on vanadium-based alloys, room temperature compression tests were performed and microhardness measurements were carried out by the Vickers indentation method. The lattice parameters were determined by X-ray diffraction analysis and they were calculated using the density functional theory. By carrying out these experiments, the data available for vanadium-based alloys were extended being now available for further optimization of novel V-Si-B high temperature alloys.

1 Introduction

Vanadium base alloys as new structural lightweight materials are of interest for high temperature applications, e.g. in turbines for energy conversion, because of their promising stress-strain behavior in combination with a low density and high melting point of vanadium [1, 2]. In particular, V-Si-B alloys show outstanding mechanical properties at high temperatures, which are comparable to the state-of-the-art Ni-based superalloy CMSX-4 [3]. Therefore, and due to the similarities compared to the well-investigated Mo-Si-B system, which is also a promising candidate to replace Ni-based alloys, ternary V-Si-B alloys recently gained more scientific interest [4-9].

Comparing the isothermal sections available in the literature for the V-Si-B [10] and Mo-Si-B [11], both metal-silicon-boron (Me-Si-B; Me = V, Mo) systems show comparable three phase fields in their metal-rich corner. These fields consist of an A2 structured Me_{SS} metal solid solution phase (V_{SS}, Mo_{SS}), an intermetallic A15 structured Me_3Si phase (V_3Si, Mo_3Si) and a ternary Me_5SiB_2 (V_5SiB_2, Mo_5SiB_2) intermetallic phase with a $D8_l$ crystal structure [10, 12, 13]. The latter is the so called T2 phase. In context of high temperature applications and according to the literature available for Me-Si-B alloys, a microstructure consisting of a Me_{SS} matrix with homogeneously distributed intermetallic Me_3Si and Me_5SiB_2 phases is expected as the desired microstructure [3, 12, 14]. The solid solution matrix may be responsible for the fracture toughness and the intermetallic phases control the high temperature strength as well as the creep and oxidation resistance of those alloys [3, 12, 14-16]. Hasemann et al. [7] recently investigated the ternary eutectics consisting of the three phases as mentioned above for the Mo-Si-B and V-Si-B systems. They determined the volume fractions in the ternary eutectic Me_{SS}-Me_5SiB_2-Me_3Si

grains to be clearly different comparing both systems. The V-Si-B eutectic consists of about 50 vol.% solid solution phase whereas the Mo-Si-B counterpart shows only 20 vol.% of the Mo_{SS} phase [7]. A study on ternary V-Si-B alloys has also shown, that the phase fraction of V_{SS} phase truly governs the plastic deformability at room temperature in V_{SS}-V_5SiB_2-V_3Si alloys. Transmission electron microscopy (TEM) investigations on previously deformed compression samples proved that dislocations can only be observed in the V_{SS} regions [6]. In 1997, Henshall et al. [17] already demonstrated that the volume fraction of the V_{SS} phase can be used to control the fracture toughness of binary V_{SS}-V_3Si alloys. These results [6, 7, 17] allow the conclusion that for the V-Si-B system, the materials properties can rather be controlled by optimization of the solid solution phase than in the Mo-Si-B case. However, the effects of several alloying elements on the microstructure and mechanical properties of vanadium solid solutions are insufficiently examined, yet. Therefore, the present work addresses the alloying effects of various elements on vanadium-based alloys.

Properties like ambient and high temperature strength, adequate oxidation behavior, a low brittle-to-ductile transition temperature (BDTT) as well as an acceptable fracture toughness and a low density are known to define the profile of requirements for high temperature materials [2, 18]. Most of these properties are related to a high melting point, thus, several alloy systems based on refractory metals are in focus of research since several years [19-21]. Therefore, the addition of refractory metals Mo and Nb were examined in the present work with respect to their effects on the V_{SS} phase. In the context of structural materials for fusion reactors, several V-Cr-Ti alloys were investigated in the past. The literature indicates that an improvement in strengthening can also be expected by the addition of titanium. In particular, the solid solution strengthening [22-24] and the oxidation resistance can be improved by addition of Ti [25, 26]. In order to use the beneficial effects of those alloy elements (Nb, Ti, Mo) for future vanadium-based high temperature alloys, basic investigations like room temperature compression tests, microstructural investigations and hardness measurements were performed in the present study.

2 Materials and Methods

The alloys V-5Nb, V-5Ti and V-5Mo and a reference sample of plain vanadium were produced by conventional arc melting under argon atmosphere in a water cooled copper crucible with a diameter of 25 mm (Arc Melter MAM-1; Edmund Buehler GmbH). Prior to melting, the starting materials vanadium (min. 99.9 % purity; HMW-Hauner GmbH), niobium (min. 99.9 %; HMW-Hauner GmbH), titanium (min. 99.8 %; HMW-Hauner GmbH) and molybdenum (min. 99.9 %; HMW-Hauner GmbH) were carefully weighed out (10 g per alloy) according to the alloy compositions. To ensure a high degree of homogeneity over the sample size, the button-shaped samples were re-melted for five times and flipped between each melting step. The mass losses measured after melting were negligible for all samples (< 0.1 wt.%). All investigations within the present study were carried out under as-cast conditions. Inductively coupled plasma optical emission spectroscopy (ICP-OES) was used to examine the chemical alloy compositions. Sample pieces of each alloy were dissolved in HCl + HNO_3 + HF and analyzed using a Thermo Scientific iCAP6500 spectrometer (measuring accuracy < 3 %). The oxygen concentrations (in wt.ppm) were determined in accordance with the carrier gas procedure via melt extraction with subsequent infrared detection using a Bruker G8 Galileo System. Table 1 summarizes the results of the chemical analyses.

Samples for metallographic preparation were embedded in hot mounting epoxy resin (Struers Poly Fast) and subsequently grinded from 180 - 2200 grit, followed by mechanical polishing with 3 μm and 1 μm diamond suspension and finished using colloidal silica. Microstructures (as-cast) were investigated using optical microscopy (OM; Zeiss Axiovert 200M) and a FEI (Thermo Fisher Scientific) Scios DualBeam scanning electron microscope (SEM). SEM images were typically performed in the backscattered electron (BSE) mode. Energy dispersive X-ray spectroscopy (EDS, EDAX) were used to measure the concentration of alloying element in the V_{SS} (vanadium solid solution) phase. The phase analyses of V-5X (X = Nb, Ti, Mo) alloys was performed with X-ray diffraction analysis (XRD) in a θ/θ reflection geometry with a 2θ range from 20° to 160° using a PANalytical X'Pert Pro Bragg-

Brentano diffractometer and Co-Kα_1/α_2 radiation. XRD measurements were carried out on previously ground samples. The obtained diffraction patterns were analyzed by the Rietveld technique using Topas Academic 5 program package [27] in order to determine the phase fractions in the alloys. The lattice parameters were determined by the whole powder pattern fitting (WPPF) approach with the Pawley fit of the XRD data using Topas 5 and evaluated against the values determined by the density functional theory (DFT).

To mimic the random distribution of atoms in the V-5X (X = Nb, Ti, Mo) solid solution, special quasi-random structures (SQS) [28, 29] with 128 atoms per cell made via the Monte Carlo SQS (mcsqs) code as implemented in the alloy theoretic automated toolkit (ATAT) were used [30]. The SQS and the simple unit cells of V (100% V) were then used as starting structures for the first-principles structural relaxation carried out with Quickstep [31] as implemented in the CP2K version 5.1 program package [32]. Using the Gaussian plane wave method (GPW) [33], for V, Nb, Ti and Mo the DZVP-MOLOPT-SR-GTH basis set [34] for the atomic centered Gaussian functions were chosen, while for the interatomic part the GTH-pseudopotentials were used [35-37]. Exchange and correlation in this DFT-based method were treated with the generalized gradient approximation (GGA) functional as parameterized by Perdew, Burke and Ernzerhof (PBE-GGA) [38]. Cell shape and volume variations were allowed during the structural optimization until a total energy self-consistency of 10^{-6} Ha and until the self-consistency for the forces and maximum geometry change of 10^{-5} Ha/Bohr and 10^{-5} Bohr, respectively, were achieved. The energy cut-off for the plane waves on the grid were 600 Ha and the k-mesh sampled via the Monkhorst-Pack algorithm [39] were 4x4x4 during the cell optimization.

Table 1. Chemical compositions of as-cast V-5X alloys measured by ICP-OES and melt extraction.

Nominal composition	Nb [at.%]	Ti [at.%]	Mo [at.%]	Oxygen [wt.ppm]	Nitrogen [wt.ppm]
V-5Nb	4.96	-	-	1080 ± 90	13 ± 1
V-5Ti	-	4.96	-	1150 ± 130	4 ± 2
V-5Mo	-	-	4.80	1000 ± 90	6 ± 2
V	-	-	-	830 ± 150	7 ± 3

Mechanical properties of the alloys were investigated using compression tests and microhardness measurements. Compression samples with a typical length of 3 mm and a diameter of 1.6 mm were cut via electrical discharge machining (EDM). Compression tests were performed at room temperature using a Zwick/Roell Z100 electro-mechanical universal testing machine at an initial (engineering) strain rate of $\dot{\varepsilon} = 10^{-3}$ s^{-1}. Five samples per alloy were tested. The yield stresses were determined by the 0.2 % offset method ($\sigma_{0.2}$). Microhardness measurements referred to DIN EN ISO 4516:2002 were performed by generating a series of 30 indents set within the grains at well-defined intervals. The tests were performed by indenting a Vickers pyramid with a force of 1 N (HV 0.1) and a hold period of 10 s in the solid solution alloys.

3 Results and Discussion

3.1 Microstructural Investigations

The as-cast microstructures of the V-5X (Nb, Ti, Mo) alloys and the reference sample consisting of arc melted vanadium are illustrated in Figure 1. In accordance to the respective binary phase diagrams of V-Nb [40], V-Ti [41] and V-Mo [42], all alloys solidified within the V_{SS} phase region. Thin deformation scales evidently formed during the metallographic polishing process which can be assumed from the irregularities obtained in the SEM/OM images. Rietveld analysis performed on the measured XRD patterns (Figure 2) demonstrate, that the alloys V-5Nb, V-5Ti as well as vanadium are single phased consisting of only the V_{SS} phase. Reflexes of molybdenum, however, can be observed in the patterns of alloy V-5Mo. Results of the Rietveld analysis show a small part of ~ 1 vol.% Mo or low-concentrated Mo_{SS} in the microstructure. However, larger Mo phase regions have not be detected in the SEM analysis, which suggests that the Mo-rich phase is homogenously distributed. In order to verify the dissolved amount of alloying elements in the V_{SS} phase, EDS spot analyses were conducted. The alloy V-5Mo shows two regions of V_{SS} phases. There are grains/regions with around 5.2 at.% Mo and the others with only 1.8 at.% Mo. It cannot be explained, yet, how these deviations may be evolved during solidification. However, there are calculated results on the binary Mo-V system that show a miscibility gap at lower temperatures, which may explain the present observation [43, 44]. In case of Nb and Ti-containing alloys EDS results showed that almost 5 at.% of Nb/Ti were dissolved in the V_{SS} phase, which corresponds to the results of the ICP-OES measurements.

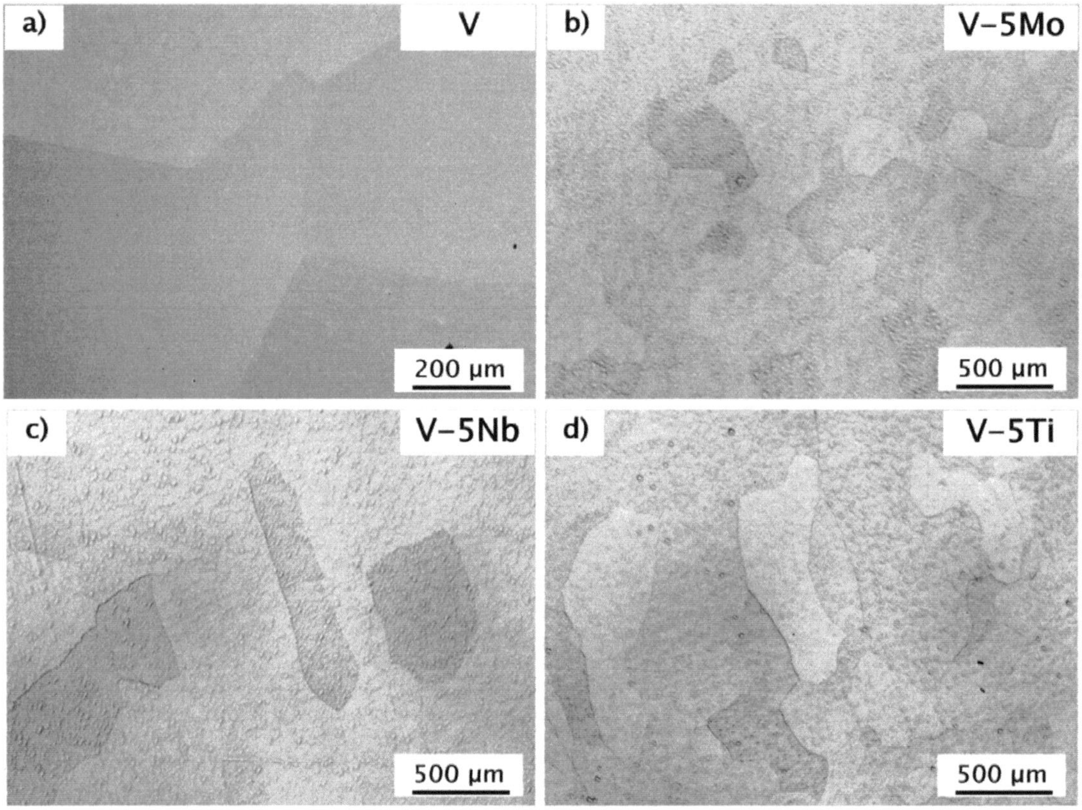

Figure 1. Micrographs of the as-cast microstructures for the alloys a) V, b) V-5Mo, c) V-5Nb, d) V-5Ti.

The lattice parameters of the as-cast alloys were determined in order to achieve additional information about the solid solution phase as well as to identify possible effects concerning the resulting mechanical properties of the alloys. The lattice parameters were calculated using the density functional theory as explained in section 2 and measured by X-ray diffraction combined with WPPF analysis. The results for both methods including their respective deviation (in brackets) compared to vanadium are displayed in Figure 2 (right side). In general, the lattice parameters of vanadium increased for all alloys by adding Ti, Nb or Mo. Due to the small amount of non-dissolved Mo in the alloy V-5Mo, the measured deviation in the lattice parameters differs by 0.1 % compared to the DFT calculations. Compared to the DFT values, the deviation of the lattice parameters measured with XRD are compared to the arc-melted

IOP Conf. Series: Materials Science and Engineering **882** (2020) 012012 doi:10.1088/1757-899X/882/1/012012

vanadium sample and thus, the values slightly differ compared to the model-based vanadium. Differences are explained by the amount of oxygen and nitrogen (see Table 1) as well as possible impurities in the as-received starting materials used for the alloy manufacturing.

All alloying elements lead to an increase of the lattice parameter. However, the qualitative increase in the lattice parameters are comparable for both techniques and can be ranked in decreasing order for Nb > Ti > Mo. Compared to V, the lattice parameter for alloy V-5Nb is increased by 0.59 %, for alloy V-5Ti by 0.43 % and for alloys V-5Mo by 0.16 %.

The DFT method seems to be promising in order to estimate lattice constants without performing elaborate experiments. Since the change in lattice parameter is known as a crucial factor in the theory of solid solution hardening [45, 46], DFT analyses can help to estimate reliable values in the future.

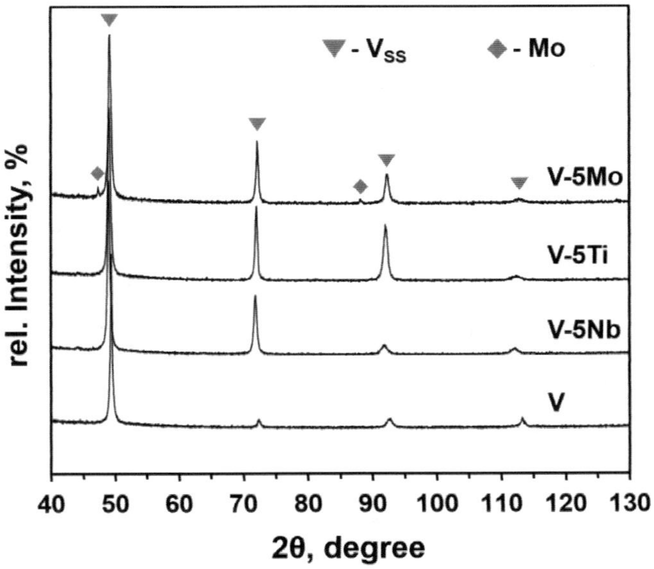

lattice parameter		
	DFT [Å]	**XRD [Å]**
V	2.9948	3.0318
V-5Nb	3.0119 (+ 0.57 %)	3.0498 (+ 0.59 %)
V-5Ti	3.0073 (+ 0.42 %)	3.0447 (+ 0.43 %)
V-5Mo	3.0026 (+ 0.26 %)	3.0365 (+ 0.16 %)

Figure 2. XRD patterns of different V-5X alloys (left) and lattice parameter with the respective deviation compared to vanadium (right). Lattice parameter were calculated via DFT and measured via XRD combined with Rietveld analysis.

3.2 Mechanical Properties

The mechanical properties of the alloys were investigated using room temperature compression tests as well as microindentation (Vickers) tests. In the engineering stress vs. strain diagram in Figure 3, the results of the compression tests are shown. For reasons of clarity, only three out of five test curves per alloy were plotted. Figure 4 displays the yield stresses, maximum compressive stresses and total compressive strains measured in compression tests. It cannot be excluded that mechanical properties in the present work were influenced by the oxygen/nitrogen content of the alloys (V_{SS} + O/N) since oxygen and nitrogen are known to embrittle V already at comparably minor portions [47]. However, the compressive strains do not indicate significant embrittlement. All binary alloys show high deformability up to 26 % total compressive strain at room temperature which is comparable to the vanadium sample. This observation is in good agreement with recently published results for different V-based alloys [4, 6]. Taking into account the compressive yield stresses ($\sigma_{0.2}$), alloying vanadium with 5 at.% Nb leads to the lowest values of 250 ± 70 MPa followed by 5 at.% Ti with 300 ± 50 MPa and 5 at.% Mo, which results in the highest yield strength of 310 ± 50 MPa. The arc-melted vanadium sample results in $\sigma_{0.2}$ of 140 ± 25 MPa. The strengthening effects of the different alloying elements can therefore be ranked in decreasing order as follows: Mo > Ti > Nb. This result is quite unexpected when taking into account the change of lattice parameter measured for alloys in this study (Figure 2). The increase in strength shows a contrary course compared to the lattice constant and contradicts the theory presented in [45] and [46].

The tendency is contrary to the mechanical data on V-based alloys already available in the literature. Harrod and Gold [48] published a comprehensive compilation and review on the mechanical properties of several V-based alloys in the 1980s. They report the alloying effects of Mo, Ti and Nb on the room temperature yield strength of vanadium in decreasing order as follows: Nb > Mo > Ti. It has to be noted that they obtained their data from various sources. Sample fabrication and purity as well as test parameters were not comprehensively reported. However, the results of tensile tests on V-10X sheet samples (X = Nb, Mo, Ti) by Rajala and van Thyne [49] show the same tendency with regard to the effect of the alloying elements on the yield strength (Nb > Mo > Ti) as reported in [48]. A final ranking with respect to the strength contributions of the respective alloying elements (Mo, Ti and Nb), however, does not seem to be expedient due to the equality of compression data measured in the present work (Figure 4).

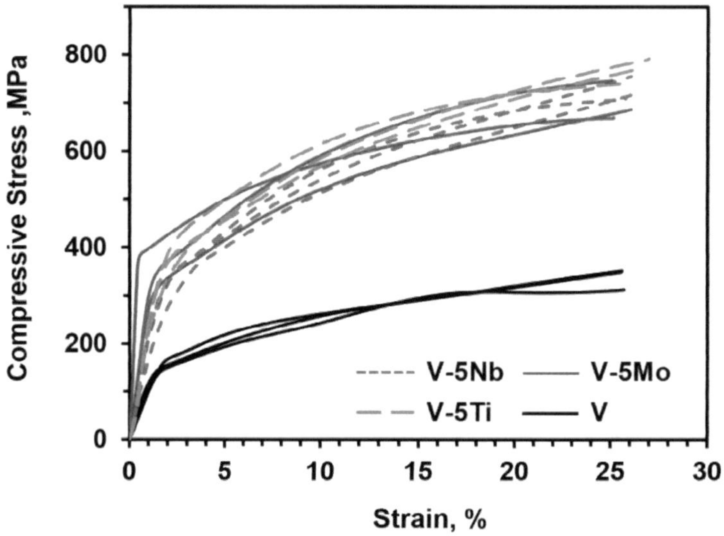

Figure 3. Compressive engineering stress vs. strain curves at room temperature of alloys V-5Nb, V-5Ti, V-5Mo and arc-melted vanadium.

The hardness values obtained from Vickers microindentaion tests are summarized in Table 2. The vanadium sample has a microhardness of 93 HV0.1. For the V-5X samples, alloy V-5Ti has the highest hardness value with 190 ± 7 HV0.1 followed by V-5Nb with 180 ± 3 HV0.1 and V-5Mo with 174 ± 5 HV0.1. As was also apparent from the results of the compression tests, none of the alloying elements (Mo, Ti and Nb) can particularly be recommended regarding their potential to optimize the room temperature mechanical properties of the V_{SS} phase. All alloying elements have led to a significant increase in strength and hardness.

Table 2. Microhardness (HV0.1) values for V-5X alloys compared to vanadium.

Nominal composition	Microhardness HV0.1
V-5Nb	180 ± 3
V-5Ti	190 ± 7
V-5Mo	174 ± 5
V	93 ± 1

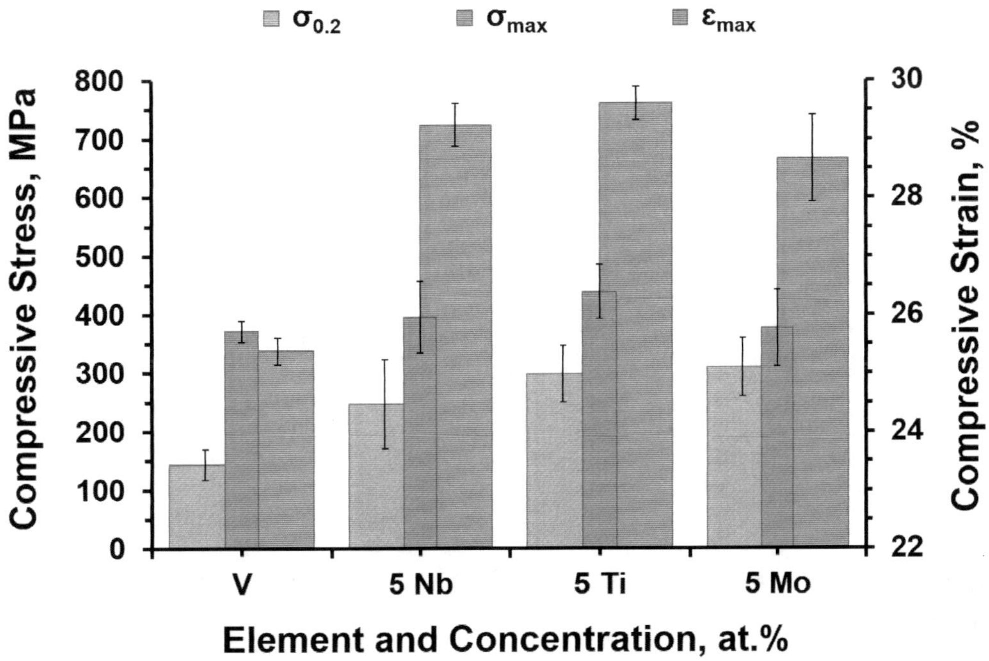

Figure 4. Comparison of the yield stress ($\sigma_{0.2}$), maximum compressive stress (σ_{max}) and total compressive strain (ε_{max}) for V and V-5X alloys (X = Nb, Ti, Mo); all data were generated at room temperature.

In context of the promising V-Si-B alloys, it has to be noted that the strengths measured for alloys in this study are significantly lower than those of a comparably fabricated V-5Si sample. Hasemann et al. [6] recently published compression strength data for arc-melted binary as-cast V-Si alloys. The V-5Si sample in their study reached a yield stress ($\sigma_{0.2}$) of 580 ± 78 MPa [6], which is about 250 MPa higher than stresses measured for alloys in the present work (Figure 4). However, alloying with 5 at.% Si exceeds the solubility limit at room temperature, which will result in a certain amount of V_3Si and therefore, this effect has to be considered in order to maintain balanced strength and ductility in technical alloys [2, 5, 6]. Data in the present paper showed that additions of 5 at.% Nb, Ti or Mo increase the room temperature strengths and does not lead to a decrease of deformability. Thus, they may help to optimize the V_{SS} phase in future V-Si-B alloys. However, a more detailed description can be made when the influence on Nb, Mo and Ti on the microstructure formation and the liquidus projection (cooling paths) in the respective V-Si-B-X systems has been studied more intensively.

4 Summary and Conclusions

The alloying effects in three different V-5X (X = Nb, Ti, Mo) vanadium-based alloys were investigated in order to estimate their opportunity to optimize novel V-Si-B high temperature alloys. All alloys are located in the V_{SS} phase region in their respective phase diagrams. However, while V-5Nb and V-5Ti solidified as single phase alloys, the alloy V-5Mo includes different solid solution phases. The present study reports on mechanical data of the as-cast alloys determined by room temperature compression tests and microhardness measurements. Additions of 5 at.% Nb, Ti or Mo lead to an increase of the compression strengths compared to arc-melted vanadium and does not cause any embrittlement. All alloys showed a high ductility up to 26 % total compressive strain in room temperature compression tests. XRD analysis and calculations using the density functional theory (DFT) were used to examine the alloys lattice parameters. All alloying elements increased the lattice parameter, compared to plain vanadium. A comparison between XRD and DFT values showed that the deviation of the lattice parameters with respect to vanadium is approximately the same for both methods. It can therefore be concluded that the DFT calculation provides reliable values and can minimize experimental effort to estimate the lattice constants. Based on these experimental results, the data available for V-based alloys were clearly extended.

Acknowledgments
The present research was conducted within the context of the International Graduate School MEMoRIAL at Otto-von-Guericke University (OVGU) Magdeburg, Germany, kindly supported by the European Structural and Investment Funds (ESF) under the program "Sachsen-Anhalt WISSENSCHAFT Internationalisierung" (project no. ZS/2016/08/80646). The authors kindly thank Henrike Thärichen, Hermann Scholz, Kilian Kirsten and Han Chen for supporting the experimental work in the context of their student project at the Otto-von-Guericke University Magdeburg. Special thanks goes to Oliver Michael for supporting the metallographic sample preparation. Financial support of the Methodisch-Diagnostisches Zentrum Werkstoffprüfung (MDZWP) e.V., Magdeburg, Germany is greatly acknowledged. Rachid S. Touzani thanks the URZ OVGU Magdeburg for calculation power and time.

References

[1] Krüger M, Bolbut V, Gang F and Hasemann G 2016 Microstructure variations and creep properties of novel high temperature V-Si-B materials *JOM-J Min. Met. Mat. S.* **68** 2811-16
[2] Krüger M 2017 *Innovative metallische Hochtemperaturwerkstoffe* Habilitation at Otto-von-Guericke University Magdeburg Germany
[3] Krüger M 2016 High temperature compression strength and oxidation of a V-9Si-13B alloy *Scripta Mater.* **121** 75-78
[4] Müller C, Hasemann G, Regenberg M, Betke U and Krüger M 2020 Microstructure and compression properties of VSS-V3B2 eutectic alloys in the V-Si-B system *Materials* **13** 2100
[5] Hasemann G 2020 Experimental study of the liquidus surface in the V-rich portion of the V-Si-B system *J. Alloy. Compd.* **824** 153843
[6] Hasemann G, Müller C, Grüner D, Wessel E and Krüger M 2019 Room temperature plastic deformability in V-rich V-Si-B alloys *Acta Mater.* **175** 140-47
[7] Hasemann G, Krüger M, Palm M and Stein F 2018 Microstructures of ternary eutectic refractory Me-Si-B (Me = Mo, V) alloy systems *Mater. Sci. Forum* **941** 827-32
[8] Schmelzer J, Baumann T, Dieck S and Krüger M 2016 Hardening of V-Si alloys during high energy ball milling *Powder Technol.* **294** 493-97
[9] Krüger M, Schmelzer J and Helmecke M 2016 Similarities and differences in mechanical alloying processes of V-Si-B and Mo-Si-B powders *Metals* **241** 2-16
[10] Nunes C A, De Lima B B, Coelho G C and Suzuki P A 2009 Isothermal section of the V-Si-B system at 1600°C in the V-VSi2-VB region *J. Phase Equilibria Diffusion* **30** 345-50
[11] Nowotny H, Dimakopoulou E and Kudielka H 1957 Untersuchungen in den Dreistoffsystemen Molybdän-Silizium-Bor, Wolfram-Silizium-Bor und in dem System: VSi2-TaSi2 *Mh. Chem.* **88** 180-92
[12] Rioult F A, Imhoff S D, Sakidja R and Perepezko J H, 2009 Transient oxidation of Mo-Si-B alloys: effect of the microstructure size scale *Acta Mater.* **57** 4600-13
[13] Colinet C and Tedenac J-C 2014 First principles calculations of the stability of the T2 and D88 phases in the V-Si-B system *Intermetallics* **50** 108-16
[14] Krüger M, Jain P, Kumar K S and Heilmaier M 2014 Correlation between microstructure and properties of fine grained Mo-Mo3Si-Mo5SiB2 alloys *Intermetallics* **48** 10-18
[15] Schneibel J H, Kruzic J J and Ritchie R O 2003 Mo-Si-B alloy development *Proc. 17th Annu. Conf. Foss. Energy Mater.* Baltimore/Maryland pp 3-9
[16] Williams J and Akinc M 1998 Oxidation behavior of V5Si3 based materials *Intermetallics* **6** 269-75

[17] Henshall G A, Sturm M J, Bewlay B P and Sutliff J A 1997 Ductile-phase toughening in V-V3Si in situ composites *Metall. Mater. Trans.* A **28** 2555-64
[18] Wright I G and Gibbons T B 2007 Recent developments in gas turbine materials and technology and their implications for syngas firing *Int. J. Hydrogen Energ.* **32** 3610-21
[19] Bewlay B P, Jackson M R, Subramanian P R and Zhao J-C 2003 A review of very-high-temperature Nb-silicide-based composites *Metall. Mater. Trans.* A **34** 2043-52
[20] Wenderoth M, Völkl R, Vorberg S, Yamabe-Mitarai Y, Harada H and Glatzel U 2007 Microstructure, oxidation resistance and high-temperature strength of γ' hardened Pt-base alloys *Intermetallics* **15** 539-49
[21] Jain P and Kumar K S 2010 Dissolved Si in Mo and its effects on the properties of Mo-Si-B alloys *Scripta Mater.* **62** 1-4
[22] Pink E and Arsenault R J 1972 Solid-Solution strengthening and weakening of vanadium-titanium alloys *Metal Science Journal* **6** 1-6
[23] Li Z D, Li Q, Li Y, Lin C G, Cui S and Ma T D 2017 Mechanical properties and precipitation evolution behavior of V-5Cr-5Ti alloy after cold rolling *Fusion Eng.Des.* **114** 76-83
[24] Li Z D, Li Q, Li Y and Ma T D 2018 Microstructure and properties of V-5Cr-5Ti alloy after hot forging *Fusion Eng. Des.* **127** 83-90
[25] Chen J M, Muroga T, Qiu S Y, Nagasaka T, Huang W G, Tu M J, Chen Y, Xu Y and Xu Z Y 2004 The development of advanced vanadium alloys for fusion applications *J. Nucl. Mater.* **329-333** 401-05
[26] Natesan K and Uz M 2000 Oxidation performance of V-Cr-Ti alloys *Fusion Eng. Des.* **51-52** 145-52
[27] Rietveld A B 1969 A profile refinement method for nuclear and magnetic structures *J. Appl. Crystallogr.* **2** 65-71
[28] Zunger A, Wei S H, Ferreira L G and Bernard J E 1990 Special quasirandom structures *Phys. Rev. Lett.* **65** 353-56
[29] Wei S H, Ferreira L G, Bernard J E and Zunger A 1990 Electronic properties of random alloys: Special quasirandom structures *Phys. Rev.* B **42** 9622-49
[30] Van De Walle A, Tiwary P, De Jong M, Olmsted D L, Asta M, Dick A, Shin D, Wang Y, Chen L Q and Liu Z K 2013 Efficient stochastic generation of special quasirandom structures *Calphad* **42** 13-18
[31] VandeVondele J, Krack M, Mohamed F, Parrinello M, Chassaing T and Hutter J 2005 Quickstep: Fast and accurate density functional calculations using a mixed Gaussian and plane waves approach *Comput. Phys. Commun.* **167** 103-28
[32] Hutter J, Iannuzzi M, Schiffmann F and VandeVondele J, 2014 cp2k: atomistic simulations of condensed matter systems *WIREs Comput. Mol. Sci.* **4** 15-25
[33] Lippert G, Hutter J and Parrinello M 1997 A hybrid Gaussian and plane wave density functional scheme *Mol. Phys.* **92** 477-88
[34] VandeVondele J and Hutter J 2007 Gaussian basis sets for accurate calculations on molecular systems in gas and condensed phases *J. Chem. Phys.* **127** 114105
[35] Goedecker S, Teter M and Hutter J 1996 Separable dual-space Gaussian pseudopotentials *Phys. Rev.* B **54** 1703-10
[36] Hartwigsen C, Goedecker S and Hutter J 1998 Relativistic separable dual-space Gaussian pseudopotentials from H to Rn *Phys. Rev.* B **58** 3641-62
[37] Krack M 2005 Pseudopotentials for H to Kr optimized for gradient-corrected exchange-correlation functionals *Theor. Chem. Acc.* **114** 145-52
[38] Perdew J P, Burke K and Ernzerhof M 1996 Generalized gradient approximation made simple *Phys. Rev. Lett.* **77** 3865-68
[39] Monkhorst H J and Pack J D 1976 Special points for Brillouin-zone integrations *Phys. Rev.* B **13** 5188-92
[40] Smith J F and Carlson O N 1983 The Nb-V (Niobium-Vanadium) system *Bull. Alloy Phase*

IOP Conf. Series: Materials Science and Engineering **882** (2020) 012012 doi:10.1088/1757-899X/882/1/012012

Diagrams **4** 46-49
[41] Murray J L 1981 The Ti-V (Titanium-Vanadium) system *Bull. Alloy Phase Diagrams* **2** 48-55
[42] Smith J F 1992 The Mo-V system (Molybdenum-Vanadium) *J. Phase Equilib.* **13** 50-53
[43] Centre for Research in Computational Thermochemistry, SGTE 2017 Alloy Phase Diagrams (1176), Mo-V, available online: http://www.crct.polymtl.ca/fact/phase_diagram.php?file=Mo-V.jpg&dir=SGTE2017 (accessed Jan 18th 2019)
[44] Bale C W *et al.* 2016 FactSage thermochemical software and databases, 2010-2016 *Calphad* **54** 35-53
[45] Labusch R 1970 A statistical theory of solid solution hardening *Phys. Status Solidi* B **41** 659-69
[46] Fleischer R L 1961 Solution hardening *Acta Metall. Mater.* **9** 996-1000
[47] Jo M-G, Madakashira P P, Suh J-Y and Han H N 2016 Effect of oxygen and nitrogen on microstructure and mechanical properties of vanadium *Mater. Sci. Eng* A **675** 92-98
[48] Harrod D L and Gold R E 1980 Mechanical properties of vanadium and vanadium-base alloys *Int. Mater. Rev.* **4** 163-221
[49] Rajala B R and van Thyne R J 1961 The development of improved vanadium-base alloys for use at temperatures up to 1800°F *J. Less Common Met.* **3** 489-500

IOP Conf. Series: Materials Science and Engineering **882** (2020) 012013 doi:10.1088/1757-899X/882/1/012013

Simulative and experimental investigations on single-sided resistance spot brazing of sheet metal tube connections as an alternative to single-sided resistance spot welding

M Najuch[1,2], S Jüttner[1] and D Schmicker[2]

[1] Otto-von-Guericke-University Magdeburg, Faculty of Mechanical Engineering, Institute of Materials and Joining Technology, Universitätsplatz 2, 39106 Magdeburg, Germany

[2] IFA Technologies GmbH, Industriestr. 6, 39340 Haldensleben, Germany

marcel.najuch@ovgu.de

Abstract. In order to reduce the deficits in single-sided resistance welding (SSSW), single-sided resistance brazing is a promising alternative. The reason for this is a process-related reduced thermal and mechanical component impact compared to SSSW. In the underlying article, the single-sided resistance brazing process will therefore be presented. In addition to the process architecture, the process advantage is presented and the joining characteristics are clarified. Finally, the presumed process advantage is confirmed by means of a practical example.

1 Introduction

In order to achieve lightweight construction and thus ensure a reduction in CO2-emission, hollow sections made of high-strength steel are used in many industrial sectors, including the automotive sector. In contrast to individual sheet metal bodies, which are produced using cost-intensive forming processes and then being joined together to form closed structures, steel sections can be manufactured using simple and cost-effective production methods such as roll forming. In addition to car body construction, there is another application in the area of the powertrain. Here, thin-walled tubes are used to transmit torque (e.g. propshaft). A major challenge is the limited possibility of joining these profiles with sheet metal. Therefore, the so-called single-sided spot welding (short: SSSW), i.e. resistance welding with one-sided accessibility, was developed for the production of such sheet metal hollow section joints [1], [2]. Figure 1 a) and Figure 1 b) show the process structure schematically.

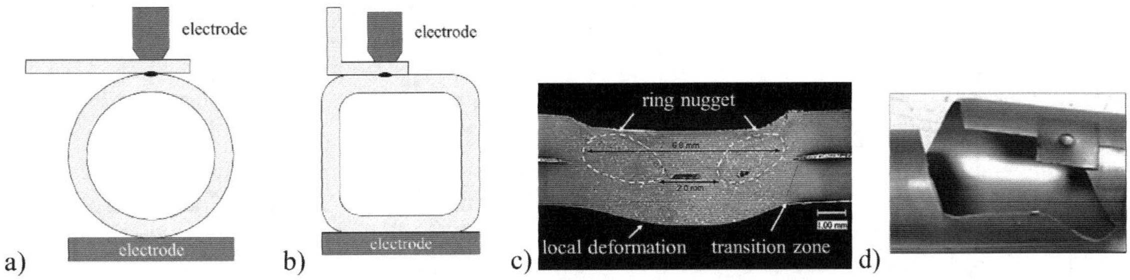

Figure 1. Schematic illustration of single-sided resistance spot welding a) sheet metal - tube, b) sheet metal - hollow section, c) characteristic welding nugget, d) crack in tube after dynamic torsional stress; [2], [3].

However, if one considers the previous investigations and findings presented in the state of research and development, the deficits of the so-called SSSW become clear: Due to the lack of accessibility and thus supporting effect of the electrode, considerable mechanically and thermally activated plastic deformation occurs during the process (local and global deformation) [2], [4], [5], [6], [7], [8]. When, for example, comparing the electrical current process window in conventional resistance welding (RP = 2 - 3 kA) with the one of single-sided resistance welding (SSSW = 1 - 2 kA), the difference becomes obvious. [1], [9] In case of too low contact pressure, splashing also occurs in the process [1], [5], [6], [9] whereby the splash limit is sometimes below the lower quality limit [2]. If one further compares the microsections of a conventional resistance weld with a one-sided joint, a curved solidification line is visible due to a change in heat conduction [3]. The characteristic deformation processes in the SSSW also leads to an increased tendency to blowholes in the welding nugget [4] and to the formation of a so-called ring nugget [2], [3], [5], [6]. In addition, the hollow profile stiffness significantly limits the process window. With a profile wall thickness $t < 1.25$ mm, for example, no satisfactory results could be obtained due to very limited welding parameter ranges with splashes and strong deformation. This is due to the limited stiffness of the hollow profile, which is why only low electrode forces could be selected. [6] Finally, the high heat input during welding reduces the fatigue strength of the joined components. The reason for this is the formation of a sharp-shaped hardening zone together with the characteristic local deformation in the welding area. [7], [8] A welding nugget with the characteristic deficits is shown in Figure 1 c).

A practical example from the field of propshaft production will illustrate the problem. Here, sheets with a defined weight (so-called balancing sheets) are applied to the outer diameter of the shaft to compensate imbalances. Due to the limited accessibility, caused by the tube cross section, the single-sided resistance welding is used as application method. The process-related thermal and mechanical damage mechanisms already described lead to a loss of strength in the propshaft. Under dynamic loading, tube failure occurs in the area of the balancing sheet weld (Figure 1 d)). To counteract this cause of failure, the tubes used are currently oversized. This limits a possible reduction in the cross-section or wall thickness of the tube and consequently the achievement of lightweight construction goals through the joining process. For this reason, the development is being pushed ahead in order to develop thermally minimally invasive alternative processes for the application of balancing sheets.

A promising alternative to resistance welding is resistance brazing. In [10], [11], [12], [13] investigations into resistance spot brazing of thin sheets have already been described. In comparison to conventionally resistance welded sheet metal-sheet metal joints, an increased fatigue strength (150 %) could be proven by the reduction of thermally and mechanically caused stress [13]. In [8] a brazing process for the application of sheet metal to tubes is described. However, investigations on single-sided resistance brazing are not known. From the described findings, the conclusion is obvious to use the listed advantages of resistance brazing to minimize the deficits of SSSW. The aim is therefore to develop a process that causes a lower thermal and mechanical load on the joining partners compared to single-sided resistance welding. In the underlying contribution, therefore, the single-sided resistance brazing with mechanically introduced solder is to be investigated more closely. In addition to the process architecture, the advantage of single-sided resistance brazing is presented and confirmed with the aid of a practical example.

2 Process- and simulation setup

As in single-sided resistance welding, single-sided resistance brazing involves joining a sheet to a hollow profile or tube. Figure 2 a) schematically shows the process arrangement for single-sided resistance brazing. However, as this is a brazing process, the solder application and the choice of a suitable solder material are decisive. Based on [10], [11], [12] a copper solder was chosen, which is mechanically inserted into a 1.5 mm thick DC01 plate. Due to the mechanical insert of the solder, the sheet metal to be joined is deformed, so that a hump is formed. Care has been taken to ensure that a gap s is created between the solder and the tube in the initial contact state to allow indirect heating of the solder depot (Figure 2 b)). The sheet metal used has a radius of 30 mm. The outer diameter of the tube is 55 mm with a wall thickness of 1.5 mm. The tube material selected is 26MnB5. In [14] a simulation model was built and validated with the software SORPAS 2D in order to effectively design the process development and to gain an understanding of the process (Figure 2 c)). The simulation model is also used in the underlying investigations.

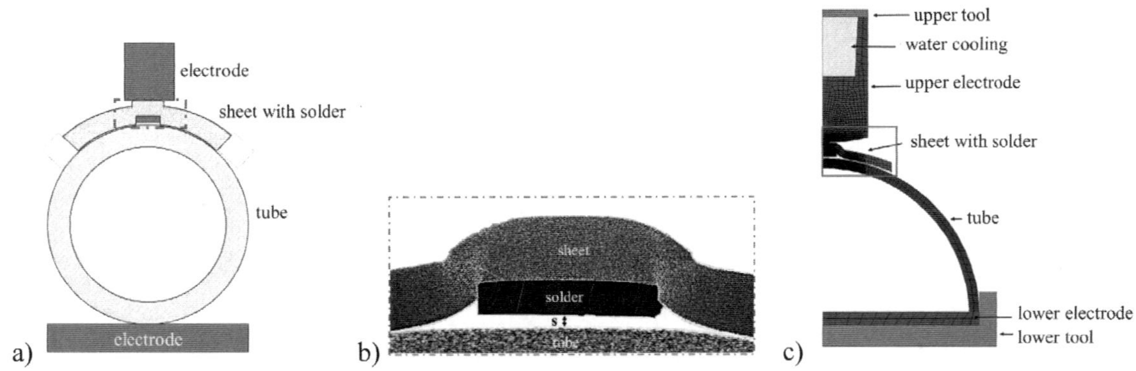

Figure 2. a) schematic process arrangement for single-sided resistance brazing; b) detailed representation of the solder depot with gap s between solder and tube; c) representation of the simulation model [14].

3 Numerical and experimental investigations on single-sided resistance brazing

3.1 Numerical comparison of the process flow for single-sided resistance brazing with single-sided resistance welding

In order to be able to make a process comparison between brazing and welding, a simulation model for single-sided resistance welding was also built using the SORPAS 2D software. If the simulation results are compared with experimental results, it becomes clear that the simulation achieves valid results. If, for example, the nugget diameter is considered, the simulation deviates by only 3.80 % from the metallographically determined nugget diameter. Based on the two simulation models, the process flow of single-sided resistance welding was compared with the process of single-sided resistance brazing. In order to obtain the best possible analogy, the geometrical and material parameters were chosen to be as identical as possible. However, the electrode shape of the working electrode differs due to the process characteristics. Here, a conical electrode shape was used for welding. In each case, a parameter set was selected in which an acceptable joint was produced in experimental tests. The heating process was derived from the calculated temperature distributions. Figure 3 shows the heating mechanism of both processes. In single-sided resistance welding, generally higher current intensities and shorter process times must be selected. The initial heating takes place in the interface between the sheet metal and the tube. The heat propagation is further multidirectional (Figure 3 a)). If one considers the corresponding peak temperatures of the joining partners, the described process becomes clear again (Figure 3 b)). Both the tube and the sheet are heated equally strongly after the squeeze phase (here 100 ms).

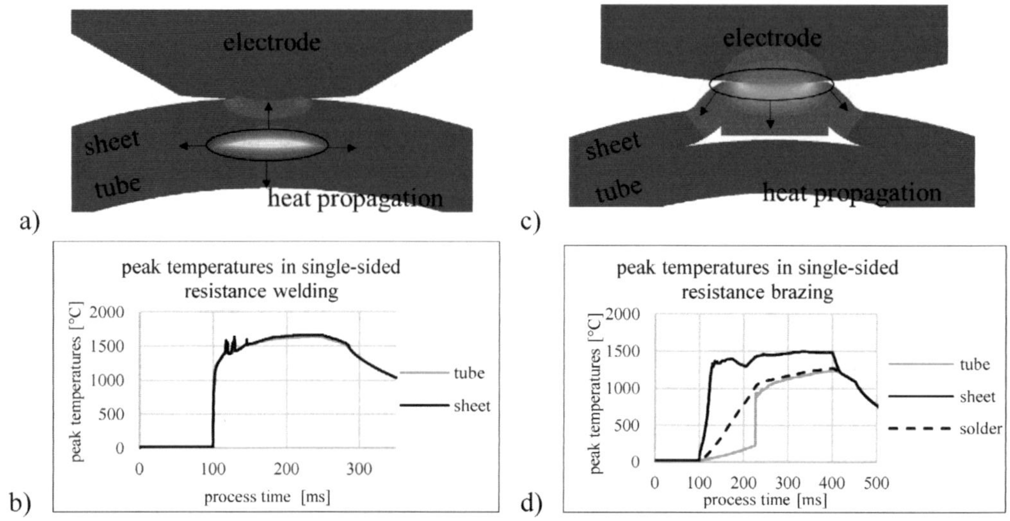

Figure 3. a) schematic heating process in single-sided resistance welding; b) peak temperatures in single-sided resistance welding; c) schematic heating process in single-sided resistance brazing; d) peak temperatures in single-sided resistance brazing.

In comparison, the heating process in single-sided resistance brazing differs significantly. Here the initial heating takes place in the interface between electrode and sheet metal. The heat propagation is in the direction of the tube and the solder (Figure 3 c)). The calculated temperature curve reflects the process again (Figure 3 d)). After the squeeze phase (100 ms) there is a sudden increase in the sheet metal temperature due to the initial heat development in the sheet metal hump area. The heat propagation causes the solder to heat up continuously. Due to the gap s between solder and tube, the tube remains almost unaffected by heat in the first process section (approx. 200 °C). After approx. 230 ms a sudden rise in tube temperature is visible. From this point on, the initial gap s is completely closed so that contact can be established between the strongly heated solder (approx. 949 °C) and the tube. This results in a heating of the tube in the contact area between solder and tube. In the further course, the tube temperature follows the maximum solder temperature. As soon as the solder has reached the melting temperature (T_{solder} = 1080 °C), a solid connection can be assumed. If one ultimately compares the thermal influence on the joining partners of both processes, the difference becomes clear. In resistance welding, the peak temperature of the tube is 1635 °C. In comparison, a maximum tube temperature of 1240 °C was determined for resistance brazing. The sheet metal is also heated less ($T_{max,sheet\ metal,welding}$ = 1662 °C; $T_{max,sheet\ metal,brazing}$ = 1496 °C). In order to clarify the heating process described for single-sided resistance brazing, experimental process flow tests were carried out. Looking at the microsections, it can be seen that the simulated process sequence corresponds to the real process (Figure 4). With increasing process time, the expansion of the heat-affected zone (dotted line) reflects the heating process.

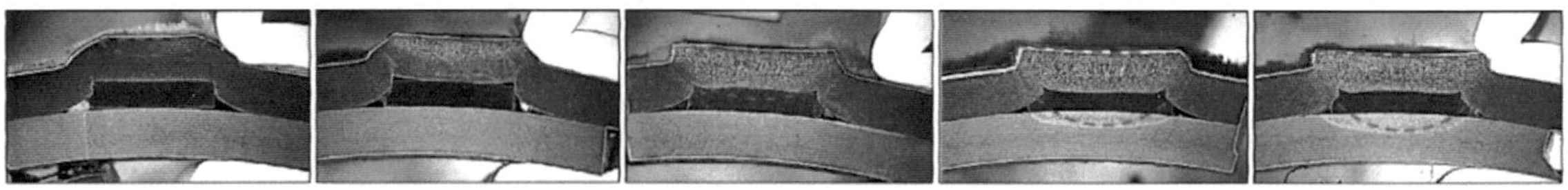

Figure 4. Joining result for single-sided resistance brazing with increasing process time.

It was shown that the single-sided resistance brazing process represents a thermally minimal process alternative to single-sided resistance welding. Due to the lower temperatures, the thermally activated plastic tube deformation can be avoided, especially with smaller tube wall thicknesses.

IOP Conf. Series: Materials Science and Engineering **882** (2020) 012013 doi:10.1088/1757-899X/882/1/012013

3.2 Characterization of the connection for single-sided resistance brazing

If one further compares the resulting joining zone of a resistance-welded specimen with a resistance-brazed specimen, a clear difference can be seen. Table 1 shows possible criteria for differentiating between the two joining methods. In the case of resistance welding, the joint formation is realized by the formation of a welding nugget. In order to be able to make an evaluation, the minimum nugget diameter is generally determined in the microsection. As a further quality criterion, the plug diameter can be measured after a peel test accompanying series production (Figure 5 a)). The characteristic heating process described above results in a high thermal influence on the sheet metal and the tube. An increased risk of thermally activated local deformation is the result.

Table 1. Criteria of a resistance brazed and a resistance welded joint.

	Single-sided resistance welding	**Single-sided resistance brazing**
microsection	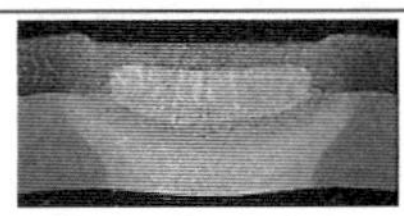	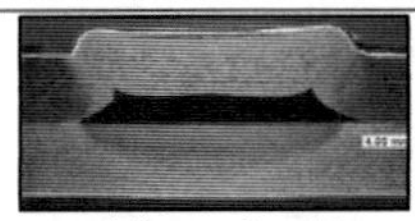
connection	forming of weld nugget	melting of solder
metallographic evaluation	minimum nugget diameter	contact lenght solder-tube
destructive evaluation during series production by peel test	measurement of plug diameter	evaluation of fracture type measurement of fracture diameter
initial heating	interface sheet metal - tube	interface electrode - sheet metal
thermal influence sheet metal	high	medium
thermal influence tube	high	low
risk of local deformation	high	low

In comparison, the connection in resistance brazing is mainly achieved by melting the solder. Due to the heating process a lower thermal influence on the joining partners can be achieved. The danger of thermally activated local deformation is reduced. Since no welding nugget is produced during resistance brazing, the contact length between the solder and the tube can be used as a comparative value in the context of metallographic evaluation. However, further investigations are necessary to determine a minimum length depending on the load case. For the following investigations, a minimum contact length $d_{bd} \geq 3.5 \cdot \sqrt{t_{sheet}}$ is determined based on [15]. A further difference becomes clear when looking at the evaluation accompanying series production by means of the peel test. Because there is no plug after the test, alternative criteria must also be selected here. Depending on the process parameters, characteristic fracture surfaces occur during resistance brazing. Figure 5 b) shows the development of the fracture surface with increasing process time.

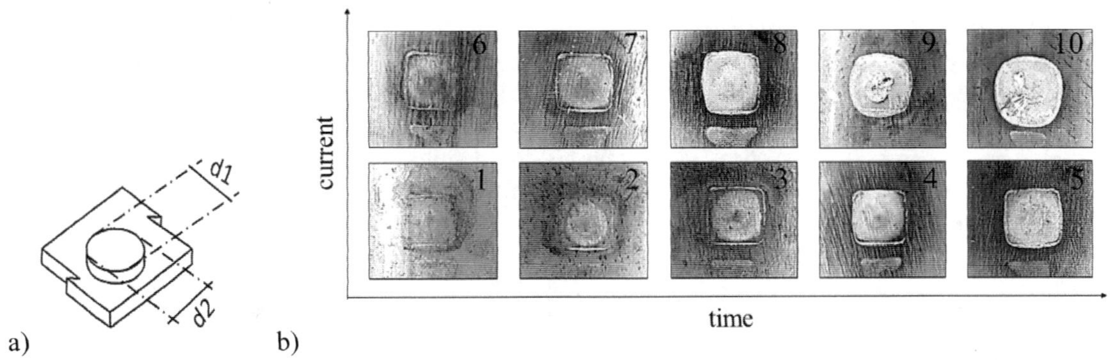

Figure 5. Result of the peel test: a) plug evaluation during resistance welding b) development of the fracture surface on the tube during resistance brazing.

If the energy input is too low (Figure 5 b), example 1) a purely adhesive failure occurs between the solder and the tube without a clear copper-colored solder imprint being visible on the tube. As time and/or current increases, the size of the visible solder imprint increases (Figure 5 b), example 2 - 8). If the energy input is large enough, a cohesive fracture is also visible (Figure 5 b), example 9). A further increase of the energy results in a purely cohesive fracture (Figure 5 b), example 10). In addition to the different fracture types, the size of the fracture surface visible on the tube also increases. Due to the lack of guidelines, further investigations must be carried out in order to enable a uniform evaluation of the resistance brazed joint. The aim of the following investigations is to produce an adhesive-cohesive or purely cohesive fracture (Figure 5 b), examples 9 - 10).

In order to get an impression of the joint strength in relation to the fracture surfaces, shear tests were carried out. The test setup and the results are shown in Figure 6. The joined sheet metal-tube connection was axially loaded to failure by means of a tube with an inner diameter of 55 mm. The maximum force at the time of failure was measured three times and the average was determined. Furthermore, the resulting fracture type and the size of the fracture surface were analyzed (Figure 6 b)). Adhesive failure was found in the test specimen group 1. As no clear copper-colored solder imprint was available, the fracture surface was not measured after the shear test. An average force of 0.87 kN was determined. In comparison, the process time was increased for test specimen group 2. Due to the increased energy input, an increase in shear strength to 4.0 kN was determined. The fracture has a pronounced adhesive character. An average fracture surface of 5.5 mm was measured. Only when the process time was increased again, a purely cohesive fracture with an average area of 6.3 mm was detected. A renewed increase in shear strength to 5.5 kN was determined. It is thus clear that the type of fracture and the fracture area can be used as an evaluation parameter for single-sided resistance brazing. However, more detailed investigations are necessary to establish precise guidelines.

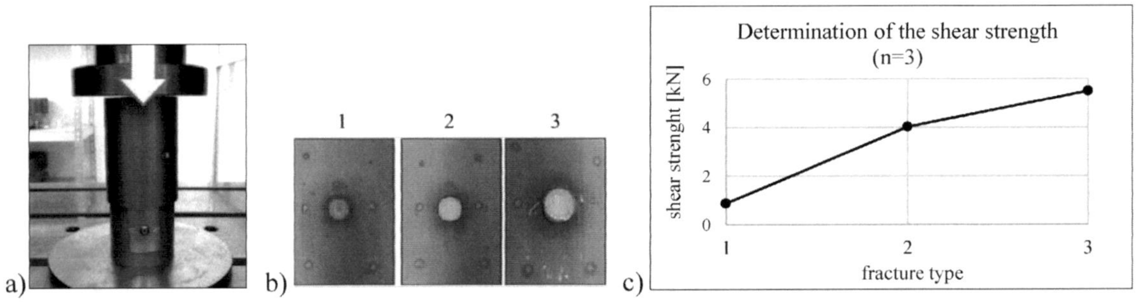

Figure 6. Determination of shear strength as a function of fracture type: a) test set-up; b) characteristic fracture surfaces; c) results.

3.3 Presentation of the process advantage using a practical example

As already mentioned at the beginning, the welding of balance sheets in propshaft production leads to a significant loss of strength under dynamic loads. This is caused by a failure of the tube in the area of the heat-affected zone of the weld (Figure 1 d)). In order to verify the process advantage of the single-sided resistance brazing process in practice, 30 test shafts with identical geometry and material data were set up. The necessary sheets were brazed onto 15 of each of these shafts and welded onto the remaining 15 shafts. This test setup reflects the balancing process necessary for propshaft production. Care was taken to ensure that the parameter set used for each process corresponds to an acceptable joining result. Subsequently, the test shafts were subjected to a dynamic torsion test under alternating loads until failure. Three identical load horizons were tested for each application method, so that five test shafts were tested per load horizon and application process. Subsequently, a process-specific woehler curve was determined in each case. The test setup and the test results are shown in Figure 7.

a)
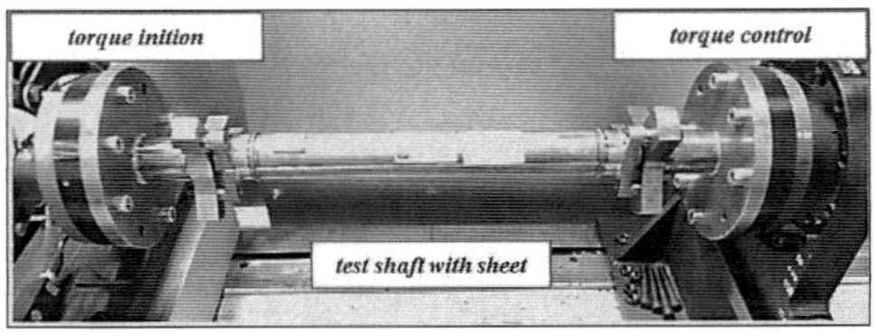

b)
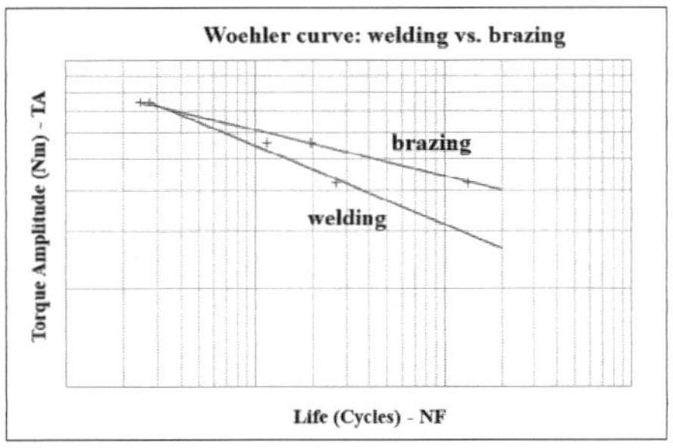

Figure 7. Comparison of brazed shafts with welded shafts: a) presentation of the test setup; b) test results of the dynamic torsion test

If the results of the individual load levels are compared more precisely, the most significant difference was found in the lowest load case. Here an average number of load cycles of 263841 was determined for the welded version. By brazing as a minimally invasive process alternative, the time to failure could be increased by 503 %. The average load cycle here was 1326377. At higher load amplitudes only smaller increases become visible. In the intermediate load case the percentage improvement by brazing is still 170 %. At the highest load case almost identical results are visible. The reason for the reduction of the improvement effect at higher load amplitudes is the characteristic failure cause of the shaft. At low loads and the associated high number of load cycles, a crack of the tube as a result of the sheet metal application was consistently detected in the welded version. The higher the load amplitude, the higher the probability that other causes of failure (e.g. plastic deformation of the tube, fracture of the friction weld) will occur. Thus, the sheet metal application process mainly influences the lowest load horizon. An increase in performance under dynamic loading has been demonstrated. This is due to a process-related reduced thermal and mechanical influence on the components.

4 Summary

In order to minimize the deficits in single-sided resistance welding, single-sided resistance brazing was examined more closely. In addition to the process architecture, a simulation model was also presented in order to build up an understanding of the process. In a comparison between single-sided resistance welding and single-sided resistance brazing, the advantage of the new process becomes clear. While the initial heating in the SSSW takes place in the contact area of the sheet metal tube and then spreads multidirectional, in resistance brazing the tube is only subjected to very low thermal stress in the initial stage of the process. This is due to the specific indirect heating process of the solder. As soon as the constructive gap s between solder and tube is closed, the tube is also heated for a short time. Comparing the maximum temperatures in the tube, a difference of 400 °C was found. In the following, the differences between a welded and a brazed tube-sheet metal joint were shown and the joint created by resistance brazing was characterized. Due to the dissimilar connection mechanisms, alternative evaluation criteria have to be chosen for the brazing process. In addition to the contact length between solder and tube which can be measured in the metallographic analysis, the fracture surface provides information about the joint quality. However, in order to determine more precise quality parameters, further detailed investigations are necessary. Finally, the process advantage was demonstrated. Using the example of propshaft production, test shafts were equipped with a sheet metal. In addition to the SSSW, the sheet metal was also applied using single-sided resistance brazing. In the following dynamic torsion test it was shown that the fatigue strength of the component increases significantly. The reason for this is the reduced thermal and mechanical component influence of the brazing process. It was thus proven that single-sided resistance brazing has high potential as an alternative to single-sided resistance welding.

5 References

[1] Nielsen C, Chergui A and Zhang W, "Single-sided sheet-to-tube spot welding investigated by 3D numerical simulations," in 7th International Seminar on Advances in Resistance Welding, Busan, Republic of. Korea, 2012.

[2] Rudolf H, Widerstandspunktschweißen von Blech-Hohlprofil-Verbindungen an verzinkten Stahlblechen, Berlin: TUDpress, 2006.

[3] Schiebahn A, "Abschlussbericht IGF-Vorhaben 17.685N: Untersuchung und Qualifizierung der verfahrensspezifischen Merkmale beim einseitigen Widerstandspunktschweißen ohne Gegenlage," Institut für Schweißtechnik und Fügetechnik - RWTH Aachen University, Aachen, 2016.

[4] Liang C, Lin Z, Chen G and Li Y, "Numerical analysis of single sided spot welding process used in sheet to tube joining," *Science and Technology of Welding an Joining*, no. 11, pp. 606 - 617, 2006.

[5] Yang H, Hu S, Zhang Y, Li Y and Lai X, "Experimental study of single sided sheet to tube resistance spot welding," *Science and Technology of Welding and Joining,* vol. 6, no. 12, pp. 530-535, 2007.

[6] Dilthey U and Böhme D, "Abschlußbericht AiF 12.739 N - Einseitiges Widerstandsschweißen von Stahl-Hohlprofilen," Institut für Schweißtechnische Fertigungsverfahren der Rheinisch-Westfälischen Technischen Hochschule Aachen, Aachen, 2003.

[7] Papke H, "Wuchtgewicht für Kreuzgelenkwellen". Deutschland Patent DE3931055A1, 18 09 1989.

[8] Pullen T and Cremerius R, "Hohlwelle mit mindestens einem Wuchtgewicht sowie Verfahren zu deren Herstellung, wobei das mindestens eine Wuchtgewicht mittels Löten an der Hohlwelle befestigt wird". Deutschland Patent WO2006/027012 A1, 10 09 2004.

[9] Li Y. B, Liang C, Zhang Y and Lin Z, "Application of electrode force change in single sided resistance spot welding using servo gun," *Science and Technology of Welding and Joining,* vol. 13, no. 7, pp. 671 - 678, 2008.

[10] Sitte G, Koppe K and Hindorf H, "Punktförmiges flussmittelfreies Widerstandslöten," Blech, Rohre, Profile: Herstellen, Verarbeiten und Veredeln von Band, Blech, Rohren und Profilen; CIRP related magazine; Referatsorgan der Internationalen Forschungsgemeinschaft für Produktionstechnik, vol. 9, no. 45, p. 57 pp, 1998.

[11] Sitte G, „SCHWEISSTECHNIK UND MEHR - Punktförmiges Widerstandslöten - Die Alternative zum Widerstandspunktschweißen," *Der Praktiker: Das Magazin für Schweißtechnik und mehr*, Bd. 5, Nr. 53, pp. 196 - 199, 2001.

[12] Koppe K and Sitte G, "Umformwiderstandsschweiß- und Lötprozesse - neue Verbindungstechnik für Feinbleche," in 2. Fachtagung: Fügen und Konstruieren im Schienenfahrzeugbau, Halle, 1997.

[13] Sitte G and Brund C, "Punktlöten und Punktlötkleben," in Hart- und Hochtemperaturlöten und Diffusionsschweißen: Vorträge und Posterbeiträge des 8. Internationalen Kolloquiums in Aachen, Aachen, 2007.

[14] Najuch M, Jüttner S, Schmicker D, Krüger A and Haverland C, "Simulation und Implementierung eines thermisch minimalinvasiven Fertigungsprozesses für die Applikation von Wuchtgewichten," in 5. Fachtagung Sensitive Fertigungstechnik, Magdeburg, 2019.

[15] Merkblatt DVS 2916:1978-07, Prüfen von Punktschweißungen.

IOP Conf. Series: Materials Science and Engineering **882** (2020) 012014 doi:10.1088/1757-899X/882/1/012014

Microstructure-property relations of eutectic V-Si and V-B alloys

M Regenberg, G Hasemann, C Müller and M Krüger

Otto-von-Guericke University Magdeburg, Faculty of Mechanical Engineering, Institute of Materials and Joining Technology, Universitätsplatz 2, 39106 Magdeburg, Germany

maximilian.regenberg@ovgu.de

Abstract. Subject of the present work are principle studies on V-Si and V-B alloys with hypo- and hypereutectic compositions, regarding the effects of primary phase solidification and the formation of binary microstructures on the resulting mechanical properties. For this purpose, button shaped samples were produced by arc-melting, metallographically prepared and analyzed by scanning electron microscopy. Furthermore, room temperature compression tests have been carried out and the resulting deformations were subsequently analyzed by using electron backscatter diffraction. As a conclusion, the materials' failure behavior was characterized and discussed with respect to the developed microstructure. The main goal of the experiments conducted is to critically discuss previously published works on the V-Si and V-B alloying systems and to obtain more profound insights on microstructure-property relations with respect to the strengthening effects and plastic deformability.

1 Introduction

Due to the variety of attractive offers in civil aviation, the airplane has become one of the most popular means of transportation in recent decades. According to a statistic of the German VuMA (Arbeitsgemeinschaft Verbrauchs- und Medienanalyse), about 56.5% of residents used the airplane (scheduled flights, charter flights and low-cost airlines combined) to reach their (holiday) destination in the years 2016-2018 [1]. Furthermore, a rising trend in the number of transported passengers (private and business travelers) in air traffic can be stated [2]. These statistic numbers illustrate the growing importance and popularity of air traffic as means of mass transport, while simultaneously the question regarding the environmental compatibility of flying also gains increasing attention. Although the airplane often is the most time-efficient way of traveling, a drawback can be seen if the ecological aspects are considered.

Table 1. Comparison of average emissions from individual means of passenger transport (reference year 2017, collected by the german Federal Environment Agency; emissions are expressed in grams per passenger kilometer (g/Pkm); emissions from the provision and conversion of energy sources into electricity, petrol, diesel and kerosene are included [3].

		Passenger car	Coach[1]	Railway, long-dist.	Airplane	Line bus	Railway, local	Tram, subway
Greenhouse gases[2]	(in g/Pkm)	139	32	36[3]	201[4]	75	60	64
Carbon monoxide	(in g/Pkm)	0.6	0.04	0.02	0.13	0.05	0.04	0.04
Volatile hydrogen[5]	(in g/Pkm)	0.14	0.01	0	0.04	0.03	0.01	0
Nitrogen oxide	(in g/Pkm)	0.34	0.17	0.04	0.51	0.28	0.18	0.06
Particular matter[6]	(in g/Pkm)	0.004	0.003	0	0.004	0.002	0.002	0
Capacity utilisation		1.5 ppl/car	60%	56%	82%	21%	27%	19%

[1] Category "Coach" includes buses on irregular shedules (e. g. school trips and coffee cruises) and long-distance bus services

[2] CO_2, CH_4 and N_2O are stated in CO_2-equivalents

[3] Stated emission factors for railways are based on the average energy-mix in Germany

[4] In consideration of all climate-impacting effects of air traffic (EWF = Emission Weighting Factor = 2)

[5] Methane not included

[6] Abrasion not included

Regarding the emissions of greenhouse gases and nitrogen oxides, aircraft turbines show the highest values. Additionally, the numbers of emitted carbon monoxide, volatile hydrogen and particulate matter are comparably high, as can be seen in Table 1. In order to reduce emissions during the operation of flight, the efficiency of the aircraft engine (turbine) has to be increased [4]. To fulfill these requirements the development and application of novel structural materials is inevitable.

Reports on low alloyed vanadium-based materials with additions of Cr, Ti or Si for the use in fusion reactor applications (f.e. in self cooled Li blankets) have already been published since the 1990's [5-7]. Besides that, more recent findings have shown that, due to their low density and high mechanical strength at elevated temperatures up to 1000 °C, V-based materials show a high potential for the use as high temperature structural materials [8, 9]. In particular, the material class of eutectic vanadium-silicon alloys, with operating temperatures up to 700 °C and simultaneously good thermal shock resistance, have proven increased suitability for the potential use as structural components in the field of stationary gas- and aircraft turbines [10, 11]. Extensive research has been carried out by the working group of Krüger et al. in 2016 [8], concerning the mechanical properties of a V-9Si-13B alloy. The samples investigated were produced via a powder metallurgical route and were then compared to the state-of-the-art single-crystal nickel-based superalloy CMSX-4®. The results showed that the V-based alloy can compete with the superalloy, especially in terms of yield stress and the higher specific strength, due to a lower density [8].

In principle, the present study aims to re-evaluate the eutectic compositions of vanadium-silicon and vanadium-boron alloys as they have been stated in the phase diagrams by Smith [12] and more recently by Lima-Kühn et al. [13] (V-Si system), respectively by Spear et al. [14] and Nunes et al. [15] (V-B system) in an experimental way. In addition, the aforementioned authors reported on the obtained microstructure-property relations of the vanadium-based alloys, with regard to present volume fractions of eutectic in the produced samples.

IOP Conf. Series: Materials Science and Engineering **882** (2020) 012014 doi:10.1088/1757-899X/882/1/012014

2 Experimental procedures

The alloying systems vanadium-silicon and vanadium-boron were investigated in the presented study. A total of six button-shaped samples with different hypo- and hyper-eutectic alloying compositions (shown in Table 2) were melted in a water-cooled copper crucible, using a non-consumable tungsten electrode in a conventional vacuum-arc-furnace (comparable to Arc Melter MAM-1, Edmund Buehler GmbH, Bodelshausen). The raw materials consisted of vanadium chippings (99.7%), silicon granules (99.99%) and boron granules (99.0%), which were carefully weighed in. During arc-melting, the buttons were flipped and re-melted five times to ensure good homogeneity. The actual chemical alloy composition after melting was verified using inductively coupled plasma optical emission spectroscopy (ICP-OES), see Table 2.

Table 2. Nominal and by Inductively Coupled Plasma-Optical Emission Spectrometry (ICP-OES) measured composition of the V-Si and V-B alloy.

Nominal composition (at.%)	Si (at.%)	Deviation of Si (%)
V-9.75Si	10.73	1.10
V-12.5Si	13.00	1.04
V-15.75Si	16.95	1.08

Nominal composition (at.%)	B (at.%)	Deviation of B (%)
V-5.5B	5.78	1.05
V-11B	11.54	1.05
V-14.5B	14.18	0.98

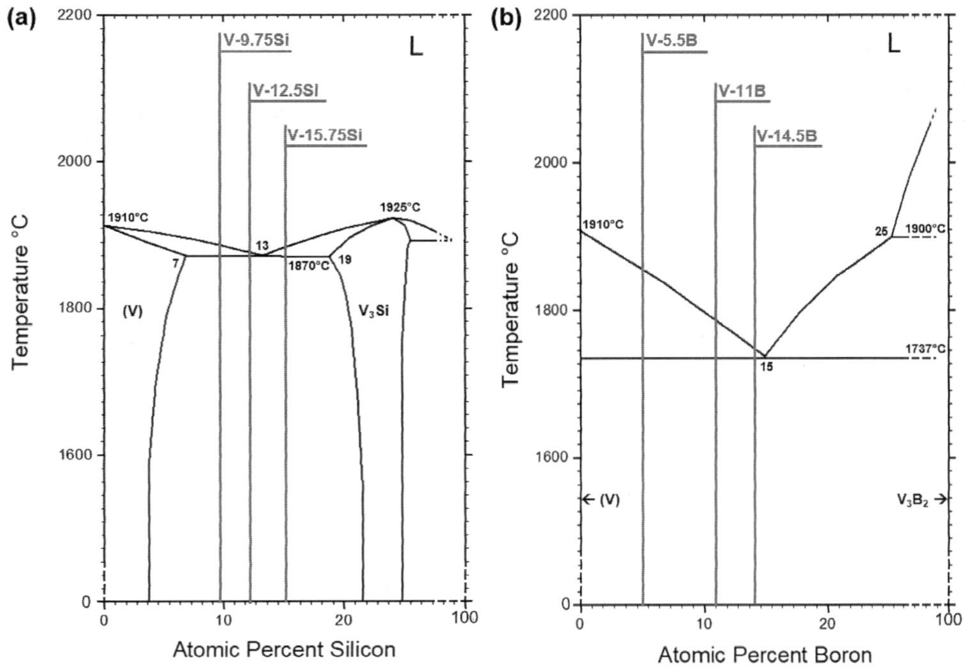

Figure 1. Hypothetical compositions of the target alloys in the respective parts of the (a) V-Si phase diagram by Smith et al. [12] and (b) V-B phase diagram by Massalski [16].

In order to determine the composition of the samples considering the eutectic and hypo-, respectively hyper-eutectic fractions, the lever rule was used in the vanadium-rich proportions of the V-Si phase diagram according to Smith [12] and the V-B phase diagram compiled by Massalski et al. [16], as can be seen in Figure 1. It must be stated that, according to the displayed vanadium-boron phase diagram, the V-14.5B alloy is expected to solidify at a temperature slightly below the eutectic point, thus developing a hypoeutectic microstructure. In fact this is not the case, since recent investigations carried out by the Nunes work group [13, 14] have shown a more accurate prediction of the invariant reactions, which were taken into account for the present calculations.

The button-shaped samples were cut with a diamond coated saw blade on a cutting machine (Brillant 220, ATM Qness GmbH, Mammelzen) and embedded in epoxy resin. In the course of metallographic preparation, the specimens were subsequently grinded on a semi-automatic grinding and polishing machine (Saphir 550, ATM Qness GmbH, Mammelzen) and polished on a single-specimen polishing machine (MiniMet 1000, Buehler, Lake Bluff), using 6 μm, 3 μm and 1 μm polycrystalline diamond suspension consecutively, finishing with a colloidal silica suspension (OP-S). Microstructural investigations were carried out with a scanning electron microscope (SEM; Merlin, Carl Zeiss AG, Oberkochen). In preparation for the SEM analysis, the samples were coated with a thin layer of iridium and afterwards scanned, using the backscattered electron mode (BSE) and wavelength-dispersive X-ray spectroscopy (WDS, Oxford Instruments, Abingdon). The phase fractions of the different microstructures were determined with the optical image analysis software FIJI, using different grey scale values of the images.

Standard cylindrical compression samples [19] of 3 mm in length and 1.8 mm in diameter were prepared by electro discharge machining (EDM) and surface-cleaned (removal of a burr and the surface tarnish) subsequently, using 800 and 1200 grit SiC paper. In addition, a 3 h lapping process with sandblasting granules (glass-beads) in ethanol was carried out, to avoid any impact of the surface condition on the subsequent mechanical tests. Room temperature compression tests were performed on a universal testing machine (Type 1381, Instron, Norwood) with a maximum load capacity of 30 kN and a load cell with an operating range of ± 10 kN. According to DIN 501006 [19], a constant cross-head speed corresponding to an initial (engineering) strain rate of $\dot{\varepsilon} = 10^{-3}\ s^{-1}$ was used. As results of the compression tests, the compressive yield stress functions were determined using the 0.2% offset method and displayed in engineering compressive stress-strain curves. After testing, macroscopic images of the samples were obtained on a digital microscope (Keyence VHX-5000, Osaka). Additionally, the specimens were embedded in epoxy resin, grinded down to about half their cylindrical diameter and prepared afterwards, following the same steps as described above (see preparation of the button-shaped samples). Subsequentially, the mechanical deformation was determined by means of local misorientation maps, using electron backscatter diffraction (EBSD, Oxford Instruments, Abingdon).

3 Results and discussion

The as-cast microstructure of the V-Si alloys is displayed in Figure 2. Significant differences between the primary phases and the eutectic areas were obtained in the alloying system, which are emphasized by different grey values. The stated values for the phase fractions represent the results of the optical image analysis. In Figure 2 (a) the V-9.75Si microstructure, consisting of primary solidified vanadium solid solution (V_{SS}) phase (≈ 60 vol.% phase fraction) and surrounding eutectic matrix is present, whereas Figure 2 (b) displays the eutectic arrangement of the V-12.5Si alloy and Figure 2 (c) depicts the hypereutectic V-15.75Si microstructure, containing primary solidified V_3Si dendrites (≈ 40 vol.% phase fraction) and eutectic microstructure. The results of the SEM analysis, regarding the V-B microstructures are displayed in Figure 3. The hypoeutectic V-5.5B alloy (Figure 3 (a)) shows fractions of V_{SS} phase (≈ 60 vol.% phase fraction) and eutectic matrix, whereas the V-11B alloy in Figure 3 (b) is considered as near-eutectic since small proportions of V_3B_2 phase (< 3 vol.% phase fraction) were obtained in these specimens. The hypereutectic alloy V-14.5B (Figure 3 (c)) depicts a fraction of primary V_3B_2 (≈ 14 vol.% phase fraction) and a respectively larger fraction of the eutectic matrix, compared to alloy V-5.5B. It is noteworthy, that the vanadium-boron alloys formed a non-facetted-facetted eutectic in contrast to the vanadium-silicon alloys which form a lamellar eutectic. This can be explained by deviating solid-liquid interfacial energies of the phases, different melting entropies of the eutectic phases and kinetic coefficients for coarse surfaces or different local supercooling effects [20-22]. The present deviations from the calculated compositions regarding both alloying systems can be explained by the rapid cooling conditions of the crucible, which prevents the diffusion of the elements and thus the formation of phases under equilibrium condition (as predicted by the lever rule) and the apparent inaccuracies in the published phase diagrams.

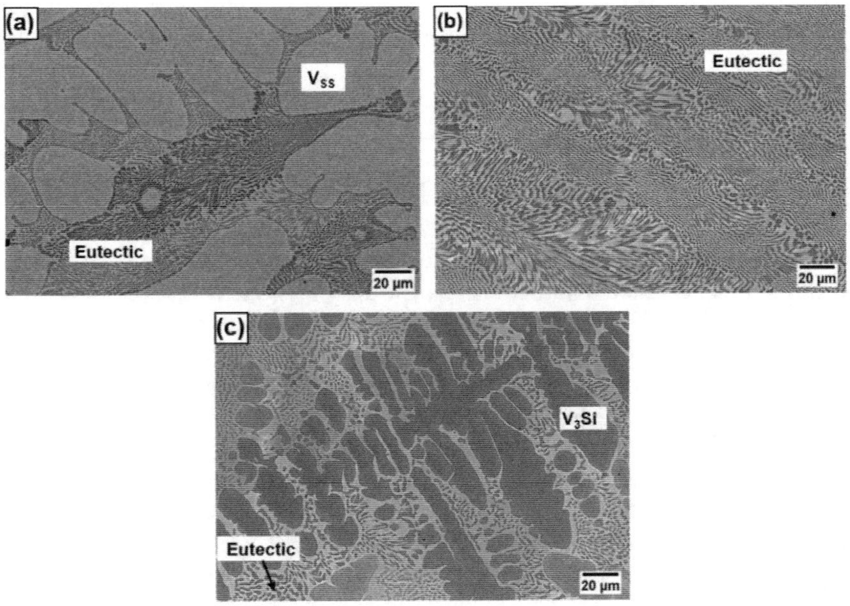

Figure 2. SEM-BSE images of the (a) hypoeutectic alloy V-9.75Si, (b) eutectic alloy V-12.5Si and (c) hypereutectic alloy V-15.75Si microstructures.

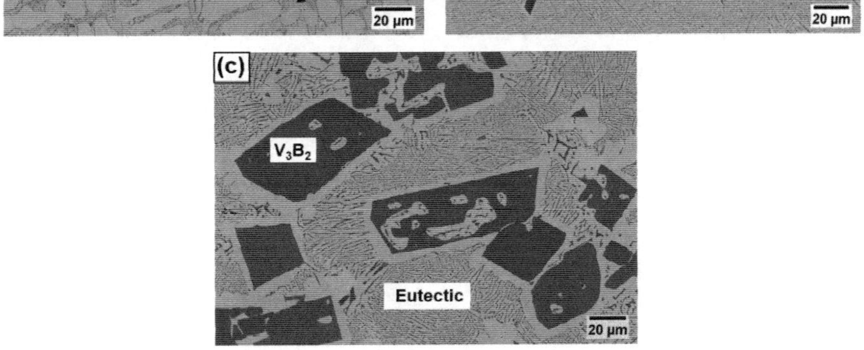

Figure 3. SEM-BSE images of the (a) hypoeutectic alloy V-5.5B, (b) near-eutectic alloy V-11B and (c) hypereutectic alloy V-14.5B microstructures.

The results of the room temperature compression tests are displayed in Figure 4. In order to provide a statistical evaluation, 4-6 samples of each alloy were tested, which provided a high reproducibility of the results. Considering the V-Si alloys (Figure 4 (a)), a strong correlation between the Si content and the compressive strength, as well as the maximum plastic deformation until failure was observed. This is in good agreement with the varying degrees of bulging of the samples. The hypoeutectic alloy V-9.75Si shows the highest grade of plastic deformation ($\varepsilon_{P\ max} \approx$ 14%), respectively lowest strength ($\sigma_{0.2} \approx$ 849 MPa), which is attributed to the high fraction of ductile V_{SS} phase in the alloy composition [23]. It is stated that the sample failed in a ductile regime, which is indicated by shear failure along the diagonal axis of the cylindrical specimens during the compression tests. The eutectic alloy V-12.5Si shows an increase in strength ($\sigma_{0.2} \approx$ 914 MPa) and reduced plastic deformability ($\varepsilon_{P\ max} \approx$ 7.5%), compared to V-9.75Si, which is caused by higher fractions of the intermetallic V_3Si phase. Additionally, less bulging and shear plastic deformability of the eutectic specimen was found. The highest mechanical strength ($\sigma_{0.2} \approx$ 1202 MPa) and the lowest plastic deformability ($\varepsilon_{P\ max} \approx$ 3 %) values were measured for the hypoeutectic alloy V-15.75Si, which possesses the highest percentage of the V_3Si phase compared to all alloys tested. As mentioned before, the increasing phase fraction of intermetallic silicides may be interpreted as the origin for the increase in strength and embrittlement of the samples. Regarding the failure mechanism, it must be noted that there is significantly less ductility, even though the crack propagates in a diagonal direction, no measurable bulging of the cylinder and no shear behavior as such has been observed.

Considering the results of the compression-loaded V-B alloys (Figure 4 (b)), the dependency between alloyed B content and resulting mechanical properties cannot be derived as clear as those from the V-Si samples. Specifically the compressive stress-strain curves of the eutectic alloy V-11B and the hypereutectic alloy V-14.5B exhibit overlays and show minor differences concerning the compressive strength. The lowest strength ($\sigma_{0.2} \approx$ 272 MPa) values were obtained from the V-5.5B samples, and an almost ideal plastic behavior was determined. The samples of the near eutectic V-11B alloy exhibited comparatively high strength values ($\sigma_{0.2} \approx$ 460 MPa) and good plastic deformability ($\varepsilon_{P\ max} \approx$ 20%). Again, it is noteworthy that the results are to be viewed in a more differentiated way, since the present material failure is still close to ideal plastic behavior and thus, crack characterization is not applicable. The bulging of the V-11B specimen is also distinctive and a deformation alongside the diagonal axis indicates failure by shear plastic deformation of the compression-tested samples. In comparison with the hypo- and hypereutectic vanadium-boron alloys, the results of the V-14.5B tests have shown higher compressive yield strength values ($\sigma_{0.2} \approx$ 548 MPa) at high plastic deformations ($\varepsilon_{P\ max} \approx$ 22%). Taking the corresponding results of the compression tests into account, it can be stated that the graphs show less similarity to an ideal-plastic curve. Additionally, bulging of these samples is less significant and the crack induced by the deformation along the diagonal points out to a minor brittle fracture behavior. Based on the observations, it can be concluded that a tendency towards lower deformability with increasing boron content (hence an increasing content of intermetallic vanadium boride phase) is present.

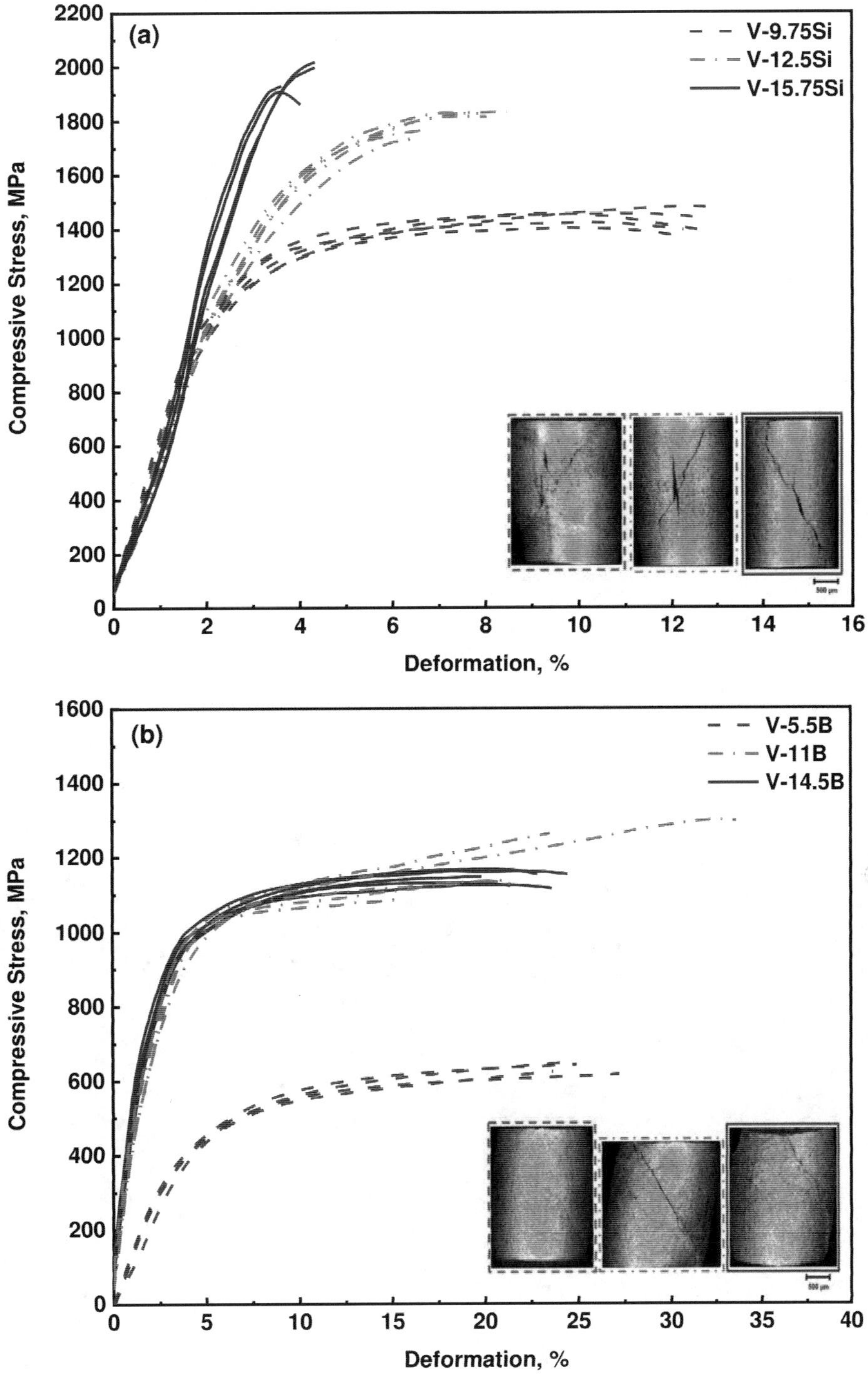

Figure 4. Engineering compressive stress-strain curves of the examined V-Si (top) and V-B (bottom) alloys and photographs of typical samples after compression testing, see inset.

The EBSD misorientation maps, see Figure 5, are in good agreement with the results obtained from the compression tests. Regarding the low alloyed V-9.75Si sample (Figure 5 (a)), it can be stated that the energy introduced during compression is almost entirely transferred via the deformation of the vanadium-rich fraction of the eutectic (green color-coded areas). In contrast, the V_{SS} primary dendrites show almost no deformation (blue areas), merely a slight green veil can be noted around the grains which is an indicator for piled-up dislocations at the grain boundaries. Further attention must be paid to the emerging thin green lines on the primary dendrites, which are presumed to be dislocation slip bands caused by dislocation pile-ups and local stress peaks at the grain boundaries. On the one hand, the origin of the present cracks in the two-phase eutectic region can be described as a result of stress-peaks in the specimens, induced by the strong deformation of the V_{SS}-rich matrix (leading to the aforementioned piled-up dislocations at the grain boundaries). On the other hand, the cracks in the V_3Si grains indicate a stress relief mechanism, determined by a break-up of the hard, intermetallic phase [11]. The misorientation map of the V-12.5Si alloy (Figure 5 (b)) shows an evenly distributed deformation behavior of the V_{SS} over the entire sample. The present fraction of intermetallic V_3Si is larger compared to the hypoeutectic V-Si alloy and the proportion of vanadium solid solution is smaller, respectively, resulting in an increase of the strength and a decrease of the ductility. Moreover, it is mentioned that an irregularity in the distribution of misorientations can be seen in areas where the V_3Si phase increasingly solidified in larger, rod shaped structures. The deformation obtained in the regions between the dendrites is smaller, compared to the sections with a higher ratio of the globular silicide phase. This observation supports the assumption, that the alloy's mechanical properties may be adjusted by a combination of the strength enhancing V_3Si phase and the ductile V_{SS} matrix. According to the previous results, the deformation behavior of the V-15.75Si sample is in good agreement with the literature [11,24,25]. Due to the higher silicon content in the hypereutectic specimen, the fraction of intermetallic V_3Si phase is larger in comparison with the hypo- and fully eutectic alloys. The EBSD image of alloy V-15.75Si (Figure 5 (c)) shows a similar correlation regarding the distribution of the force through deformation in the matrix phase and the breaking up of the silicide phases in case of stress overload, as stated before [11].

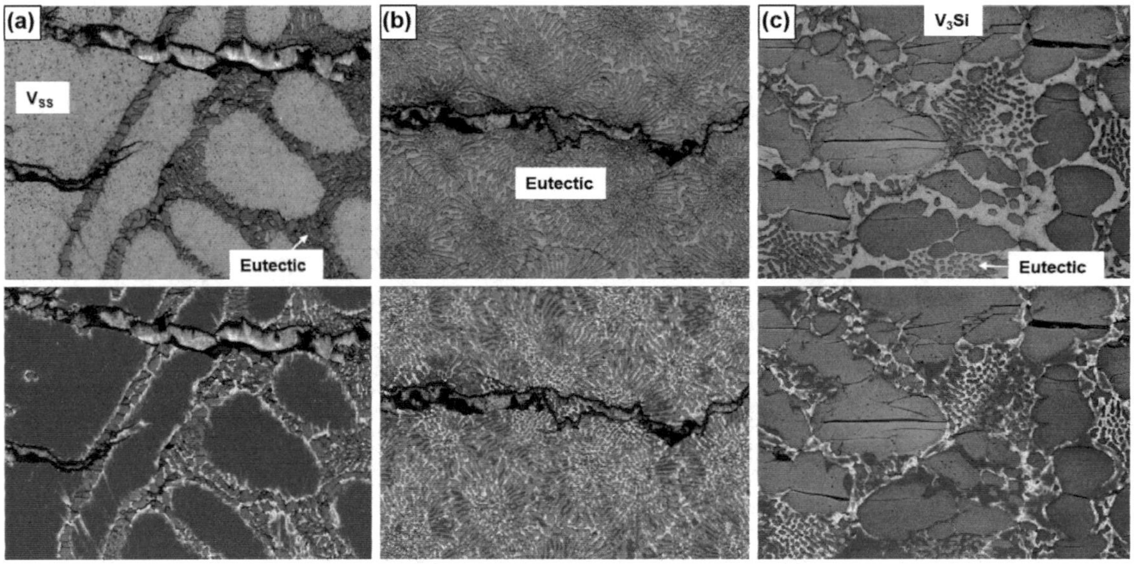

Figure 5. SEM images (top) and color-coded EBSD misorientation maps regarding vanadium (bottom); From left to right:(a) V-9.75Si, (b) V-12.5Si, (c) V-15.75Si.

IOP Conf. Series: Materials Science and Engineering **882** (2020) 012014 doi:10.1088/1757-899X/882/1/012014

The misorientation maps of the vanadium-boron alloy are displayed in Figure 6. The hypoeutectic V-5.5B sample (Figure 6 (a)) shows a homogeneous distribution of the introduced deformation via the eutectic phase fraction. At high magnitudes, small amounts of cracked, rod-shaped V_3B_2 crystallites were obtained in the microstructure. This supports the assumptions stated above (concerning the failure mechanism of the V-Si alloys), that a stress relief mechanism, determined by a break-up of the hard, intermetallic phase is also present. At this point, the attention is drawn to a special feature, present in Figure 6 (a). A primary grain (which can hardly be identified as such in the band contrast image on the left side) in the upper section of the image seems strongly permeated by the eutectic structure. This can be attributed to a residual amount of melt that was accumulated during cooling, and it solidified partly as primary grain and subsequently as eutectic microstructure. Furthermore, the green veils on the blue colored V_{SS} grains are noticeable: It is challenging to differentiate whether these are dislocation belts (similar to the V-9.75Si sample) or sample preparation defects, induced by grinding and polishing. The eutectic microstructure of the V-B alloys is present in a fine non-faceted-faceted arrangement. Due to the resolution limitation of the EBSD detector, a more precise characterization of this area was not possible. The images of the V-11B compression samples (Figure 6 (b)) display a regular distribution of deformation across the vanadium-rich fractions of the eutectic microstructure, comparable to that of the eutectic V-Si specimens. The present vanadium boride facets and the primary crystal remains almost undeformed. The V_3B_2 crystal depicted in Figure 6 (b) shows a green veil at its phase boundaries, indicating a pile-up of dislocations. Based on the fine eutectic microstructure and the strong deformability, conclusions were drawn regarding the high levels of plastic deformation present in some of the tested samples: Similar to the V-15.75Si alloy, the misorientation map of the hypereutectic V-14.5B sample (Figure 6 (c)) indicates a failure behavior, determined by a distribution of the deformation between the V-rich V_{SS} phase and the intermetallic phase. The plastic deformation is illustrated in green (V-rich phase), while the grey (V_3B_2 phase) fractions show cracks, induced by stress-induced overload.

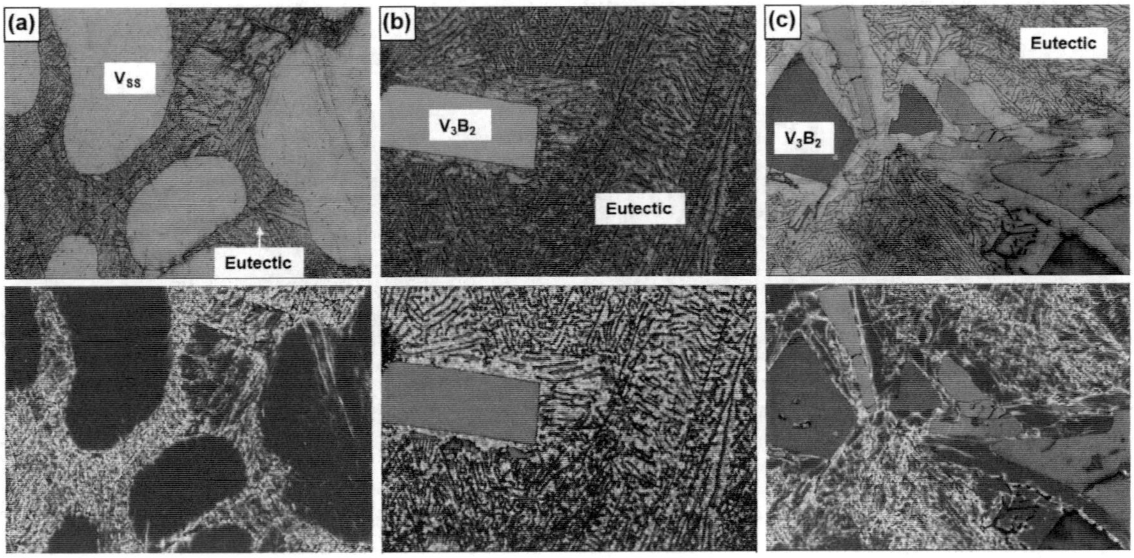

Figure 6. SEM images (top) and color-coded EBSD misorientation maps regarding vanadium (bottom); From left to right: (a) V-5.5B, (b) V-11B, (c) V-14.5B.

In addition to the alloy characteristics discussed above, a bar chart is presented in Figure 7, showing the experimentally obtained compression yield strengths. When taking the values, displayed in Table 3 into account, a clearly differentiated increase with respect to the Si and B contents in the alloy composition was observed. In comparison, the $\sigma_{0.2}$(Si) values are generally higher than the $\sigma_{0.2}$(B) numbers, resulting in a difference in the overall (compressive) strength of ~430 MPa, regarding the hypoeutectic alloys, ~440 MPa, referred to the eutectic alloys and ~520 MPa, concerning the hypereutectic alloys. The displayed results, again, indicate, that the mechanical properties of both alloying systems are determined by the present fraction of eutectic (ductile) and intermetallic (brittle but high strength) microstructure in proportion by volume. Furthermore, the compressive strength of the samples was not taken for evaluation because of the partly present ideal plastic failure behavior, leading to intersections with the fracture stresses and therefore resulting in misleading or imprecise conclusions.

Table 3. $\sigma_{0.2}$ values of the V-Si and V-B alloys tested.

Alloy	V-9.75Si	V-12.5Si	V-15.75Si	V-5.5B	V-11B	V-14.5B
$\sigma_{0.2}$ (MPa)	~745	~840	~1015	~272	~400	~490
Standard deviation (MPa)	65	41	105	43	59	33

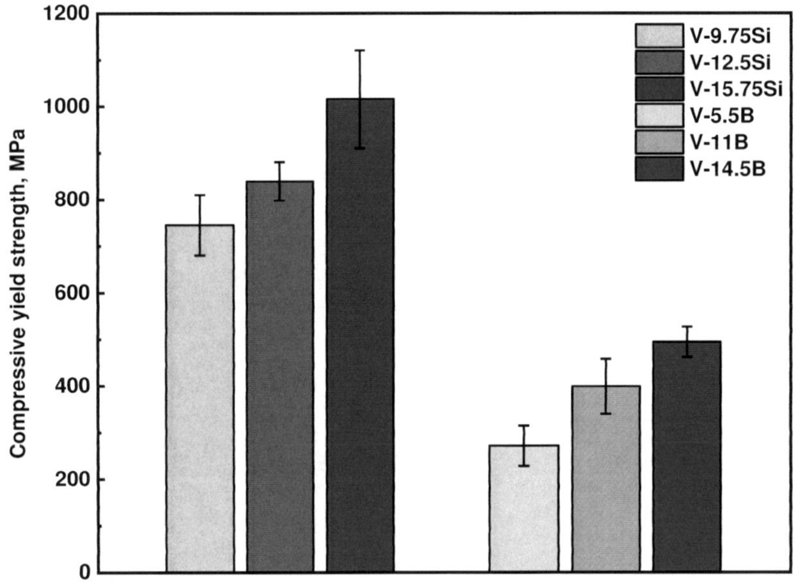

Figure 7. Comparison of the compressive yield strengths of the V-9.75Si, V-12.5Si, V-15.75Si, V-5.5B, V-11B and V-15.5B samples.

4 Summary and conclusion

In the present work, different alloys from the system vanadium-silicon and vanadium-boron were produced by an arc-melting process. The microstructures were designed to contain either ~50% of the primary phases in comparison to the respective eutectic alloy composition, or a 100% eutectic microstructure, evaluated with computer aided image analysis. Room temperature compression tests on cylindrical samples were carried out and the mechanical properties such as compressive yield stress and deformation, were investigated. Based on the compression tests, the room temperature plasticity was determined for all samples, which was influenced by the different alloying elements and the different volume fractions of the primary phases. Grain misorientation maps, displaying orientation changes by calculating the misorientation between each pixel and the surrounding pixels, thus highlighting strongly deformed regions, were generated with EBSD analysis. The maps indicate that the applied mechanical stress is absorbed by deformation of the V_{SS} phase, whilst the high-strength intermetallic silicide and boride phases act as strength enhancing components. Dislocation pile-ups on the phase boundaries and resulting cracks were stated as the criteria of failure in almost all samples. An exception is alloy V-5.5B, which exhibits an almost ideal-plastic deformation behavior and thus, no cracking in the investigated range of deformation was observed. It can be noted, that the present findings are in good agreement with previous investigations on comparable alloy systems.

The results of the present work indicate a high potential of vanadium-based materials with respect to structural applications. To gain a more profound understanding of the deformation mechanisms, and/or on the dislocation movement in those alloys, transmission electron microscopic analysis (TEM) may be a helpful tool, and more V-B samples with higher fractions of boron should be investigated.

Acknowledgements
The present research was funded by the German Research Foundation (DFG) under the grant number 410338871 and was in part conducted within the context of the International Graduate School MEMoRIAL at Otto-von-Guericke-University (OVGU) Magdeburg, Germany, kindly supported by the European Structural and Investment Funds (ESF) under the program "Sachsen-Anhalt WISSENSCHAFT Internationalisierung" (project no. ZS/2016/08/80646). Financial support of the Methodisch-Diagnostisches Zentrum Werkstoffprüfung (MDZWP) e. V., Magdeburg, Germany is greatly acknowledged. We kindly thank Dr. E. Wessel (IEK-2, FZ Jülich) for providing the SEM images and the EBSD analysis.

References

[1] IFAK; GFK Media and Communication Research; forsa marplan 2019 Meistgenutzte Verkehrsmittel zum Erreichen des Urlaubsziels in Deutschland in den Jahren 2017 bis 2019 1
[2] Statistisches Bundesamt 2019 Anzahl der beförderten Personen im Luftverkehr in den Jahren 2004 bis 2018 in Deutschland 1
[3] Umweltbundesamt 2018 *Vergleich der durchschnittlichen Emissionen einzelner Verkehrsmittel im Güterverkehr - Bezugsjahr: 2018* (Umweltbundesamt)
[4] Schesky E and Kral M 2003 *Flugzeugtriebwerke: Kolben- und Gasturbinentriebwerke ; Aufbau, Wirkungsweise und Betriebsverhalten* (Rhombos-Verlag)
[5] Smith D L, Chung H M, Loomis B A, Matsui H, Votinov S and Van Witzenburg W 1995 Development of vanadium-base alloys for fusion first-wall-blanket applications *Fusion Eng. Des.* **29** 399–410
[6] Zinkle S J, Matsui H, Smith D L, Rowcliffe A F, Osch E van, Abe K and Kazakov V A 1998 Research and development on vanadium alloys for fusion applications *J. Nucl. Mater.* **258–263** 205–14
[7] Ehrlich K 2003 Die Entwicklung von Strukturmaterialien für die Kernfusion *Materwiss. Werksttech.* **34** 39–48
[8] Krüger M 2016 High temperature compression strength and oxidation of a V-9Si-13B alloy *Scr. Mater.* **121** 75–8
[9] Krüger M and Köppe-Grabow B 2017 Prozessabhängige mikrostrukturausbildung von Vss-V3Si-V5SiB2-werkstoffen *Prakt. Metallogr. Metallogr.* **54** 293–307
[10] Krüger M 2016 *Innovative metallische Hochtemperaturwerkstoffe* (Otto-von-Guericke-Universität Magdeburg)
[11] Bei H, George E P, Kenik E A and Pharr G M 2004 Microstructures and mechanical properties of V-V3Si eutectic composites *Zeitschrift für Met.* **95** 505–12
[12] Smith J F 1989 Si-V (Silicon-Vanadium) *Phase Diagrams of Binary Vanadium Alloys* (ASM International) pp 261–7

[13] De Lima-Kühn B B, Da Silvaa A A A P, Suzukia P A, Coelho G C and Nunes C A 2016 Microstructural characterization of as-cast V-Si alloys and reevaluation of the invariant reactions involving the liquid phase of the V-Si system *Mater. Res.* **19** 1122–6

[14] Spear K E, Liao P K and Smith J F 1989 B-V (Boron-Vanadium) *Phase Diagrams of Binary Vanadium Alloys* ed J F Smith (Metals Park, Ohio, United States of America: ASM International) pp 20–6

[15] Nunes C A, De Lima B B, Coelho G C, Rogl P and Suzuki P A 2004 On the stability of the V5B6-phase *J. Alloys Compd.* **370** 164–8

[16] Massalski T B, Okamoto H, Subramanian P R and Kacprzak L 1990 *Binary Alloy Phase Diagrams* (Materials Park Ohio: ASM International)

[17] de Lima B B, Coelho G C, Suzuki P A, Nunes C A and Rogl P 2004 Evaluation of the invariant reactions of the V-B system *J. Phase Equilibria Diffus.* **25** 134–9

[18] Pinto da Silva A A A, Chaia N, Ferreira F, Carvalho Coelho G, Fiorani J M, David N, Vilasi M and Nunes C A 2017 Thermodynamic modeling of the V-Si-B system *Calphad Comput. Coupling Phase Diagrams Thermochem.* **59** 199–206

[19] DIN 50106 2016 Prüfung metallischer Werkstoffe – Druckversuch bei Raumtemperatur 1–10

[20] Elliott R 1984 Eutectic solidification *Metall. Trans. A* **65** 85–92

[21] Karma A and Plapp M 2004 New insights into the morphological stability of eutectic and peritectic coupled growth *JOM* **56** 28–32

[22] Sahm P R, Egry I and Volkmann T 1999 *Schmelze, Erstarrung, Grenzflächen* ed P R Sahm, I Egry and T Volkmann (Berlin, Heidelberg: Springer Berlin Heidelberg)

[23] Bauer G, Güther V, Hess H, Otto A, Roidl O, Roller H, Sattelberger S, Köther-Becker S and Beyer T 2017 Vanadium and Vanadium Compounds *Ullmann's Encycl. Ind. Chem.* **38** 1–22

[24] Hasemann G, Müller C, Grüner D, Wessel E and Krüger M 2019 Room temperature plastic deformability in V-rich V–Si–B alloys *Acta Mater.* **175** 140–7

[25] Müller C, Hasemann G, Regenberg M, Betke U and Krüger M 2019 Microstructure and Compression Properties of V SS -V 3 B 2 Eutectic Alloys in the V-Si-B System *Materials (Basel).* **13** 1–12, 2100.

Hydrogen diffusion and local Volta potential in high- and medium-entropy alloys

M Rhode[1,3], A Wetzel[2], O Ozcan[2], J Nietzke[1,3], T Richter[1] and D Schroepfer[1]

[1] Department 9 - Component Safety, Bundesanstalt für Materialforschung und -prüfung (BAM), Unter den Eichen 87, 12205 Berlin, Germany
[2] Department 6 - Materials Chemistry, Bundesanstalt für Materialforschung und -prüfung (BAM), Unter den Eichen 87, 12205 Berlin, Germany
[3] Institute for Materials and Joining Technology, Faculty of Mechanical Engineering Otto-von-Guericke University, Universitätsplatz 2, 39106 Magdeburg, Germany

michael.rhode@bam.de / michael.rhode@ovgu.de

Abstract. High-entropy alloys (HEAs) are characterized by a solid solution of minimum five and medium-entropy alloys (MEAs) of minimum three principal alloying elements in equiatomic proportions. They show exceptional application properties, such as high-strength and ductility or corrosion resistance. Future HEA/MEA-components could be exposed to hydrogen containing environments like vessels for cryogenic or high-pressure storage where the hydrogen absorption and diffusion in these materials is of interest. In our study, we investigated the HEA $Co_{20}Cr_{20}Fe_{20}Mn_{20}Ni_{20}$ and the MEA $Co_{33.3}Cr_{33.3}Ni_{33.3}$. For hydrogen ingress, cathodic charging was applied and diffusion kinetic was measured by high-resolution thermal desorption spectroscopy using different heating rates up to 0.250 K/s. Peak deconvolution resulted in high-temperature desorption peaks and hydrogen trapping above 280 °C. A total hydrogen concentration > 40 ppm was identified for the MEA and > 100 ppm for HEA. This indicates two important effects: (1) delayed hydrogen diffusion and (2) considerable amount of trapped hydrogen that must be anticipated for hydrogen assisted cracking phenomenon. Local electrochemical Volta potential maps had been measured for the hydrogen free condition by means of high-resolution Scanning Kelvin Probe Force Microscopy (SKPFM).

1 Introduction

1.1 High-entropy alloys (HEAs)

High-entropy alloys (HEA) represent a new class of materials consisting of single-phase and disordered solid solutions of alloying elements (in accordance to the taxonomy initially suggested in [1]). They usually have a chemical composition of minimum five alloying elements, whereas so-called medium-entropy alloys (MEAs) comprise two to four alloying elements. In a narrow meaning, HEA or MEA encompasses equiatomic amounts of the specific alloying elements a solid solution in a single-phase microstructure (like fcc) without further precipitates or inclusions. These materials are fundamentally different from conventional alloys that are based on one principal element such as Fe in case of steels, or Al- or Ni-based superalloys. The face-centered cubic (fcc) CoCrFeMnNi material system and its derivatives have been intensively studied and possess excellent mechanical properties, i.e. high ductility, ultimate tensile strength and fracture toughness [2-4]. Among the equiatomic fcc alloys, CrMnFeCoNi-HEA and CrCoNi-MEA are considered as potential structural alloys and get more and more attention in materials science and engineering [5,6]. For that purpose, HEA and MEA application properties are of great interest.

1.2 Hydrogen absorption and diffusion in HEAs

One of the application properties characteristics of HEAs could be the resistance against hydrogen related degradation of the mechanical properties (some authors also referred to as "hydrogen embrittlement"). The literature study shows the excellent mechanical properties of CoCrFeMnNi-HEA and CoCrNi-MEA alloy in hydrogen containing environments [7-10]. This demonstrates their potential for the use in hydrogen environments as found in vessels for cryogenic or high-pressure storage and is characterized by their outstanding resistance to hydrogen assisted degradation of the mechanical properties. Even at very high hydrogen concentrations (above 30 to 40 ppm) the degradation is low, considering that comparable concentrations are already high enough to remarkably decrease the mechanical properties of other fcc-alloys like Fe-based and high-alloyed austenitic steels [7,9]. For that reason, the hydrogen absorption, diffusion and trapping in these materials is of high interest. Hydrogen can interact with crystal lattice defects (vacancies, dislocations, etc.) and can be bound to so-called traps [11]. This interaction can be expressed by the delayed diffusion compared to not-affected lattice diffusion in pure metals [12]. Generally, a critical combination of a susceptible microstructure, mechanical load and hydrogen concentration is responsible for hydrogen assisted cracking/hydrogen embrittlement [13,14]. First publications on CoCrFeMnNi-HEA showed that hydrogen diffusion at room temperature is comparable to low-carbon version of austenitic stainless steel 316L (X2CrNiMo17-12-2) at 27 °C [15]. Nonetheless, it can be concluded that hydrogen diffusion in HEA and austenitic steel are comparable, the values of the diffusion coefficients have been in the order of 10^{-6} mm²/s, which is contrary to other studies ranging from 10^{-9} to 10^{-8} mm²/s [16,17]. The hydrogen release temperature was also comparable to austenitic stainless steels, which lies between 200 to 450 °C with a maximum peak around 300 °C, whereas the CoCrFeMnNi-HEA trapped significantly more hydrogen (up to 70 ppm). Systematic literature studies on diffusion characteristics and hydrogen release temperatures are rare so far for CoCrFeMnNi-HEA and entirely missing for CoCrNi-MEA. These parameters are important for service conditions of components, namely hydrogen leakage rates by permeating hydrogen at different temperatures.

1.3 Corrosion of HEAs

HEA systems composed of passivating elements, such as Cr, Mo, Ni, Ti, tend to form protective passive oxide layers which show a positive impact on the pitting potential. Even though most HEAs (in case of more than one phase also considered as chemically complex alloys / CCAs) have adequate resistance to general corrosion, microstructural heterogeneities may also result in high susceptibility to localized corrosion [18]. It has been reported that the addition of Al, Cu, Mo and B in HEAs leads to the formation of non-uniform passive films with low corrosion resistance of HEAs [19,20]. Systems including additional Al improve the mechanical properties of CCAs [21], but also lead to the formation of bcc phases and segregation of elements. CoCrFeMnNi-HEA is assumed to be more resistant to pitting corrosion, since it is Al-free and has single-phase fcc-structure [22,23]. Nonetheless, existing literature studies show contradicting results. In [24], the CoCrFeMnNi was compared to 316L steel during immersion tests in NaCl solution with CO_2. A lower corrosion resistance compared to 304 stainless steel in H_2SO_4 was reported [25]. Hence, it is of interest to identify the local corrosion behavior of the HEA and MEA materials depending on their microstructure and chemical composition. In addition, hydrogen can be generated during corrosion in acids as a result of the cathodic part of the corrosion reaction, especially in acidified pits or crevices [26,27] and be introduced into the microstructure and stimulate, or initiate hydrogen assisted stress corrosion cracking (if a sufficiently high mechanical load is applied). Continuous hydrogen charging was applied on CoCrFeMnNi tensile samples but did not result in any remarkable degradation of the mechanical properties [9]. But it must be anticipated that due to the applied strain rate (10^{-4} 1/s) breakdown of the passive layer occurred but test speed was perhaps too high for stress corrosion cracking as hydrogen could not diffuse fast enough.

The presented brief literature review demonstrates that HEA/MEA behavior in the presence of hydrogen is still an open question and demands further investigations. This is particularly the case for MEAs. For that reason, we examined and compared two different equiatomic alloys and their behavior towards absorbed hydrogen, high-temperature diffusion and local electrochemical potentials.

2 Materials and methods

2.1 Investigated materials

In our study, we investigated two different materials, the common five-elements CoCrFeMnNi-HEA (so-called "Cantor"-alloy and a three-elements CoCrNi-MEA. We used vacuum melted ingots that have been manufactured by the group of Prof. G. Laplanche (Ruhr-University Bochum). Both alloys had been homogenized after casting and recrystallized for 1 h (CoCrFeMnNi-HEA at 1020 °C and CoCrNi-MEA at 1060 °C) resulting a single-phase fcc-microstructure with an average grain size of 50 µm. More details on alloy preparation and microstructure features can be found elsewhere in [28]. Table 1 shows the chemical composition (in at.-%) of the investigated alloys as determined by electron microprobe analysis (EMPA, with experimental error of about 1 at.-%).

Table 1. Chemical composition (in atomic %) of the CoCrFeMnNi-HEA and CoCrNi-MEA.

Alloy	Co	Cr	Fe	Mn	Ni
HEA	20.7	20.1	19.6	19.7	19.9
MEA	34.3	-	-	33.0	32.7

Specimens with the shape of a circular sector had been used for the experiments. For that purpose, the material rods had been sliced to disks with a thickness of 2 mm and subsequently cut to the final size by electrical discharge machining. The reason was the necessary economic use of the expensive raw materials and high-sophisticated processing of the experimental materials as the sliced disks had been additionally used for milling experiments vs. degradation of surface properties additionally for experiments presented in [29]. The dimensions are described in figure 1 (a).

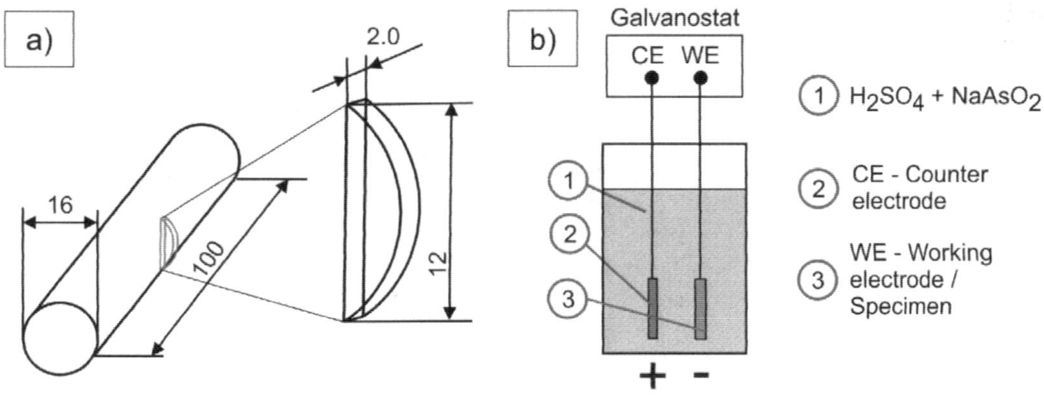

Figure 1. (a) Cut position of flat sample in material rod, dimensions of examined samples and (b) cathodic hydrogen charging set-up.

2.2 Hydrogen charging, measurement and thermal desorption spectroscopy

The samples have been cathodically charged with hydrogen in 0.1 M H_2SO_4 aqueous solution with addition of 0.05 M $NaAsO_2$ as a recombination inhibitor to increase hydrogen absorption in the materials, see figure 1(b). The sample was the working electrode (WE) and the counter electrode (CE) was a Pt 1800 electrode (SI Analytics), connected to a Wenking Potentiostat/Galvanostat TG94. Before charging, each specimen surface was manually ground by 500 SiC grit paper, cleaned in ethanol using an ultrasonic bath and rinsed in dry nitrogen gas flow. To ensure an almost homogenous hydrogen distribution, different combinations of charging current density and charging time have been tested. As a result, a galvanostatic charging current of 50 mA/cm² was applied for 240 hours (10 days) to both CoCrFeMnNi and CoCrNi, assuming a low hydrogen diffusivity like in case of austenitic stainless steels. For each material and heating rate two samples have been charged and tested (due to the limited amount of available sample material). The trapping and desorption characteristics were analyzed by BRUKER G8 Galileo using the so-called carrier gas hot extraction technique (CGHE). This analyzer uses an infrared radiation furnace placed around a quartz glass tube. In this tube the sample is placed, heated and the hydrogen desorbs from the sample. The hydrogen is collected in an inert gas (nitrogen) flow and carried to the coupled quadrupole mass spectrometer (MS). The desorbing hydrogen corresponds to a monitored ion current in A (see figure 3 to figure 6). This ion current is integrated within the monitored time range and corresponds to the absolute hydrogen amount that was absorbed by the sample. Via a previously determined calibration factor and the sample weight, the corresponding hydrogen concentration can be calculated in ppm. Further details on CGHE can be found in [30-33]. The temperature was measured using samples with same dimensions and attached type-K thermocouple at the surface. These investigations ensured that the surface temperature corresponds to the programmed heating rate. Deviation of the adjusted and measured temperature is less than 2 K (and therefore ignored in the data analysis). For that reason, the quasi-linear temperature profiles could be idealized and are used for the correlation of the peak temperatures and the corresponding temperature at the sample surface (see figure 3 to figure 6).

Thermal desorption spectroscopy (TDS) can be used to determine effects of microstructure and temperature on hydrogen trapping. In our study, we determined the corresponding desorption spectra using heating rates of $\Phi = 0.125$ and 0.250 K/s. The starting temperature was 25 °C and the final temperature was 450 °C. Subsequently, the samples were hold at 450 °C for at least 10 minutes to potentially desorb very deeply trapped hydrogen (which was not the case). These temperature dependent desorption spectra are the sum of overlapped single peaks with assumed Gaussian distribution. These had been identified and analyzed individually by a peak deconvolution technique using the Gaussian peak fit function of the software Origin 9.7. The result is that each peak can be correlated to a specific peak temperature T_P (maximum effusion), which indicates an activation energy for hydrogen diffusion, i.e. release of trapped hydrogen. In general, the desorption peak shifts to higher temperatures with increasing Φ. Ideally, each Gaussian peak represents a type of a major trap-site in the material (2D or 3D-lattice defects).

2.3 Scanning Kelvin Force Probe Microscopy (SKPFM) analysis

Atomic Force Microscopy (AFM) based Scanning Kelvin Force Probe Microscopy (SKPFM) allows the analysis of Volta-Potential differences on heterogeneous alloy surfaces and at buried metal-polymer interfaces [34,35]. In the Kelvin mode, the surface potential of the sample is measured in addition to the sample topography. Alternating line scans in topography and SKPFM lift mode are carried out, where the tip follows the measured height profile with a constant distance to the surface at a constant DC bias and a superimposed AC bias. The AC bias generates an alternating charging between SKPFM tip and sample surface, which induces an alternating modulation of the force on the tip, which is used for the determination of the surface potential, see figure 2.

As the potential values are not referenced to a standard potential as in the classical macroscopic Scanning Kelvin Probe (SKP) technique, the Volta-potential maps obtained by means of SKPFM deliver Volta-potential differences on the heterogeneous surface and not absolute potential values. The measurements were performed under ambient conditions in air with a NanoWizard 4 (JPK Instruments, Bruker Nano GmbH) operating with a resolution of 1024×1024 pixels. For the topographic scan, the contact mode was used, whereas for the Kelvin Probe scan the tip was lifted to a predefined height of 40 nm and an AC voltage of 0.8 V was applied to the tip to determine the surface potential following the topography profile. The AFM and SKPFM images were collected at a scan frequency of 0.5-1.0 Hz using a silicon cantilever with a conductive Cr/Pt-coating (Tap300-G tip, Budget Sensors) with a nominal spring constant of 40 N/m. The reported surface potentials values are relative to the potential of the tip. The software JPK Data Processing Suite 6.1.88 was used for offline analysis of the topographies. To remove tilt and the vertical z-offset between line scans, image flattening was performed with the first order least-square polynomial function.

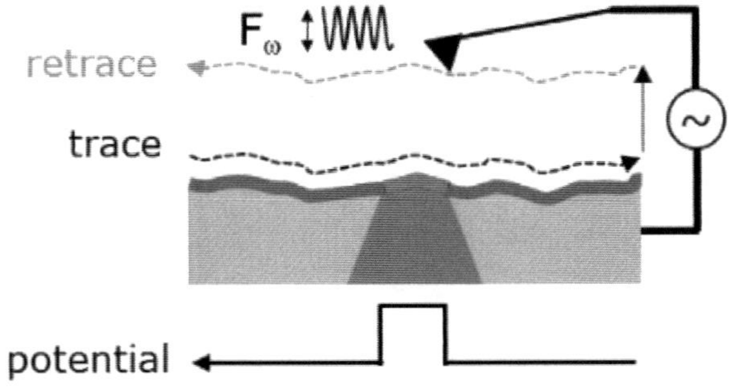

Figure 2. Work principle of SKPFM.

In contrast to the TDS-samples, we only investigated hydrogen charged CoCrNi-samples by AFM. For that purpose, the working electrode was the CoCrNi-sample, counter was graphite rod and reference electrode was Ag/AgCl (for potential measurement during hydrogen charging). The electrolyte had the same composition as used for the TDS-samples. The volume was 10 ml (cm^3) and charging current density was 1 mA/cm^2 applied for 1 and 2 hours.

3 Results and discussion

3.1 Temperature dependent hydrogen diffusion trapping by TDA peak deconvolution

The figures in this subsection show the different hydrogen diffusion and effusion behavior dependent on both the chemical composition and applied Φ. For that purpose, figure 3 shows the obtained different hydrogen effusion spectra for (a) CoCrFeMnNi and (b) CoCrNi for the applied Φ of 0.125 K/s and 0.250 K/s.

Figure 3. Hydrogen effusion spectra for Φ = 0.125 K/s and 0.250 K/s: (a) CoCrFeMnNi, (b) CoCrNi.

Both materials had quite different hydrogen diffusion behavior. This is expressed by the different maximum ion current and the total effusion time that was necessary to release the hydrogen. In case of the CoCrFeMnNi, a delayed hydrogen diffusion was identified that is expressed by an increased total desorption time (e.g. CoCrFeMnNi with approximately 2000 s compared to 1700 s for CoCrNi at Φ = 0.250 K/s). Both applied heating rates of 0.125 K/s and 0.250 K/s resulted in a shift of the maximum peak value and time of appearance. That corresponds to different peak temperatures and indicates that

IOP Conf. Series: Materials Science and Engineering **882** (2020) 012015 doi:10.1088/1757-899X/882/1/012015

e.g. the different atoms within the crystal lattice of the CoCrFeMnNi-HEA and CoCrNi-MEA have significant impact on the hydrogen diffusion, i.e. the predominant trap sites are different. Considering that the integrated spectra correspond to the total desorbed hydrogen, it is obvious that CoCrFeMnNi-HEA (max. ion current $1.19*10^{-12}$ A) had a higher absorbed hydrogen concentration compared to the CoCrNi-MEA ($5.17*10^{-12}$ A). Detailed hydrogen concentration values are shown in table 3.

The obtained spectra had been subjected to peak deconvolution with an increasing number of peaks. Figure 4 shows the results for the peak deconvolution with two and three peaks for CoCrFeMnNi (parts a, b) and CoCrNi (parts c, d).

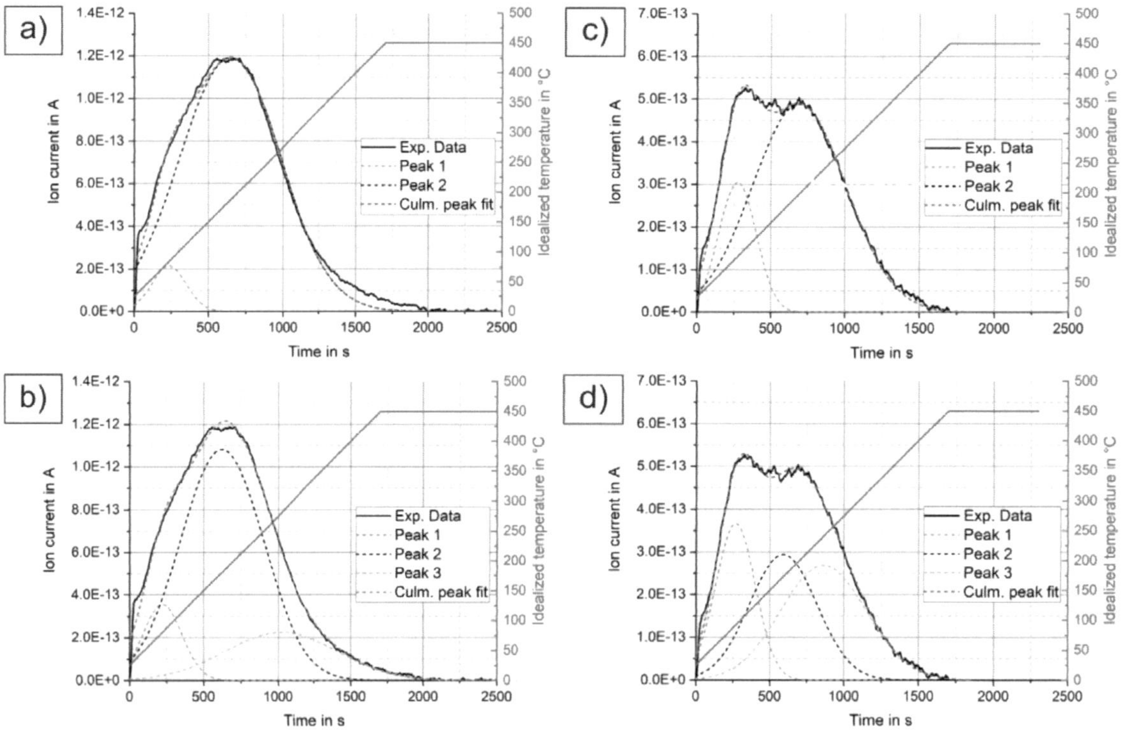

Figure 4. Peak deconvolution at $\Phi = 0.250$ K/s for CoCrFeMnNi-HEA with 2 peaks (a), 3 peaks (b) and CoCrNi-MEA with 2 peaks (c) and 3 peaks (d)

If the TDS spectra is deconvoluted in two peaks (figure 4 a, c), the first peak always appears in a relatively low temperature range (around 75 °C). A major trap site was identified for both CoCrFeMnNi and CoCrNi at approximately 200 °C. This could already correspond to certain microstructure effects. This interpretation changes if the spectra are deconvoluted into three peaks, see figure 4 (b, d). As a result, the previously mentioned high peak is split into smaller peaks, which results in the appearance of an additional desorption peak ("peak no. 3" in figure 4 (b, d) at higher temperatures.

This indicates that the hydrogen diffusion, i.e. trapping behavior in HEA and MEA materials is generally more complex than already reported [7,15]. For that reason, the number of deconvoluted peaks was increased to four and five as shown in figure 5 (a, b) for CoCrFeMnNi-HEA and (c, d) for CoCrNi-MEA.

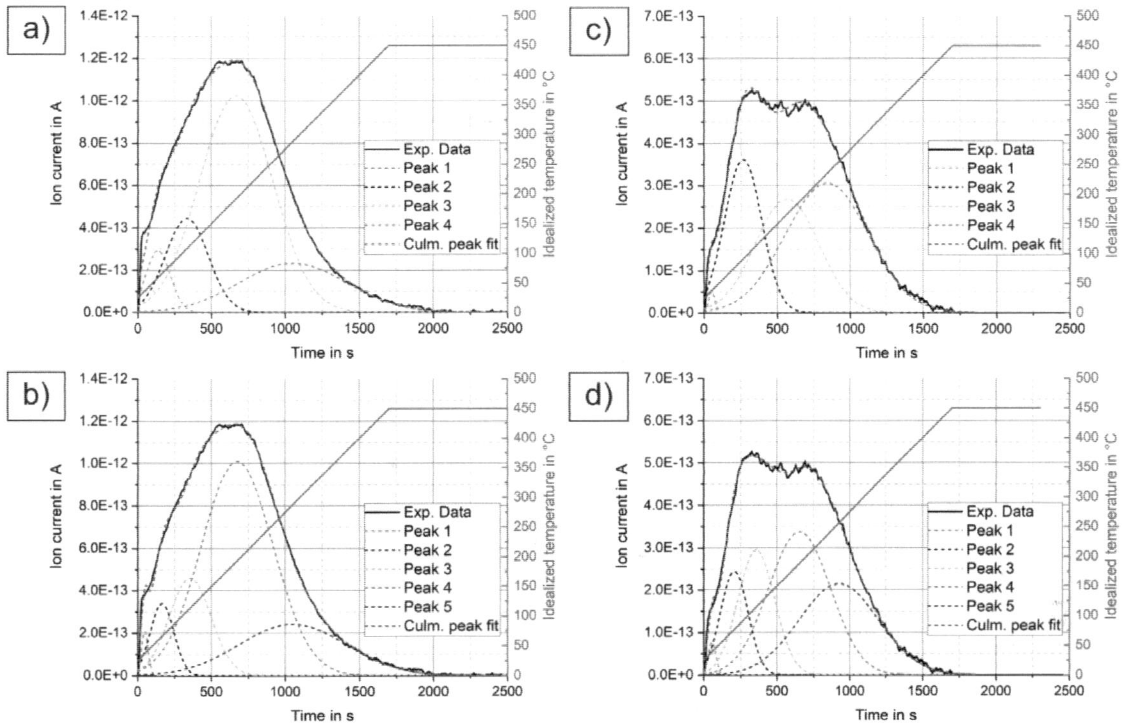

Figure 5. Peak deconvolution at Φ = 0.250 K/s
CoCrFeMnNi-HEA with 4 peaks (a), 5 peaks (b) and CoCrNi-MEA with 4 peaks (c) and 5 peaks (d)

Figure 4 and figure 5 demonstrate the challenges that are anticipated with reliable deconvolution of measured TDA spectra. With the increasing number of deconvoluted peaks to four or five, the quality of the cumulative peak fit increases (R^2-values see table 2). This is of course a mathematical effect due to the higher number of terms of the fit equation. But there is virtually no difference between the quality of four or five peak fit functions to three peaks, see figure 4 (b, d). The main high temperature peaks do not further change, especially for the CoCrFeMnNi-HEA. But the low temperature peak, $T < 70$ °C, see figure 4 (b, d), can be further expressed as a sum of smaller peaks (see figure 5).

Table 2. Peak deconvolution - model quality.

Amount of peaks	Model quality (R^2)	
	CoCrFeMnNi-HEA	CoCrNi-MEA
2	0.99217	0.99435
3	0.99609	0.99591
4	0.99755	0.99672
5	0.99885	0.99813

This raises the question what microstructure or experimental condition influences the obtained results that justify the number of chosen peaks. As example for experimental condition, it must be considered that the G8 Galileo analyzer is equipped with an infrared furnace. For that reason, the sample surface is heated, and the bulk material is heated by the heat conduction. Hydrogen at the surface near region is activated and desorbs first [30]. This results in apparent experimental-design influenced desorption peaks that have no direct relation to microstructure or chemical composition. If the number of peaks for the spectra deconvolution is insufficient (i.e. the number of the mathematical terms of the regression model), these apparent peaks are "blurred". Nonetheless, it is important to identify and to consider those peaks as they have an impact on the further desorption peaks with real microstructural background in terms of changing the peak temperatures, which could influence the further calculation of activation energies for hydrogen traps [30,36]. In that connection, the five-peak-deconvolution was chosen because it offered the highest model quality (described by R^2-value, see table 2). For comparison to effusion spectra for $\Phi = 0.250$ K/s, see figure 5 (b, d), figure 6 shows the corresponding five-peak-deconvolution of for $\Phi = 0.125$ K/s for (a) CoCrFeMnNi and (b) CoCrNi.

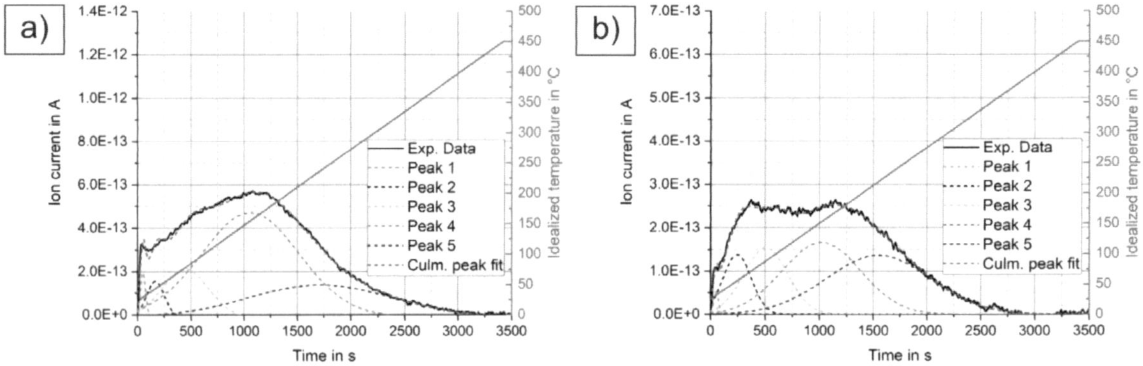

Figure 6. Peak deconvolution with five peaks at 0.125 K/s:
(a) CoCrFeMnNi / $R^2 = 0.99634$, (b) CoCrNi / $R^2 = 0.99166$.

Especially, at lower heating rate of 0.125 K/s a reliable peak deconvolution is necessary. As mentioned, the peaks at low temperatures (peak 1 and 2 for CoCrFeMnNi below 50 °C and for CoCrNi below 60 °C) could be unique experimental effects of the CGHE. Nonetheless, both materials showed a significant hydrogen effusion for nearly 3,500 s (CoCrFeMnNi) and 3,000 s (CoCrNi) despite the continuous increasing temperature up to 450 °C. Considering the small sample thickness of 2.0 mm (see figure 1), the hydrogen diffusion in both materials is very slow. This was already reported for CoCrFeMnNi-HEA [7,15], but is new for CoCrNi-MEA material, especially the significant hydrogen release at elevated temperatures above 100 °C.

3.2 Corresponding peak temperatures and hydrogen concentration

This subsection presents the data for the 5-peak-devonvolution for both materials and applied heating rates Φ of 0.125 K/s and 0.250 K/s with respect to the hydrogen concentration of each peak (HD_{PX}, where "PX" is the peak no. 1 to 5) and the total hydrogen concentration (HD_{tot}). The corresponding hydrogen release temperature encompasses the peak temperature (T_P), i.e. the temperature of the maximum effusion rate for the respective peak. The data are presented and summarized in table 3. The corresponding peak temperatures are presented (for presentation purposes) as integer values. The reason is the negligible difference for 1 °C temperature increment if the underlying Φ-dependent time interval is considered (0.125 K/s corresponds to 8 s increment time vs. 3,400 s total heating time and for 0.250 K/s corresponds to 4 s vs. 1,700 s).

Table 3. Peak dependent HD_{PX} and total hydrogen concentration HD_{tot} with corresponding release/desorption temperatures.

	Φ	Value	Unit	Peak number HD_{PX} 1	2	3	4	5	Total
CoCrFeMnNi-HEA	0.125 K/s	HD	[ppm]	2.3 ± 0.9	4.8 ± 1.0	13.3 ± 0.5	41.1 ± 15.1	52.5 ± 25.1	114.0 ± 11.3
		HD	[%]	2 ± 1	4 ± 1	12 ± 0.5	36 ± 13	46 ± 22	100 ± 10
		T_P	[°C]	31 ± 1	43 ± 3	70 ± 11	128 ± 27	200 ± 28	-
	0.250 K/s	HD	[ppm]	1.1 ± 0.2	10.8 ± 3.3	25.7 ± 7.3	70.5 ± 0.9	20.1 ± 4.4	128.4 ± 7.0
		HD	[%]	1 ± 1	8 ± 1	20 ± 1	55 ± 13	16 ± 22	100 ± 5
		T_P	[°C]	38 ± 1	73 ± 6	121 ± 8	197 ± 3	286 ± 1	-
CoCrNi-MEA	0.125 K/s	HD	[ppm]	0.4 ± 0.2	4.4 ± 0.2	8.7 ± 1.3	16.6 ± 2.2	14.9 ± 1.1	45.0 ± 2.8
		HD	[%]	1 ± 0.5	10 ± 0.5	19 ± 3	37 ± 5	33 ± 2	100 ± 6
		T_P	[°C]	32 ± 1	56 ± 1	90 ± 2	154 ± 2	221 ± 2	-
	0.250 K/s	HD	[ppm]	0.7 ± 0.2	4.8 ± 0.6	10.1 ± 1.4	22.5 ± 5.8	16.0 ± 2.4	54.1 ± 9.2
		HD	[%]	1 ± 1	9 ± 1	19 ± 3	42.5 ± 11	29.5 ± 4	100 ± 17
		T_P	[°C]	39 ± 0	74 ± 3	110 ± 4	184 ± 4	257 ± 1	-

The data in table 3 indicate that a complex hydrogen trapping behavior must be anticipated in both HEA and MEA (whereas the deviation was quite high for the lower heating rate of Φ= 0.125 K/s). The CoCrFeMnNi-HEA trapped a significantly higher hydrogen concentration (mean value of 114.0 ppm for Φ= 0.125 K/s; 128.4 ppm for Φ = 0.250 K/s) compared to CoCrNi-MEA (mean value 45.0 ppm for Φ = 0.125 K/s; 54.1 ppm for Φ = 0.250 K/s), which corresponds to a factor of approximately 2.5. The measured absorbed hydrogen concentration of CoCrNi-MEA corresponds to fcc-material like austenitic stainless steels [37] with approximately 40 ppm. The CoCrFeMnNi showed absorbed hydrogen concentration that is in the same range, but is higher compared to the reported 70 ppm for CoCrFeMnNi [7,15]. The increased number of equiatomic elements of CoCrFeMnNi-HEA must offer an increased number of hydrogen traps compared to the CoCrNi-MEA. This is important if those materials are exposed to hydrogen. Nonetheless, the mechanical behavior suggests superior resistance against hydrogen assisted cracking (or hydrogen embrittlement) of CoCrFeMnNi-HEA [7,9] and CoCrNi-MEA [10].

Peak no. 1 and no. 2 appeared for both heating rates and materials in the lower temperature regions, considering figure 5 (b, d) for Φ = 0.125 K/s and figure 6 (a, b) Φ of 0.250 K/s. Especially peak no. 1 always occurred close to the initial start temperature (30-32 °C). But the second peak appeared between 46 to 74 °C. In addition, they correspond to approximately 1 to 2 % for peak no. 1 and 4 to 10 % for peak no. 2 of the totally absorbed hydrogen concentration. This results in apparent desorption peaks that have no direct relation to microstructure or chemical composition. But if these peaks are not separated, they could "blur" real microstructure dependent peaks during hydrogen measurement. The low temperatures are perhaps the result of two effects of the experimental procedure respectively the data acquisition. Effect (1): The specimen has a localized small hydrogen enriched subsurface region due to the charging, i.e. a concentration gradient. A worst-case scenario for that is when the so-called blistering occurs [38,39]. In addition, the adsorbed hydrogen could be trapped (by an increased number of dislocations) because of the mechanical surface preparation [40,41] (in our case the grinding and polishing process). In addition, a pronounced texture could have an effect or the grain orientation. In addition, it is difficult to measure this thin subsurface hydrogen enriched "layer" thickness as CGHE does not allow localized hydrogen measurement. Effect (2): The use of infrared radiation is assumed to activate at first the hydrogen in this enriched region close to the specimen surface. The reason is that the heat was coupled into the specimen via the surface. Subsequently, the bulk was heated by thermal conductivity resulting in a "delay" of the higher temperature. This supports the assumption that the sample surface near hydrogen is activated very fast and must desorb first [30].

Peaks no. 3 to 5 represent microstructure, i.e. chemical composition, influenced effusion peaks. They encompass nearly 90 % of the total absorbed hydrogen. This means that in case of the CoCrFeMnNi-HEA more than 100 ppm hydrogen can be trapped at elevated temperatures. If the different heating rates are considered, peak no. 4 and 5 encompass 71 - 83 % of the absorbed hydrogen concentration with high hydrogen desorption peak temperatures within the range from 157 to 282 °C. A similar behavior was found for the CoCrNi-MEA from 154 to 256 °C. That indicates that the general diffusion behavior of both CoCrFeMnNi and CoCrNi represent typical fcc-lattice hydrogen diffusion behavior like austenitic stainless steels. But in case of the CoCrFeMnNi-HEA, the absorbed hydrogen concentration is much higher. It is believed that due to a similar average grain diameter, the grain boundaries volume as a predominant difference between both CoCrFeMnNi-HEA and CoCrNi-MEA can be excluded. Nonetheless, grain boundaries in fcc-metals can show increased numbers of dislocations. If the number of dislocations increases, the number of hydrogen traps increases and, for that reason, the absorbed and trapped hydrogen concentration [42]. This could be an explanation for the significantly increased trapped hydrogen concentration but demands e.g. further investigations on dislocation density. The lattice distortion of HEA and MEA as a predominant difference between both alloy types can also be excluded. It was shown that CoCrFeMnNi-HEA and CoCrNi-MEA had similar distorted lattices and, for that reason, no significant differences between the fcc-lattice strains [43].

It currently remains open, which predominant trap is the most important difference between hydrogen trapping in CoCrFeMnNi-HEA and CoCrNi-MEA, which is described by the significantly different hydrogen trapping (factor 2.5). Additional experiments with Time-of-Flight secondary ion mass spectroscopy will reveal predominant hydrogen trap sites in the material. Currently, the correlation of traps, i.e. distinct microstructure effects, in terms of calculation of activation energy E_A is difficult as we only investigated two heating rates (possible shift of Arrhenius-regression functions). For that reason, a heating rate $\Phi = 0.625$ K/s will be applied to determine a third set of the peak temperatures of peak no. 1 to no. 5. This enables determination of activation energies for the corresponding hydrogen traps via an Arrhenius-relationship between the heating rate Φ and the peak temperature T_P [30,31].

3.3 Local Volta potential maps by SKPFM

SKPFM was used to determine the surface topography and Volta potential differences at the same region of interest. Figure 7 shows the results of the CoCrFeMnNi-HEA. For that purpose, parts (a) and (b) included the topographic scans and parts (c) and (d) the corresponding local Volta potential maps. The upper row presents a scan area of 100 x 100 μm and the lower row 25 x 25 μm. Figure 8 shows the corresponding CoCrNi-MEA results (for 100 x 100 μm scan area). For comparison, the corresponding topographic images were presented together with the work function map.

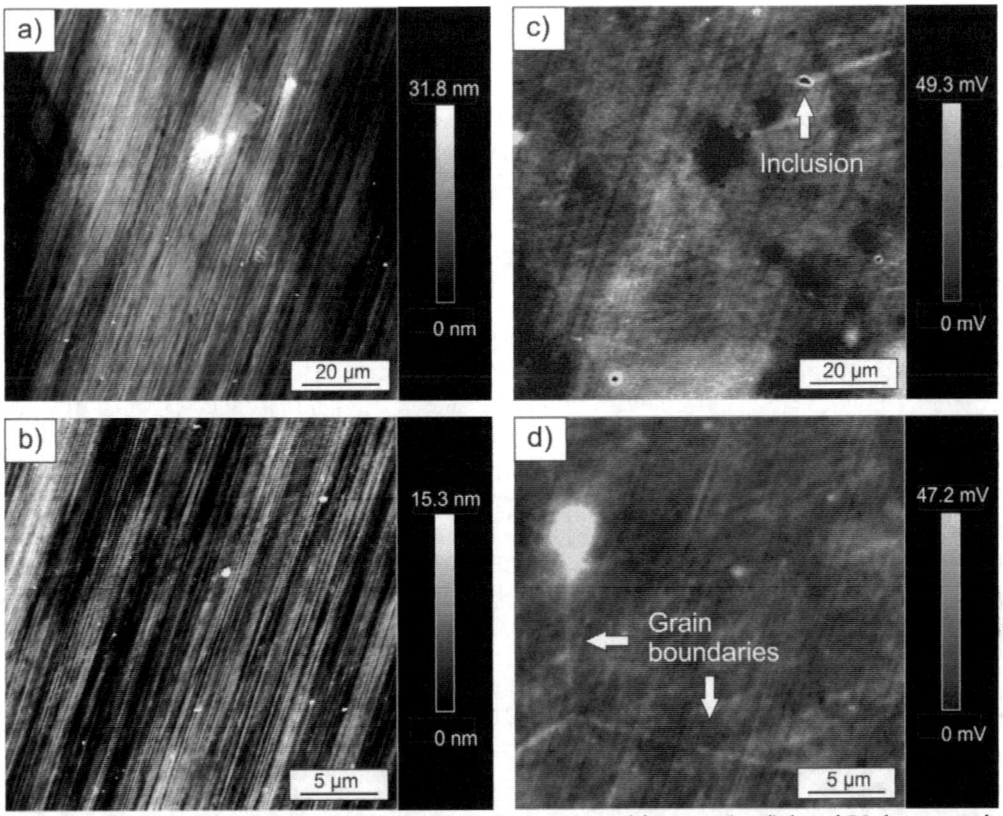

Figure 7. Uncharged CoCrFeMnNi-HEA: (a, b) topographic scan (c, d) local Volta potential.

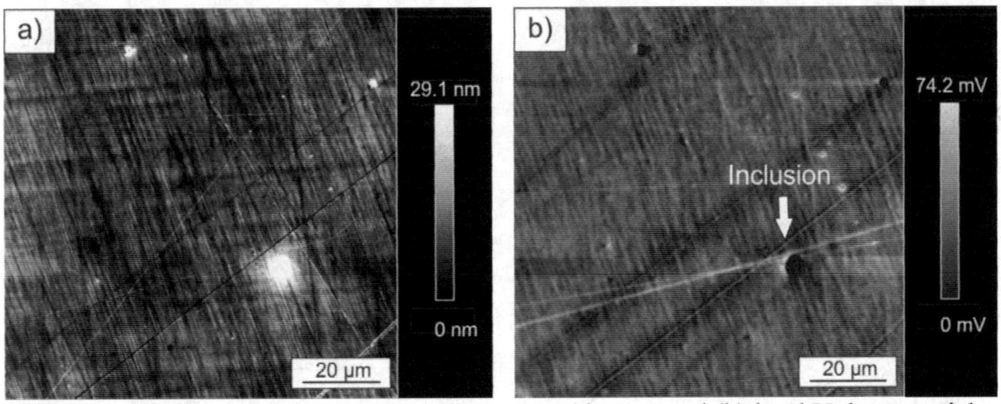

Figure 8. Uncharged CoCrNi-MEA: (a) topographic scan and (b) local Volta potential.

In all images the direction of the polishing is discernible, however the polishing lines are more pronounced in the topographic scans than in the work function / SKPFM images, as the latter tracks the topography scan during the acquisition of the potential profiles. High and low work function regions are clearly visible in the topography scans and indicate that different phases are present in the surface of both materials. This was already reported for CoCrFeMnNi-HEA [44,45]. To identify what phases these, work function areas correspond to a combination with other imaging techniques such as EBSD will be required. Another feature in case of the CoCrFeMnNi-HEA (figure 7) are grain boundaries, which are indicated by lines of higher potential (figure 7 d) that are not visible in the topographic image, see figure 7 (b). The CoCrNi SKPFM image does not show grain boundaries as the measured lines, as indicated in figure 8 (b), are of distinguishable potential but not clearly discernible from the surrounding matrix. Nevertheless, both SKFPM images of CoCrFeMnNi-HEA and CoCrNi-MEA enable the visualization of inclusions or other forms of impurities in the material that had not been observed in the topographic image (parts a). In the recorded SKPFM images this can be deduced by the spots of low potential surrounded by a rim of high potential. It is assumed that these inclusions / impurities are oxides, which are already in the material (during casting and processing) or form during the preparation of the samples for SKPFM. Considering the SKPFM work function, thick and pure oxides are a reasonable explanation as they have very low electrical conductivity, i.e. high electrical resistance and virtually suppress the measuring of any potential. This supports the result of areas with low Volta potential that are surrounded by high potential rims. These inclusions could be Al-rich (as a result of the processing route using Al-rich crucibles) [28] or further Cr- und Mn-oxides [29,46]. It remains open if inclusions generally are a considerable factor for application relevant properties like corrosion resistance (formation of galvanic elements vs. pitting). In accordance to [47], Al leads to microstructural changes, which in turn, were manifested to result in selective dissolution of Al and Al-rich phases. To identify the chemical composition and what these inclusions correspond to, a combination with other imaging techniques (like ToF-SIMS) will be required. To determine a possible hydrogen charging effect on the local Volta potential, a CoCrNi-sample was hydrogen-charged for 1 hour at 1 mA/cm² and investigated and again after 2 hours of charging time. The results are show in figure 9. Parts (a) and (b) include the topographic scans and parts (c) and (d) the local Volta potential maps (scan area of 100 x 100 μm).

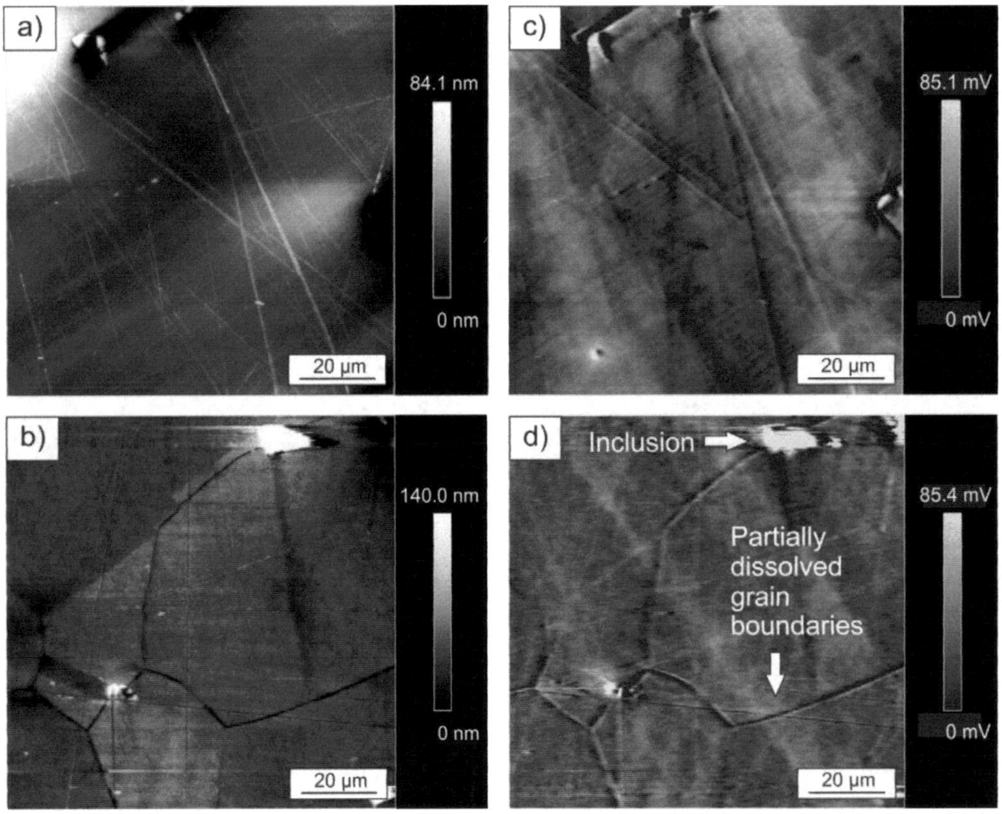

Figure 9. Hydrogen charged CoCrNi-MEA with 1 A/cm² for 1 h: (a) topographic scan, (c) local Volta potential; for 2 h: (b) topographic scan and (d) local Volta potential.

If compared to the reference condition (see figure 8), the hydrogen charging did not result in remarkable changes in the local potential (approximately 85 mV in charged condition to 78 mV in hydrogen free condition, whereas these are maximum values determined at inclusions). The homogenous potential distribution in hydrogen charged condition (parts c and d) is assumed to be a result of the limited hydrogen charging time of 1 and 2 hours. This time is rather short if it is considered that the examined MEA material has an austenitic crystal lattice and, hence, a very low hydrogen diffusion coefficient. The result is that hydrogen needs certain time to accumulate at crystal defects like inclusions or penetrate the bulk material. Hence, only a sub-surface near region could be perhaps enriched with hydrogen. Nonetheless, a significant effect of hydrogen on local potential is assumed in accordance to [48]. The hydrogen charging resulted in a surface degradation as indicated by an intergranular corrosion-like preferred dissolution of grain boundary areas in case of 2 h hydrogen charging (see figure 9 b). Currently, it remains open if the obviously dissolved areas are result of the charging parameters (as TDS-samples had been subjected too much longer exposition time at higher current density). It could be also a result of the local corrosion resistance of the MEA material. Although CoCrNi-MEA should have a homogenous distribution of the alloy elements [10], the ion concentration in the residual electrolyte was Cr-enriched and suggests that this element was preferably dissolved. But this experimental finding is still an assumption. For that reason, the combination of hydrogen charged HEA and MEA samples and corresponding SKPFM analysis must be investigated more detailed.

4 Summary and conclusions

In this study, we investigated the CoCrFeMnNi-HEA and CoCrNi-MEA. Cathodic hydrogen charging was applied and measured by high-resolution TDS with MS. For identification of temperature dependent diffusion, two different heating rates Φ = 0.125 and Φ = 0.250 K/s had been applied. Local surface topography and Volta potential scans were by SKPFM and compared for the hydrogen free condition and for CoCrNi in hydrogen charged condition. The following conclusions can be drawn from this study:

- A minimum number of peaks is needed for reasonable deconvolution of TDS hydrogen desorption spectra and five peaks had been suitable in our results. It was ascertained that at least peak no. 1 and (partially) no. 2 is perhaps a result of the applied CGHE technique (in terms of the used furnace type). This is supported by the summarized amount of max. 10 % of the total absorbed hydrogen concentration.
- High-temperature desorption peaks occurred in both materials, which indicates that high energies are necessary for the release of hydrogen from the respective trap. If the different heating rates are considered, peak no. 4 and 5 encompass 71 to 83 % of the absorbed hydrogen concentrations. This indicates two important effects: (1) delayed hydrogen diffusion and (2) a considerable amount of trapped hydrogen also at high temperatures that must be anticipated for hydrogen assisted cracking.
- The CoCrFeMnNi-HEA showed trapped hydrogen within a temperature range from 157 to 282 °C and CoCrNi-MEA from 154 to 256 °C if the respective peak temperatures are considered. Both HEA and MEA have comparable hydrogen diffusion characteristic that are additionally like other fcc-materials for example high-alloyed austenitic stainless steels. The absorbed hydrogen concentration was higher than 100 ppm for both heating rates in case of the CoCrFeMnNi-HEA, which was significantly higher than the approximately determined 40 ppm in case of the CoCrNi-MEA. It currently remains open, which predominant trap is the difference between hydrogen trapping in CoCrFeMnNi HEA and CoCrNi MEA. For calculation of activation energies for hydrogen trapping and release, further experiments with another heating rate are necessary.
- SKPFM investigations showed that both materials CoCrFeMnNi and CoCrNi contained inclusions/impurities that could be identified in the topography and Volta potential scans. The inclusions are assumed to be oxides, but further investigation is necessary to draw reliable conclusions. Their impact on possible application relevant properties such as corrosion resistance (depending on the electrolyte accompanied with hydrogen evolution and further hydrogen assisted cracking phenomena like stress corrosion cracking) is topic of ongoing research.

- Hydrogen charging of CoCrNi-MEA caused no obvious differences in the local Volta potential in case of the applied charging parameters. This is perhaps an experimental effect and suggests that hydrogen charging (as a combination of the material and the applied charging parameters) must be conducted carefully.

Acknowledgements
The authors want to thank Prof. Guillaume Laplanche (Ruhr-University, Bochum, Germany) for providing the experimental materials and further processing information. Prof. Christiane Stephan-Scherb and Dr. Wencke Schulz are thanked for materials characterization as well as Mr. Jirka Biermann for the machining of the samples (all with BAM Berlin, Germany).

References

[1] Yeh J W, Chen S K, Lin S J, Gan J Y, Chin T S, Shun T T, Tsau C H and Chang S Y 2004 *Adv. Eng. Mater.* **6** (5) 299–303
[2] Gali A and George E P 2013 *Intermetallics* **39** 74–78
[3] Gludovatz B, Hohenwarter A, Catoor D, Chang E H, George E P and Ritchie R O 2014 *Science* **345** 1153–58
[4] Wu Z, Bei H, Pharr G M and George E P 2014 *Acta Mater.* **81** 428–41
[5] Schneider M, George E P, Manescau T J, Zalezak T, Hunfeld J, Dlouhy A, Eggeler G and La planche G 2020 *Int. J. of Plast.* **124** 155–69
[6] Laplanche G, Schneider M, Scholz F, Frenzel J, Eggeler G and Schreuer J 2020 *Scripta Mater.* **177** 44–48
[7] Zhao Y, Lee D H, Seok M Y, Lee J A and Phaniraj M P 2017 *Scripta Mater.* **133** 54–58
[8] Pu Z, Chen Y and Dai L H 2018 *Mater. Sci. Eng.* A **736** 156–66
[9] Luo H, Li Z and Raabe D 2017 *Sci. Rep.* **7**, Article number 9892
[10] Soundararajan C K, Luo H, Raabe D and Li Z 2020 *Corros. Sci.* **167** 108510
[11] Pressouyre G M 1979 *Metal. Trans.* A **10** 1571–1573
[12] Grabke H J and Riecke E 2000 *Mater. Tehnol.* **34** 331–42
[13] Kannengiesser T and Boellinghaus Th 2013 *Weld. World.* **57** 3-37
[14] Rhode M, *Hydrogen diffusion and effect on degradation in welded microstructures of creep-resistant low-alloyed steels* (Berlin: BAM-Dissertationsreihe 148)
[15] Lee J, Park C, Park H and Kang N 2020 *Int. J. Hydrogen Energ.* **45** 10227–32
[16] Boellinghaus Th, Hoffmeister H and Middel C 1996 *Weld. World.* **37** 16–23
[17] Chen Y, Santos D M F and Sequeira C A C 2006 *Defect and Diffusion Forum* **258-260** 322–26
[18] Chen Y Y, Duval T, Hung U D, Yeh J W and Shih H C 2005 *Corros. Sci.* **47** 2257–79
[19] Lee C P, Chang C C, Chen Y Y, Yeh J W and Shih H C 2008 *Corros. Sci.* **50** 2053–60
[20] Hernandez-Rodriguez M A L, Laverde-Catano D A, Lozano D, Martinez-Cazeres G and Bedolla-Gil Y 2019 *Metals* **9** 307
[21] Tang Z, Gao M C, Diao H and Yang T 2013 *JOM* **65** 1848–58
[22] Otto F, Yang Y, Bei H and George E P 2013 *Acta Mater.* **61** 2628–38
[23] Liu W H, Wu Y, He J Y, Nieh T G and Lu Z P 2013 *Scripta Mater.* **68** 526–29
[24] Rodriguez A, Tylczak J H and Ziomek-Moroz M 2017 *ECS Transactions* **77** 741–52
[25] Luo H, Li Z, Mingers A M and Raabe D 2018 *Corros. Sci.* **134** 131–39
[26] N N 2003 *ASM Handbook, Vol. 13 - Corrosion* ed S D Cramer S D and Covino B S (Materials Park, Ohio: ASM International)
[27] N N 2011 *Uhlig's Corrosion Handbook* (3rd edition) ed R W Revie (New York: Wiley & Sons)
[28] Laplanche G, Berglund S, Reinhart C, Kostka A, Fox F and George E P 2018 *Acta Mater.* **161** 338–51

[29] Richter T, Schroepfer D, Rhode M and Boerner A 2020 Influence of modern machining processes on the surface integrity of high-entropy alloys *IOP Conference Series, Materials Science and Engineering* Symposium on Materials Science and Joining Technology, Magdeburg, Germany (in review)
[30] Rhode M, Mente T, Steppan E, Steger J and Kannengiesser T 2018 *Weld. World.* **62** 277–87
[31] Enomoto M, Hirakami D and Tarui T 2011 *Metall. Mater. Trans.* A **43A** 572–81
[32] Salmi S, Rhode M, Jüttner S and Zinke M 2015 *Weld. World.* **59** 137–44
[33] Rhode M, Schaupp T, Muenster C, Mente T, Kannengiesser T and Boellinghaus T, *Weld. World.* **63** 511–26
[34] Yasakau K A, Giner I, Vree C, Ozcan O, Grothe R, Oliveira A, Grundmeier G, Ferreira M G S and Zheludkevich M L 2016 *Appl. Surf. Sci.* **389** 144–56
[35] Rohwerder M, Hornung E and Stratmann M 2003 *Electrochim. Acta* **48** 1235–43
[36] Steppan E, Mantzke P, Steffens B, Rhode M and Kannengiesser T 2017 *Weld. World.* **61** 637–48
[37] Yagodzinsky Y, Todoshchenko, Papula S and Hänninnen H 2010 *Steel Res. Int.* **82** 1–6
[38] Griesche A, Dabah E, Kannengiesser T, Kardjilov N, Hilger A and Manke I 2014 *Acta Mater.* **78** 14–22
[39] Perez-Escobar D, Minambres C, Duprez L, Verbeken K and Verhaege M 2011 *Corros. Sci.* **53** 3166–76
[40] Frappart S, Oudriss A, Feaugas X, Creus J, Bouhattate J, Thebault F, Delattre L and Marchebois H 2011 *Scripta Mater.* **65** 859–862
[41] Legrand E, Oudriss A, Frappart S, Creus J, Feaugas X and Bouhattate J 2014 *Int. J. Hydrogen Energ.* **39** 1145–55
[42] Oudriss A, Creus J, Bouhattate J, Savall C, Peraudeau B and Feaugas X 2012 *Scripta Mater.* **66** 37–40
[43] Owen L R, Pickering E J, Playford H Y, Stone H J, Tucker M G and Jones N G 2017 *Acta Ma ter.* **122** 11–18
[44] Hasannaeimi, V, Ayyagari A V, Muskeri S, Salloom R and Mukherjee S 2019 *npj. Mater. Degrad.* **3** 16
[45] Shi Y, Collins L, Feng R, Zhang C, Balke N, Liaw P K and Yang B 2018 *Corros. Sci.* **133** 120–31
[46] Schulz W, Laplanche G, Schneider M, Karafiludis S and Stephan-Scherb C 2020 Insights into the high temperature oxidation behavior of the equiatomic CrMnFeCoNi and CrCoNi high- and medium-entropy alloys *Oxid. Met.* (in review)
[47] Yao Q, Thomas S, Gibson M A, Fraser H L, and Birbilis N 2017 *npj. Mater. Degrad.* **1** 15
[48] Nazarov A, Vucko F and Thierry D 2020 *Corros. Mater. Degrad.* **1** 9–20

IOP Conf. Series: Materials Science and Engineering **882** (2020) 012016 doi:10.1088/1757-899X/882/1/012016

Influence of modern machining processes on the surface integrity of high-entropy alloys

T Richter[1], D Schröpfer[1], M Rhode[1,2] and A Börner[1]

[1] Department 9 - Component Safety, Bundesanstalt für Materialforschung und -prüfung (BAM), Unter den Eichen 87, 12205 Berlin, Germany

[2] Institute for Materials and Joining Technology, Faculty of Mechanical Engineering, Otto-von-Guericke-University, Universitätsplatz 2, 39106 Magdeburg, Germany

tim.richter@bam.de

Abstract. High Entropy Alloys (HEAs) are a recent class of materials. In contrast to conventional alloys, HEAs consist of five alloying elements in equiatomic equilibrium. The high entropy effect is due, among other things, to the increased configuration entropy, which promotes solid solution formation. Many HEAs have enormous application potential due to excellent structural property combinations from very low to high temperatures. For the introduction of HEAs in real components, however, the question of the applicability of machining production technologies for component manufacture is of central importance. This has so far received little attention in global materials research. Reliable and safe processing is essential for the demand of economical component production for potential areas of application, e.g. in power plant technology. For metals, milling is the standard machining process. This article presents the results of machining analyses. It focuses on the surface integrity resulting from the milling process on a $Co_{20}Cr_{20}Fe_{20}Mn_{20}Ni_{20}$-HEA. For this purpose, investigations were carried out using ball nose end milling tools for conventional milling process in comparison to an innovative hybrid process available at BAM Berlin, Ultrasonic-Assisted Milling (USAM). USAM promises a lower degradation of the surface properties due to lower loads on the workpiece surface during machining. For this purpose, basic milling parameters (cutting speed and tooth feed) were systematically varied and cutting forces were measured during the milling experiments. The subsequent analysis of these forces allows an understanding of the mechanical loads acting on the tool and component surface. These loads cause topographical, mechanical and microstructural influences on the surface and consequently on the surface integrity. For their characterization, light and scanning electron microscopy were used, and the roughness and residual stresses via X-ray diffraction were measured. The results indicate significant advantages using USAM, especially due to reduced cutting forces compared to the conventional milling process. This causes lower mechanical loads on the tool and surface, combined with lower tensile residual stresses on and below the surface, and ultimately results in a significantly enhanced surface integrity.

1 Introduction

High Entropy Alloys (HEAs) represent a new class of alloys [1, 2], which have become a major focus of materials research. Compared to conventional alloys, they do not have one major alloying element, but at least five different alloying elements with proportions of 5 to 35 at. % [3, 4]. The special feature of HEAs is that they can form a single solid solution phase despite their high number of alloying elements [1]. HEAs offer the potential to obtain significantly improved properties and property combinations than conventional alloys [5, 6]. Therefore, HEAs are of great interest for many different applications. In this context, the $Co_{20}Cr_{20}Fe_{20}Mn_{20}Ni_{20}$-HEA, the so-called Cantor-alloy, has been investigated the longest and most intensively to date. This CoCrFeMnNi alloy exhibits excellent mechanical properties at very low temperatures. For that reason, especially cryogenic applications are conceivable here, such as pipes or tanks for liquified gases [7].

When using HEAs in real components, it must be ensured that they can be manufactured and processed safely. To date, there are very few publications dealing with the machining of HEA components, and in fact no systematic knowledge about the safe and reliable machinability of these materials [8, 9]. In this regard, machining has a great influence on the safety of components, especially the surface integrity, which is characterized by topography (roughness, defects), mechanics (residual stresses, work hardening) and metallurgy (microstructural changes). It considerably affects the resilience and service life of components in terms of corrosion resistance, crack initiation and propagation (e.g. stress corrosion cracking, fatigue cracking). For hard and brittle materials, e.g. quartz glass and ceramics, very good results have been achieved with innovative hybrid manufacturing processes, such as ultrasonic-assisted grinding, in terms of reducing cutting forces and increasing surface quality and which are already state of the art in production [10]. This has already been transferred to various metallic materials that are difficult to machine using tools with geometrically defined cutting edges. In case of ultrasonic-assisted milling (USAM) of NiCr-alloys, very good results have already been achieved in terms of increased surface integrity and reduced cutting forces, temperatures and tool wear [11-14]. Nonetheless, systematic knowledge is rare so far, especially when using radius cutters for finishing of complex components.

In this work, finish milling processes using ball nose end milling tools on a CoCrFeMnNi alloy are therefore being investigated for the first time and compared with ultrasonic-assisted milling processes. The aim is to obtain essential findings for reliable and safe machining of these innovative high-performance materials.

2 Experimental

2.1 Material preparation

The CoCrFeMnNi alloy, produced by cooperation partners of the Ruhr-University Bochum, was used for the milling tests. This alloy is already widely represented in the literature with regard to production and properties [2, 7, 15-18]. The material was melted with a vacuum induction furnace in an aluminum crucible made of elements with a purity of 99.9 wt.% and then homogenized at 1200 °C for 48 h. The diameter of the ingots could be reduced from 40 mm to 17 mm by rotary forging, followed by recrystallisation annealing at 1020 °C for 60 min, as explained in detail in [19]. As a result, the alloy shows a pure face-centered cubic phase with a grain size of ~50 µm and some Cr- and Mn-rich oxides [18, 20]. In the literature , the yield strength (273 MPa) and the ultimate tensile strength (633 MPa) are given for a similar material condition at room temperature [18]. The chemical composition was measured with an electron microprobe, cf. table 1.

Table 1. Chemical composition of CoCrFeMnNi alloy.

Element	**Co**	**Cr**	**Fe**	**Mn**	**Ni**
Atm. %	19.7	20.7	19.6	20.1	19.9

2.2 Milling experiments

For the milling experiments, the ingot was separated perpendicular to the axial direction by electrical discharge machining (EDM) and slices of approx. 2 mm thickness were produced. The experiments were conducted at the 5-axis machining center DMU 65 of Sauer GmbH/DMG MORI, modified for USAM. The experimental setup for finishing milling using solid carbide end mills with full radius (diameter 6 mm) is shown in figure 1a. The special carbide tool was designed by the manufacturer (WOLF Werkzeugtechnologie GmbH) for use on difficult-to-machine alloys in accordance to DIN 6527L with

IOP Conf. Series: Materials Science and Engineering **882** (2020) 012016 doi:10.1088/1757-899X/882/1/012016

four flutes and PVD coating with TiAlSiN [21]. The specimens were milled under dry condition (without cooling lubricant) in down milling mode and the tool was tilted in the primary (x) and secondary feed direction (y-direction), see figure 1a.

A full-factorial design of experiments was used for parameter variation, see figure 1b. In addition to the ultrasonic assistance, the cutting speed v_c and the feed rate per cutting edge f_z have been varied in accordance to the specifications of the tool manufacturer.

a)

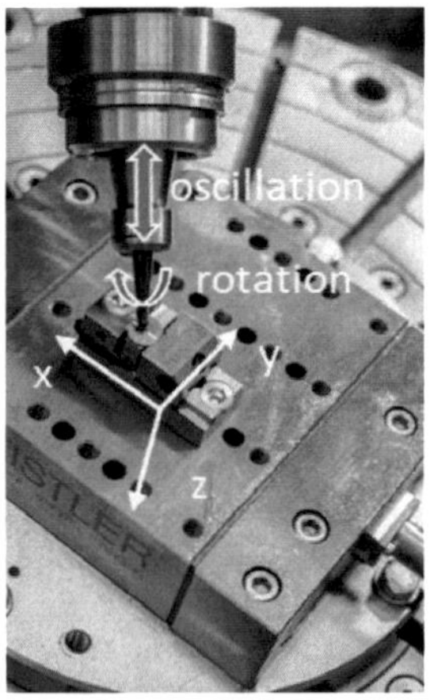

b)

angle in x-direction $\lambda = 45°$			**step over** $a_p = 0.3$ mm				**ultrasonic frequency** $f_{US} = 39.7 \pm 0.2$ kHz			
angle in y-direction $\tau = 45°$			**cutting engagement** $a_e = 0.3$ mm							
test no.	1	2	3	4	5	6	7	8	9	10
v_c **in m/min**	30	30	110	110	70	30	30	110	110	70
f_z **in mm**	0.04	0.07	0.07	0.04	0.055	0.04	0.07	0.07	0.04	0.055
USAM in %	0	0	0	0	0	100	100	100	100	100

Figure 1. (a) experimental setup, (b) experimental data.

Various investigations were carried out to characterize the surface integrity of the milled surfaces. Statements about the topography could be derived by means of light microscopic as well as SEM examinations. In addition, the roughness in x- and y- direction was measured with a contact profile meter according to DIN EN ISO 4287 [22]. The metallurgical influence and its depth could be investigated by means of SEM examinations perpendicular to the machined surface.

2.3 Force measurement

During the milling experiments the cutting forces in x- (F_f - feed force), y- (F_{fN} - normal feed force) and z-direction (F_p - passive force) were measured with a dynamometer from Kistler with a resolution of 0.1 ms, see figure 1a. To describe the acting force, the resulting cutting force F_r is calculated according to equation 1:

$$F_r = \sqrt{F_f^{\,2} + F_{fN}^{\,2} + F_p^{\,2}} \qquad (1)$$

2.4 Residual stress determination

To evaluate the mechanical effects of milling on the surface and the process related loads, principal residual stress analyses were carried out using X-ray diffraction (XRD, penetration depth approx. 5 µm, G3 Stresstech Goniometer). The determination of a residual stress profile in depth direction could be achieved by multiple electrochemical polishing. For this purpose, after each polishing step a stress measurement was carried out down to a depth of 40 µm. The electrochemical polishing was performed with a voltage of 25 V for 5 s, whereby a layer of approx. 5 µm was removed. The parameters for the XRD measurements are given in table 2. The calculation of the residual stresses is based on the material data from the literature [15] with Poisson's ratio υ (0.259) and Young's modulus E (207 GPa).

Table 2. X-ray measuring and evaluation parameters.

Measuring mode	$\sin^2\psi$	Collimator ⌀	3 mm
Radiation	Mn-Kα	Tube power	30 kV/ 6.7 mA
Detector	Linear solid-state detector	ψ – tilting	0° bis ± 45°
Diffraction line	(311)	ψ - step	9
2Θ angel	156°	Measuring time	2 s

3 Results

3.1 Forces during milling

During the milling tests, the cutting forces F_f, F_{fN}, F_p were measured and the resulting cutting force F_r was calculated. The measured cutting forces are shown exemplarily in figure 2a and b. The resulting cutting force is shown in figure 2c for a milling test and its relation to the partial forces in figure 2d. The respective peaks can be assigned to one cutting action of the tool so that four peaks occur during one tool revolution. To compare the individual force measurements, the average of the individual peak maxima of the resulting cutting force $F_{r,max}$ is used in the following.

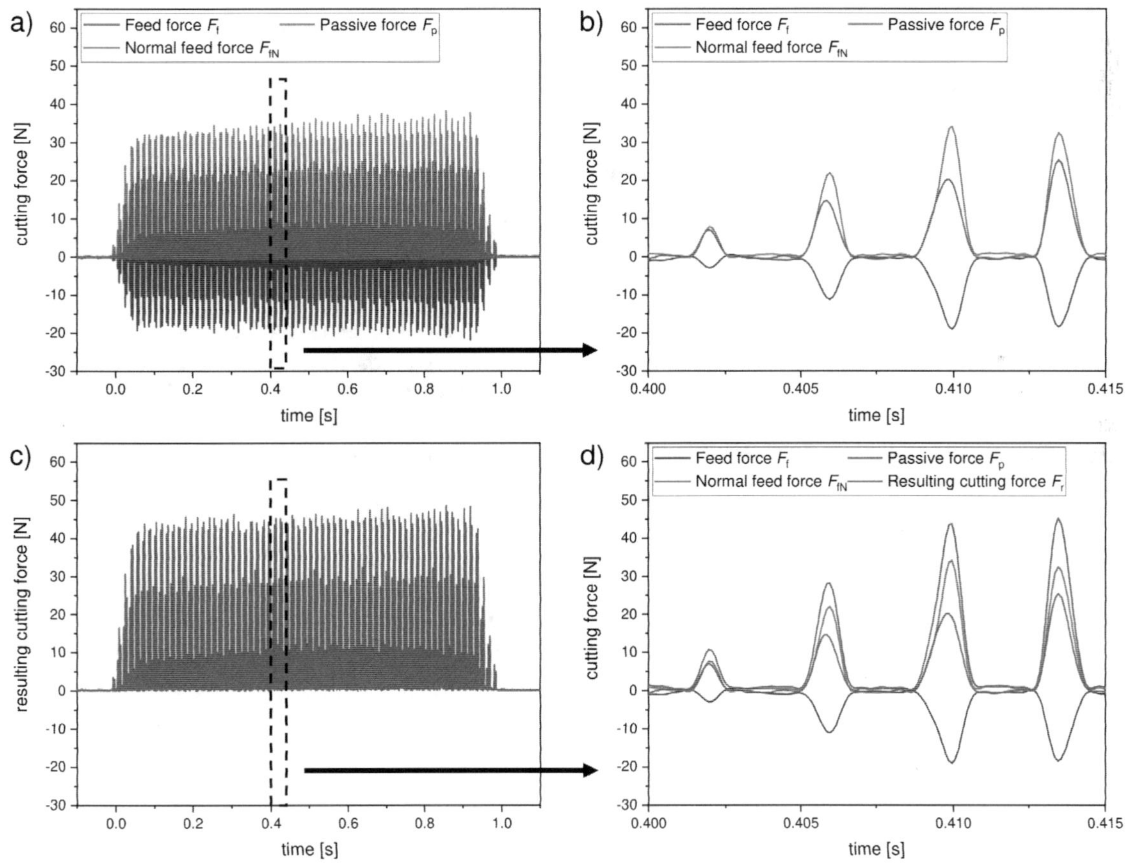

Figure 2. (a) Force measurement for one milling experiment, (b) for one tool revolution (n=1), (c) resulting cutting force and (d) calculated resulting cutting force for one tool revolution (n=1).

Differences in the cutting force maxima stem from the normal manufacturing deviation of the individual cutting edges of the newly used tool. Therefore, it should be noted that due the lower feed rate per cutting edge and cutting engagement commonly used for finish milling the cutting edges are not all equally engaged.

IOP Conf. Series: Materials Science and Engineering **882** (2020) 012016 doi:10.1088/1757-899X/882/1/012016

To show the influence of the ultrasonic assistance on the cutting forces, the force measurements of test no. 5 and test no. 10 (see figure 1b) are compared in figure 3.

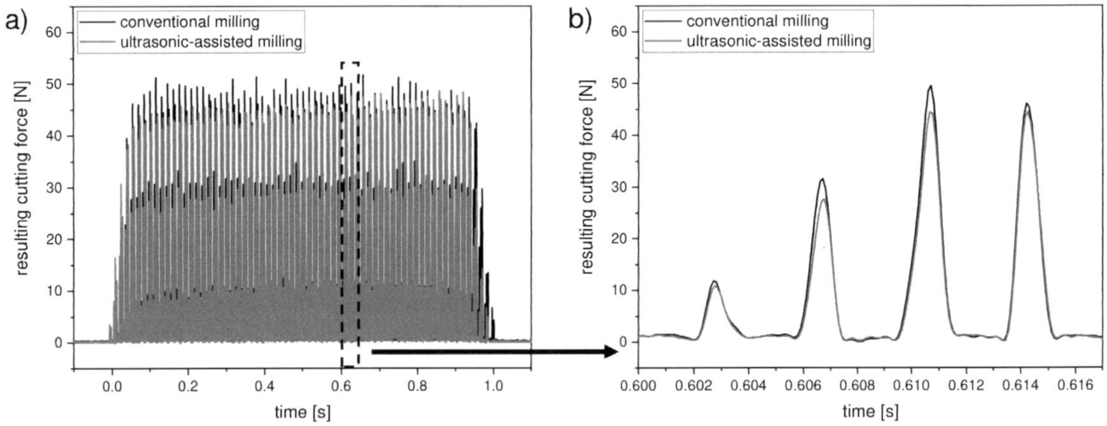

Figure 3. Comparison of the resulting cutting forces of conventional (test no. 5) and ultrasonic-assisted (test no. 10) milling of a test (a) and a tool revolution (b).

Figure 3b shows that the maximum resulting cutting force with ultrasonic assistance is lower as in conventional milling. This can be seen from the lower peak maximum of the resulting cutting force for all four cutting edges. The main reason for the reduced cutting forces is a reduced contact time due to the additional relative movement of the ultrasonic oscillation (superposition principle) with a frequency of approx. 40 kHz exceeding the cutting frequency of the individual cutting edges [23]. In addition, the coefficient of friction between tool and material surface decreases due to this high-frequency tool oscillation and causes a reduction of the cutting forces.

The influence of the milling parameters on the resulting cutting force is shown in figure 4a (f_z – feed rate per cutting edge) and b (v_c – cutting speed) for conventional and ultrasonic-assisted milling. Figure 4a shows that the resulting cutting force decreases by 38 % as the feed rate per cutting edge decreases. The cutting speed in figure 4b also has a highly significant influence on the cutting force. Within the given parameters, the cutting force can be reduced by approx. 28 % by increasing the cutting speed from 40 m/min to 110 m/min. From this, a clear influence of the feed rate per cutting edge and the cutting speed on the mechanical loads for workpiece and tool and thus on the surface integrity and service life of the tools can be concluded [24]. A comparison of conventional and ultrasonic-assisted milling shows a reduction of the resulting cutting force by approx. 5 % due to the ultrasonic assistance, for the complete parameter matrix, independently of feed rate and cutting speed, cf. figure 3. Thus, an overall positive effect of the ultrasonic assistance on the tool load can be observed. Even for such comparably small reductions of the process induced loads, distinct beneficial effects were identified considering the surface integrity, see section 3.2. From that it can be stated, that ultrasonic assistance is at least showing potential for an expansion of the working ranges in terms of machining without increasing the load.

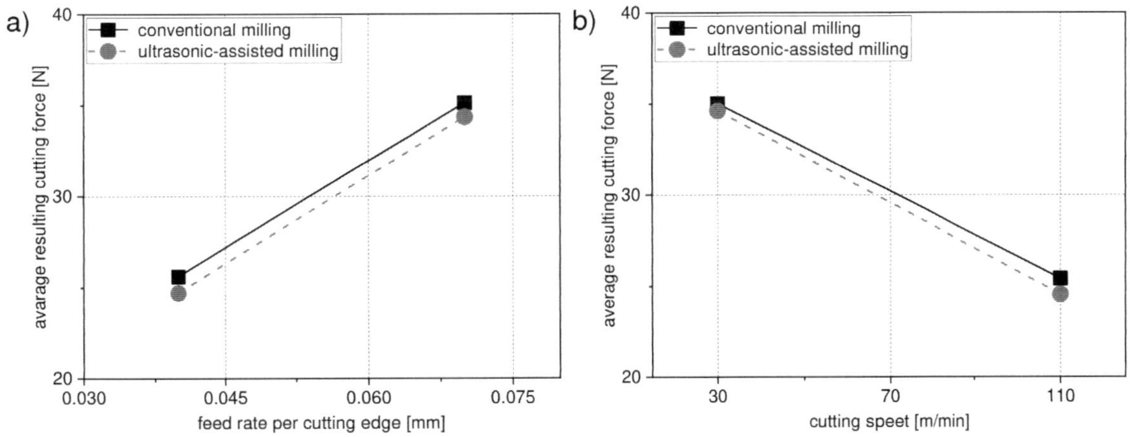

Figure 4. Influence of the milling parameters on the resulting cutting force: feed rate per cutting edge f_z (average over all v_c) (a) and cutting speed v_c (average over all f_z) (b).

3.2 Surface integrity

In order to discuss the influence of ultrasonic assistance on the surface integrity in detail, the specimens of milling tests no. 2 and no. 7 (see figure 1b) were assessed in this study, see figure 1b. These milling tests provide the highest mechanical load on the tool that was observed in the complete parameter matrix (high f_z and low v_c), cf. section 3.1. In this context, a distinct influence of USAM application was expected.

3.2.1 Topography. The topographies of conventional (a-c) and ultrasonic-assisted (d-f) machining are compared in figure 5. The light microscopic observations (a, d) show a typical milling pattern. The shape of the pattern depends mainly on the milling parameters such as the feed rate per cutting edge, the step width or the angle of inclination. In addition, figure 5a shows the influence of the manufacturing deviation of the tool (see Section 2.3) based on milling cuts of different widths. Furthermore, this figure shows milling marks which are at an angle of 45° to the feed direction x due to the double tool tilt. They are the result of the tool micro geometry and the adhesion [25]. In figures 5d, e and f, the ultrasonic pattern are visible perpendicular to the milling marks comparable to waves on a surface. These are caused by the ultrasonic oscillation of the tool. Since the ultrasonic frequency of approx. 40 kHz exceeds the cutting frequency of approximately 111 Hz (for test no. 2 and 7) by several hundred times, multiple of these ultrasonic patterns can be detected per cut, in accordance with the superposition principle.

Figures 5b and e show defects on both conventional and ultrasonic-assisted milled surfaces, which are abrasive marks and chipping of Cr and Mn oxides ($(Cr,Mn)_3O_4$). These hard oxides, present in the CoCrFeMnNi alloy [20], are ground by the cutting edge over the fresh formed surface and cause a groove due to the abrasive effect of the oxides [26]. In addition, the SEM images of conventionally milled surfaces in figure 5b and c reveal further defects. These are similar to the so-called tear defects, cf. morphology in figure 5c. Tear defects are the most common machining-related surface defects in nickel-based materials such as Inconel 718 and are usually the result of insufficient cutting speed [27, 28]. One reason for this is for instance the formation of a built-up edge in combination with unfavorable milling parameters (e. g. low cutting speed) and hard particles, such as NbC-precipitates in Inconel 718. However, the defect mechanism cannot be clearly clarified with the presented results. This kind of defects were not detected on the surface machined with USAM. For this reason, it is anticipated that such defects can effectively be avoided with USAM.

The previously described local defects can also act as initiation points for pitting corrosion or notches for crack initiation, thus reducing the surface integrity of CoCrFeMnNi alloy components [29]. Tear defects can be consistently avoided by use of ultrasonic assistance and adapted milling parameters. The oxide-related grooves, on the other hand, must be considered by adjusting the process parameters during material production to suppress such hard oxide inclusions in the microstructure.

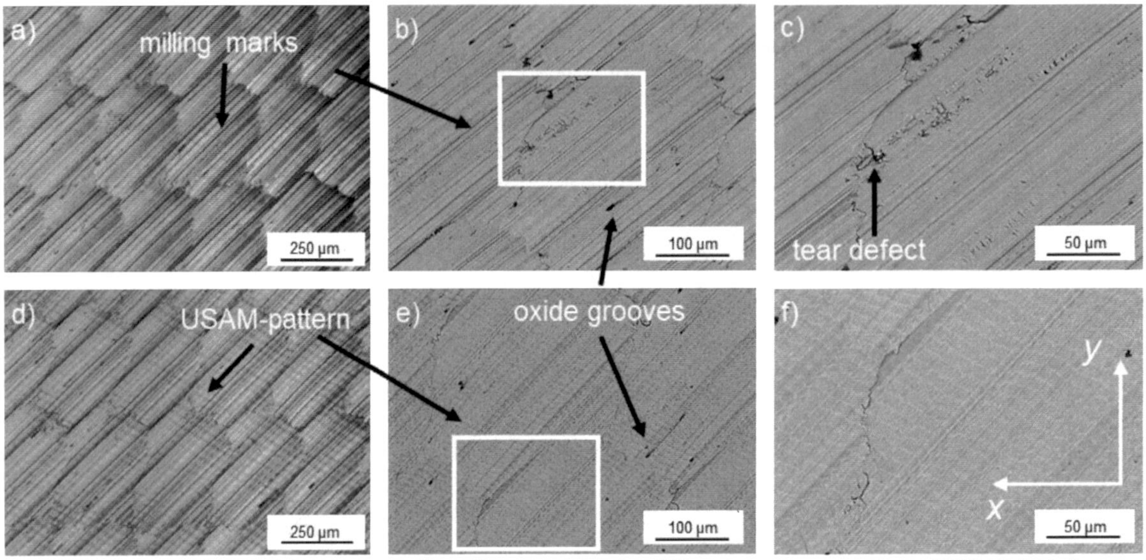

Figure 5. Comparison of the topography of conventional milling (a-c; test no. 2) and USAM (d-f; test no. 7) with light microscope (a, d) and SEM (b, c, e, f).

In comparison to the qualitative evaluation by optical light microscopy and SEM, the topography can be described quantitatively by determining the roughness values. The measured roughness values R_a (arithmetical mean deviation) and R_z (maximum profile height) are shown in table 3. There are no significant effects on the roughness values obvious within the examined test matrix if USAM is compared to conventional milling. In the secondary feed direction (y), the roughness values are higher than in the feed direction (x), which is due to the penetration kinematics or kinematic topography between the tool cutting edge and the material. This interdependency is determined by the tool macro geometry, the feed rate per cutting edge f_z, the step over a_p and the angles of inclination λ, τ [30, 31].

With an arithmetical mean deviation between 0.9 µm and 0.6 µm, the machining process can be classified in the transition between finishing and fine finishing ($R_a \leq 0.8$ µm) and thus fulfils the requirements placed on the surface of a highly corrosive and mechanically exposed component with safety relevance [32].

Table 3. Comparison of the surface roughness of conventional and ultrasonic-assisted milling using the arithmetical mean deviation R_a (a) and the maximum high of the profile R_z (b).

	Conventional milling	USAM
R_a in x – direction in µm	0.87 ± 0.10	0.88 ± 0.14
R_a in y – direction in µm	0.71 ± 0.29	0.68 ± 0.19
R_z in x – direction in µm	3.61 ± 1.12	3.60 ± 0.70
R_z in y – direction in µm	4.50 ± 0.49	4.65 ± 0.55

3.2.2 Mechanics. Process-related loads caused by milling are principally the mechanical influences, which cause residual stresses in the surface. The residual stress condition has a strong influence on the surface integrity in terms of crack initiation and propagation on the surface. For this purpose, the residual stresses were analyzed by means of X-ray diffraction at and up to 40 µm below the surface and the maximum (principal) residual stresses were determined. The results of the residual stress measurements are shown in figure 6 by means of depth profiles of the maximum residual stresses. It is obvious that for both processes a tensile residual stress level of more than 300 MPa is present at the surface. Thus, the tensile residual stresses on the surface approximately correspond to the strength of the material

(273 MPa [18]), due to plastic deformation and local temperature gradients while cutting [33]. However, the decrease of the tensile residual stresses of the USAM-specimen is comparatively greater with the distance from the surface. While the residual stresses in USAM are approximately 0 MPa at about 15 µm below the surface and change to compressive residual stresses in greater depth, a tensile residual stress level of approx. 200 MPa is present for conventional milling down to a depth of 20 µm. In addition, tensile residual stresses of about 100 MPa are still present in the conventionally milled sample down to a depth of about 40 µm below the surface.

The higher tensile residual stresses in the conventionally milled specimens can support crack formation and propagation, thus increasing the risk of component failure. From a stress-related point of view, the USAM shows a significantly higher surface integrity than the conventionally milled specimen. This makes USAM more suitable for safety-relevant components by its generation of smoother surface. That can withstand stress corrosion cracking in severe environments better or generally increased fatigue cracking resistance by avoiding mechanical notches compared to conventional milling.

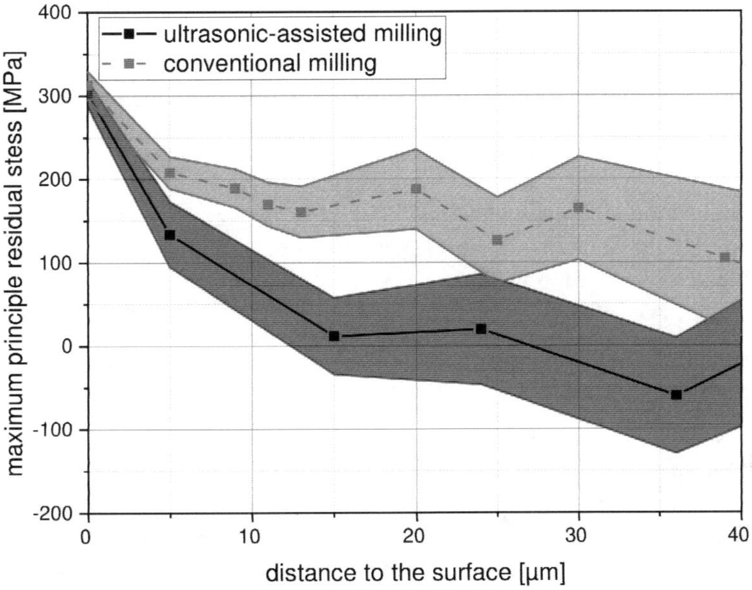

Figure 6. Influence of conventional and ultrasonic-assisted machining on the residual stresses at and below the surface.

3.2.3 Metallurgy. The metallurgical influence is mainly characterized by microstructural degradation in the area of the surface. The surface integrity highly depends on the properties and the prevailing microstructure. For this purpose, specially prepared cross sections of the milled surfaces (test no. 2 and 7) were examined by SEM, see figure 7. Shear bands in the area below the surface reveal in both the conventionally milled and the USAM-sample. A pronounced appearance of these shear bands indicates a high deformation during milling. Compared to USAM (figure 7b), the shear bands are more evident in conventional milling (figure 7a) and, at up to about 9.3 µm, extend approx. 30 % deeper under the machined surface. In these highly deformed areas, a mechanically significantly degraded microstructure can be assumed in connection with pronounced grain refinement and orientation. In this context, it can therefore be assumed that for instance the risk of crack initiation will be reduced due to the lower metallurgical influence of USAM process. These results confirm the findings of the authors [34] for the USAM of Ti-6Al-4V with higher feed rates per cutting edge and in this connection the residual stress analyses in section 3.2.2.

IOP Conf. Series: Materials Science and Engineering **882** (2020) 012016 doi:10.1088/1757-899X/882/1/012016

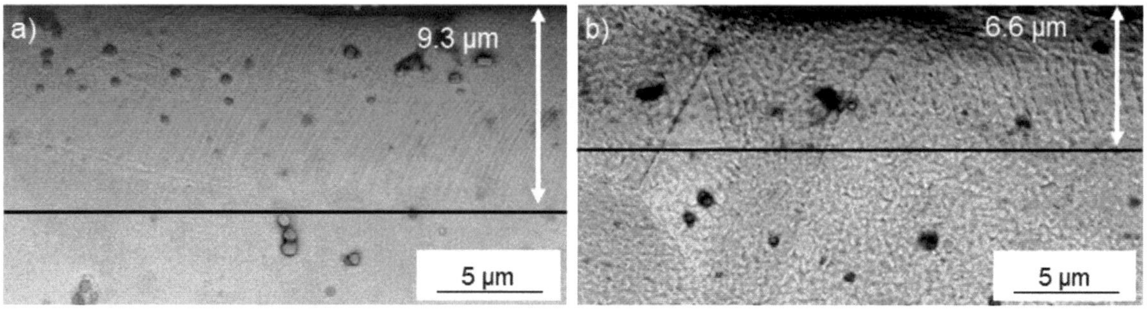

Figure 7. SEM images perpendicular to the surface from conventional (a) and ultrasonic-assisted milling (b).

The results in figure 7 reflect the results of the force evolution (sections 3.1), the topography (section 3.2.1) and the mechanical effect (section 3.2.2) with an increased surface integrity by USAM well. Here, the increased mechanical loads from conventional machining due to increased cutting forces from section 3.1 show a clear effect on the surface integrity of the CoCrFeMnNi alloy. The lower cutting forces and contact times of the tool cutting edges with the surface lead to a reduction in local overloading on the material surface in the USAM. This prevents local microdefects and increases the overall surface integrity.

Finally, the results show that the investigated CoCrFeMnNi alloy is comparatively well machinable with the carbide milling tool used. The machining process and the resulting surface integrity can be further optimized with ultrasonic assistance, respectively the working range for machining parameters can be extended if applying USAM. However, the influence of microdefects caused by oxide inclusions must be considered.

4 Summary and conclusion

In this study, the experimental results on the influence of conventional and ultrasonic-assisted finishing milling processes using ball nose end mills on a CoCrFeMnNi alloy are presented. The investigations focused on the cutting process, the cutting forces and the resulting surface integrity of the milled HEA samples. The surface integrity was characterized by the topography and the mechanical and microstructure. The following conclusions can be drawn from this study:

- The investigated CoCrFeMnNi alloy can be safely machined with the solid carbide radius cutter used. The resulting cutting force can be significantly reduced by adjusting the milling parameters. The cutting force could be reduced within the scope of the test matrix by lowering the feed rate per cutting edge by about 38 % and by increasing the cutting speed by about 28 %. Furthermore, the resulting cutting force can be reduced significantly by about 5 % on average for the entire test matrix by means of ultrasonic-assisted machining.
- The surface integrity of the CoCrFeMnNi alloy can be significantly improved compared to conventional milling with ultrasonic assistance. This is demonstrated by a reduction of microdefects on the surface, lower tensile residual stresses at or below the surface and a lower and less in-depth metallurgical influence.
- The HEA CoCrFeMnNi is comparatively well machinable with the applied milling tool and the selected boundary conditions (plane-finish milling) and milling parameters. High quality surfaces could be produced by the finishing milling processes. Further investigations will be carried out to determine the extent to which these surfaces can withstand mechanical, thermal and corrosive loads or combinations of loads. For this purpose, the influence of machining production steps on the corrosion properties and crack initiation of the alloy CoCrFeMnNi will be investigated further.

Acknowledgements
The authors want to thank Prof. Guillaume Laplanche and co-workers from Ruhr-University Bochum, Germany, for providing the experimental materials and WOLF Werkzeugtechnologie GmbH for providing milling tools.

References
[1] Yeh J W, Chen S K, Lin S J, Gan J Y, Chin T S, Shun T T, Tsau C H,Chang S Y, 2004 Advanced Engineering Materials **6** 299-303.
[2] Cantor B, Chang I T H, Knight P,Vincent A J B, 2004 Materials Science and Engineering: A **375-377** 213-218.
[3] Miracle D B,Senkov O N, 2017 Acta Materialia **122** 448-511.
[4] Gorsse S, Miracle D B,Senkov O N, 2017 Acta Materialia **135** 177-187.
[5] Pickering E J,Jones N G, 2016 International Materials Reviews **61** 183-202.
[6] Zhang Y, Zuo T T, Tang Z, Gao M C, Dahmen K A, Liaw P K,Lu Z P, 2014 Progress in Materials Science **61** 1-93.
[7] Otto F, Dlouhý A, Somsen C, Bei H, Eggeler G,George E P, 2013 Acta Materialia **61** 5743-5755.
[8] Guo J, Goh M, Zhu Z, Lee X, Nai M L S,Wei J, 2018 Materials & Design **153** 211-220.
[9] Polishetty A, Barla M M R, Littlefair G,Fabijanic D, 2015 International Conference on Manufacturing Engineering and for Manufacturing Growth Vancouver **3** 41-44
[10] Kumar M N, Subbu S K, Krishna P V,Venugopal A, 2014 Procedia Engineering **97** 1577-1586.
[11] Suárez A, Veiga F, de Lacalle L N L, Polvorosa R, Lutze S,Wretland A, 2016 Journal of Materials Engineering and Performance **25** 5076-5086.
[12] Ahmed N, Mitrofanov A V, Babitsky V I,Silberschmidt V V, 2006 Materials Science and Engineering: A **424** 318-325.
[13] Chen W, Huo D, Shi Y,Hale J M, 2018 The International Journal of Advanced Manufacturing Technology **97** 2033-2049.
[14] M'Saoubi R, Outeiro J C, Chandrasekaran H, Jr. O W D,Jawahir I S, 2008 Int. J. Sustainable Manufacturing **1** 203-236.
[15] Zaddach A J, Niu C, Koch C C,Irving D L, 2013 Jom **65** 1780-1789.
[16] Schuh B, Mendez-Martin F, Völker B, George E P, Clemens H, Pippan R,Hohenwarter A, 2015 Acta Materialia **96** 258-268.
[17] Gludovatz B, George E P,Ritchie R O, 2015 Jom **67** 2262-2270.
[18] Wu Z, David S A, Leonard D N, Feng Z,Bei H, 2018 Science and Technology of Welding and Joining **23** 585-595.
[19] Laplanche G, Berglund S, Reinhart C, Kostka A, Fox F,George E P, 2018 Acta Materialia **161** 338-351.
[20] Schulz W, Laplanche G, Schneider M, Karafiludis S,Stephan-Scherb C, 2020 Oxidation of Metals **to be published**.
[21] DIN-6527L, Vollhartmetall-Schaftfräser mit abgesetztem Zylinderschaft-Maße, Beuth publishing DIN, 2002
[22] DIN EN ISO 4287, Geometrische Produktspezifikation (GPS) - Oberflächenbeschaffenheit: Tastschnittverfahren - Benennungen, Definitionen und Kenngrößen der Oberflächenbeschaffenheit, Beuth publishing DIN, 2010
[23] Ni C, Zhu L, Liu C,Yang Z, 2018 International Journal of Mechanical Sciences **142-143** 97-111.
[24] Choudhury S K,Rao I V K A, 1999 Int J Mach Tool Manu **39** 343-353.
[25] Grzesik W, 1996 Wear **194** 143-148.
[26] Klocke F,Köning W, Fertigungsverfahren 1, Springer - Verlag, 2008.
[27] Liu C, Ren C, Wang G, Yang Y,Zhang L, 2015 Journal of Mechanical Science and Technology **29** 1723-1730.

IOP Conf. Series: Materials Science and Engineering **882** (2020) 012016 doi:10.1088/1757-899X/882/1/012016

[28] Dosbaeva G K, Veldhuis S C, Elfizy A, Fox-Rabinovich G,Wagg T, 2010 Journal of Materials Engineering and Performance **19** 1193-1198.
[29] Roberge P R, Handbook of Corrosion Engineering, McGraw-Hill, 2000.
[30] Vopát T, Peterka J, Šimna V,Kuruc M, 2015 Procedia Engineering **100** 868-876.
[31] Baccar M H, Bayraktar E, Rickert T, Boujelbene M,Katundi D, 2013 Experimental and Applied Mechanics, Volume 4 191-201
[32] Keferstein C P, Marxer M,Bach C, Fertigungsmesstechnik, Springer Vieweg, 2018.
[33] Sharma V,Pandey P M, 2016 Ultrasonics **70** 172-182.
[34] Liu J, Jiang X, Han X, Gao Z,Zhang D, 2018 The International Journal of Advanced Manufacturing Technology **101** 1451-1465.

KorroPad testing - applications from industry and research

P Rosemann, N Kauss and A Heyn

Otto-von-Guericke-University Magdeburg, Faculty of Mechanical Engineering, Institute of Materials and Joining Technology, Universitätsplatz 2, 39106 Magdeburg, Germany

paul.rosemann@ovgu.de

Abstract. The pitting corrosion resistance of stainless steels can be assessed with the KorroPad test in a very simple, quick and inexpensive way. The KorroPad is therefore an interesting alternative to time-consuming long-term tests and complex electrochemical investigations for manufacturers, processors and users of stainless steels. In this article, the basic functionality of the KorroPad is described and selected examples from industry and research illustrate it´s application. The assessment of grinding processes and post-weld treatment are first presented as relevant examples regarding industrial surface treatments on the pitting corrosion resistance. Subsequently, current research to the modification of the KorroPad test is presented, which shows the specific detection limit of KorroPads on binary Fe-Cr reference alloys. Finally, modified KorroPads are used to detect microstructure related susceptibility to pitting corrosion induced by sensitizing heat treatment. The KorroPad enables the creation of a sensitization T-t diagram based on the pitting corrosion resistance.

1 Introduction

The so-called "KorroPad" was developed in the research project "Detection of corrosion-sensitive surfaces of stainless steels by processors" [1,2]. The KorroPad consists of a gel-like electrolyte, which indicates corrosion-prone surface conditions on stainless steels by means of a visible color reaction. The principle of the KorroPad test is presented schematically in Figure 1.

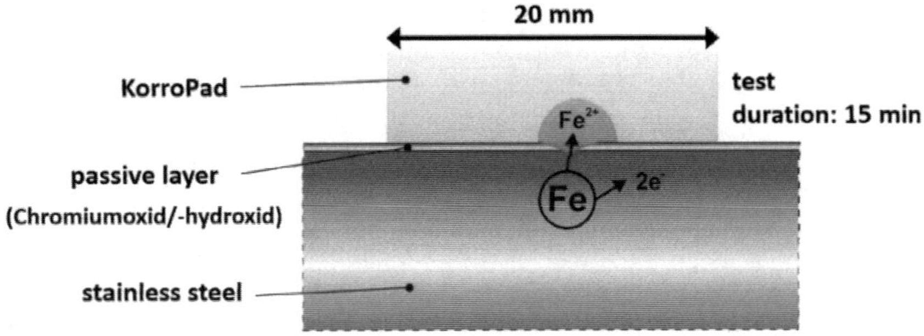

Figure 1. Test principle of the KorroPad [1].

The method was originally developed to detect an incompletely formed or instable passive layer on the austenitic stainless steels X5CrNi18-10 (1.4301). The KorroPad is a water-based agar gel, which contains 3 wt.-% agar, 1 mM potassium hexacyanoferrate III ($K_3[Fe(CN)_6]$) and 0.1 M sodium chloride

(NaCl). The agar forms a gel matrix, which holds and immobilizes the electrolyte, while the $K_3[Fe(CN)_6]$ causes the KorroPad to achieve a redox potential of up to +320 $mV_{Ag/AgCl}$ (measured on platinum) and is thus able to polarize the surface of a stainless steel up to +200 $mV_{Ag/AgCl}$. The redox potential can be shifted by increasing or decreasing the $K_3[Fe(CN)_6]$ content [1]. In addition to the anodic polarization, the NaCl dissolved in the KorroPad also ensures that pitting corrosion is initiated and eventually stabilized.

Due to this composition, the KorroPad has a defined corrosion load, which can trigger localized metal dissolution (mostly through pitting corrosion) at weak points and defects at the surface. If the critical pitting corrosion potential is reached or exceeded by the anodic polarization iron ions pass into the KorroPad at the phase boundary and react with a blue colour change with $K_3[Fe(CN)_6]$ to Prussian Blue ($Fe_4[Fe(CN_6]_3$). This insoluble pigment remains within the gel matrix of the KorroPad and the corrosion resistance can be estimated due to the permanent blue discoloration.

Three typical appearances can be observed in the KorroPad test, which are shown as examples in Figure 2. This classification enables a visual assessment of the susceptibility to corrosion.

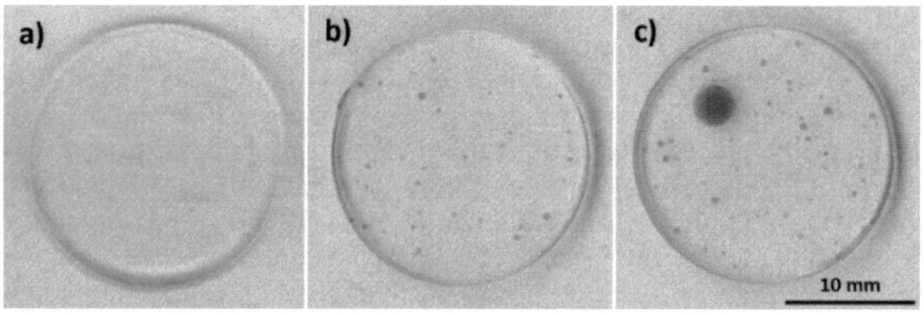

Figure 2. Typical appearances in the KorroPad test:
a) no pitting corrosion (stable passive layer) ≈ no indication;
b) metastable pitting corrosion (with repassivation) ≈ many small indications;
c) stable pitting corrosion (without repassivation) ≈ one large indication.

An assessment of the susceptibility to pitting corrosion of the stainless steels is also possible based on the percentage area of the blue indications [1, 3]. The susceptibility to pitting corrosion increases with the proportion of the coloured area. Values of more than 1 % reflect an increased susceptibility to pitting corrosion in atmospheric conditions [3]. However, the image processing required for this is significantly more complex than a visual assessment according to Figure 2.

When assessing the formation of a passive layer, which is significantly influenced by the surface treatment, it is useful to consider the proportion of the area in order to investigate the kinetics of the formation of the passive layer [4] or to compare the quality of the passive layer regarding different surface conditions. However, it should be noted that the number of metastable pitting indications is reduced by passivation (e.g. at high relative humidity) reducing especially the smaller metastable indications and the overall area fraction of indications [4]. This must be considered when comparing different surface conditions by determining the area fraction. The sensitivity of the KorroPad method regarding the characterization of the pitting corrosion resistance is comparable to electrochemical methods and correlate with standard tests such as the alternating immersion tests [5], the salt spray tests [1, 3] and long-term exposure tests in natural atmospheres [1]. Research results on austenitic stainless steels [3, 4, 6-8] and martensitic stainless steels [5, 9-14] were recently published. The KorroPad was able to characterize reduced pitting corrosion resistance caused by increased sulphur content [8] and unfavourable surface treatments [3, 7, 15].

The following four chapter present selected application examples from industry and research, in which the KorroPad test was successfully applied. The first two examples deal with the possibility of assessing the quality of industrially manufactured surfaces and the detection of corrosion-prone surface conditions. The two final chapters show how the KorroPad test can also be used to verify material and structural causes of pitting corrosion induced by sensitizing heat treatment.

IOP Conf. Series: Materials Science and Engineering **882** (2020) 012017 doi:10.1088/1757-899X/882/1/012017

2 Reduced corrosion resistance and staining of industrially ground stainless-steel surfaces
Impairment of the visual appearance and staining of industrially ground stainless steels surfaces in non-critical atmospheric exposure conditions has led to complaints in the manufacturing industry. In many of these complaints, abrasive belts with aluminium oxide (Al_2O_3) were used for the final surface treatment. This is the reason why it was suspected that Al_2O_3 abrasive belts are responsible for the problem of staining [1]. Systematic studies on the influence of the used abrasive belt on the corrosion resistance of the austenitic stainless steel X5CrNi18-10 (1.4301) were carried out in order to investigate this assumption [16, 17]. The abrasive belt (single-layer grain / granulate grain) and the grain type (Al_2O_3 / SiC) were varied and the grinding parameters like cutting speed 24 m/s, counterrotation and mineral oil-free cooling lubricant are kept constant. The aim was to create critical surface conditions, detect them with the KorroPad and identify the cause of staining. Figure 3 shows the applied abrasive belts (240 grit) in the unused condition, the ground surfaces and the associated results of the KorroPad test on these surfaces.

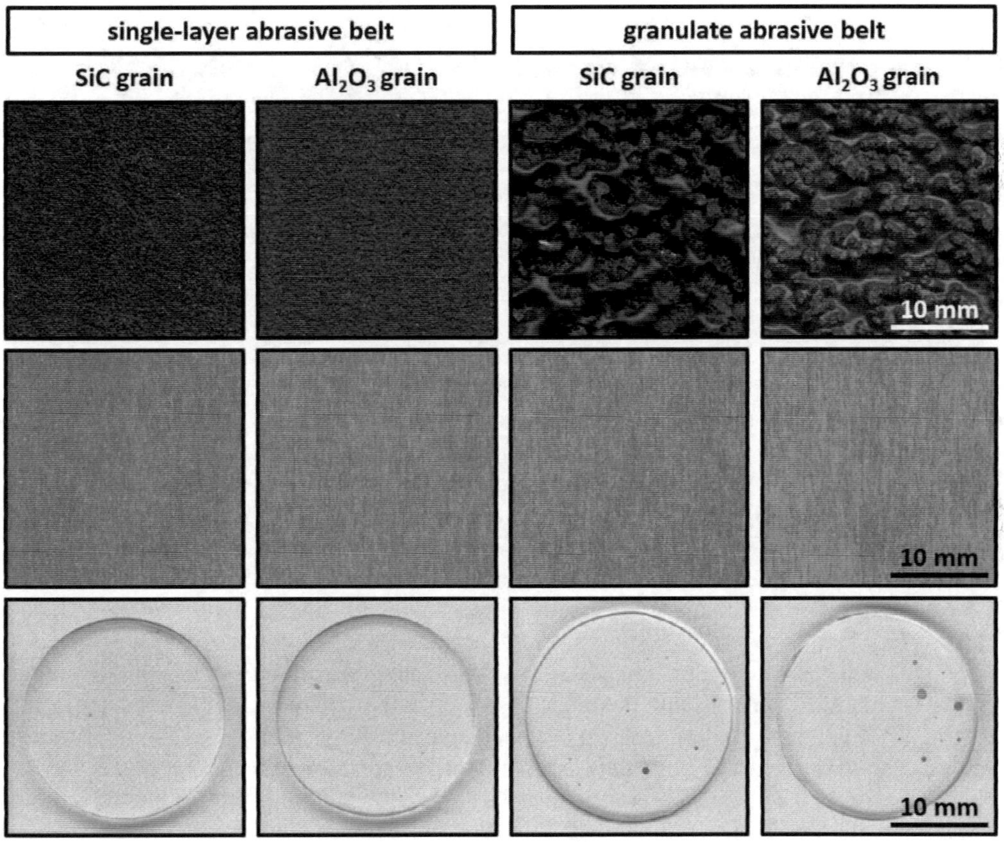

Figure 3. Appearance of abrasive belts with single-layer grain / granulate and SiC / Al_2O_3 abrasive grains, the surfaces ground with it and associated results of the KorroPad.

The surfaces ground with granulate abrasive belts show more and larger KorroPad indications than the surfaces ground with single-layer grain. This effect can be demonstrated for products of several manufacturers and no significant differences were observed with regard to the surface quality (roughness and visual appearance). In addition, surfaces ground with SiC abrasive belts show less and smaller indications in direct comparison to surfaces ground with Al_2O_3 abrasive belts.

A major advantage of the KorroPad test is the localization of critical defects at the surface, which can be detected by the position of the large blue indications. This was used to determine the cause of the

IOP Conf. Series: Materials Science and Engineering **882** (2020) 012017 doi:10.1088/1757-899X/882/1/012017

higher susceptibility of the surfaces ground with granulate to staining. Figure 4 shows SEM images (SE contrast) of the ground surface areas where the largest KorroPad indications emerged.

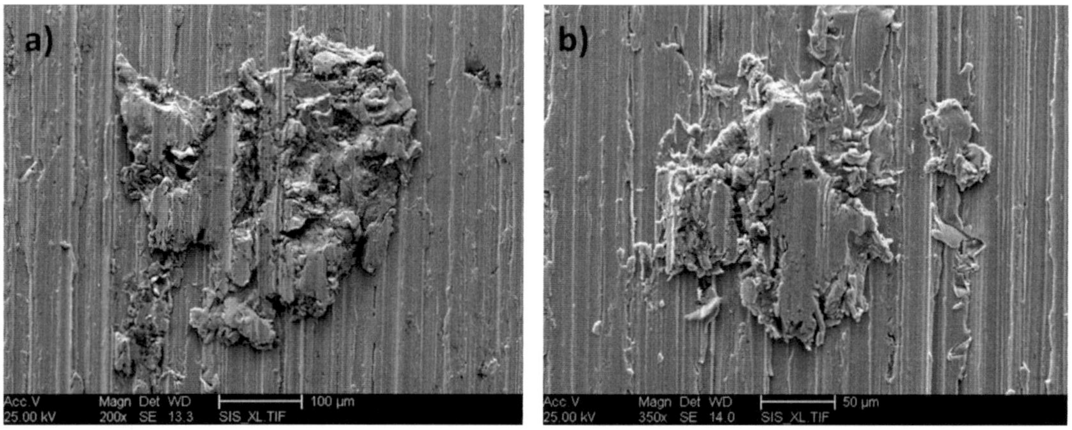

Figure 4. SEM images of the ground surfaces, where large KorroPad indications emerged a) granulate abrasive belt with corundum grain [17] b) granulate abrasive belt with SiC grain.

Large particles (200 µm to 600 µm) of the ground steel can be seen at these surfaces, which are located on the top of the surface and are firmly attached to it. These particles create a large number of geometric crevices and contain microparticles of the respective abrasive grain (Al_2O_3 or SiC). The formation of these particles can be traced back to the pronounced topography of the granulate abrasive belt, which favours the remains of chips in the abrasive belt, while this is not the case for single-layer grain. The particles are merged to the surface in the grinding process and remain on the ground surface. The abrasive chip particles represent local defects, which impairs the formation of the passive layer. In the case of atmospheric exposure, this leads to the local release of iron ions and cause the phenomenon of brownish staining, since iron ions are converted to rust in subsequent reactions. Clarifying the cause of the increased susceptibility to corrosion was significantly facilitated by the KorroPad test, which is due to the good differentiability between different surfaces and the localization of critical defects.

3 Influence of weld post-treatment on corrosion resistance

There has been a significant increase in the use of duplex stainless steels in construction [18], since low Ni and Mo steels (1.4062, 1.4162 and 1.4362) reduce material costs compared to austenitic stainless grades for different technical applications [19, 20]. At the same time, duplex steels possess higher yield and ultimate tensile strength and an equivalent or better corrosion resistance compared to austenitic stainless steels 1.4301 and 1.4404 [19, 21]. The disadvantage is that production, processing and especially welding of duplex stainless steels is more demanding and requires adaptations.

The corrosion resistance of welds is influenced by the base material, the filler metal, the welding process and the weld seam surface finish [22, 23]. The tempering colours from welding significantly reduce the corrosion resistance of welded joints and must therefore be removed by suitable surface treatment [19, 21, 23]. This requires a process-related optimization of the production considering the corrosion resistance [19, 24, 25].

The aim of these investigations was to determine the influence of different surface treatments on the corrosion resistance of duplex steel weldments using the KorroPad test. One main question was whether blasting with glass-bead is equivalent to blasting with corundum grain regarding the weldment's corrosion resistance. Weldments of 1.4062 with pickled, blasted, electropolished and plasma-polished surfaces were provided by Wilhelm Modersohn GmbH & Co.KG for different characterizations. Figure 5 shows welds in different surface states.

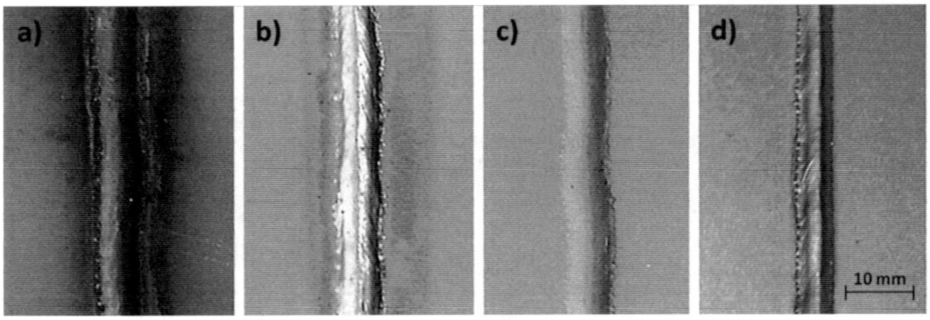

Figure 5. Weldments with different surface states: a) without treatment, b) pickled, c) blasted and d) electropolished [15].

The KorroPad test was used to check whether a stable passive layer was formed after passivation for 24 h at 95% relative air humidity on the different surfaces. The KorroPads were applied to areas directly at the weld seam (SN) and with 60 mm distance to the weld seam to additionally detect the sole effect of surface treatment. Figure 6 shows the results of the KorroPad test for the various conditions.

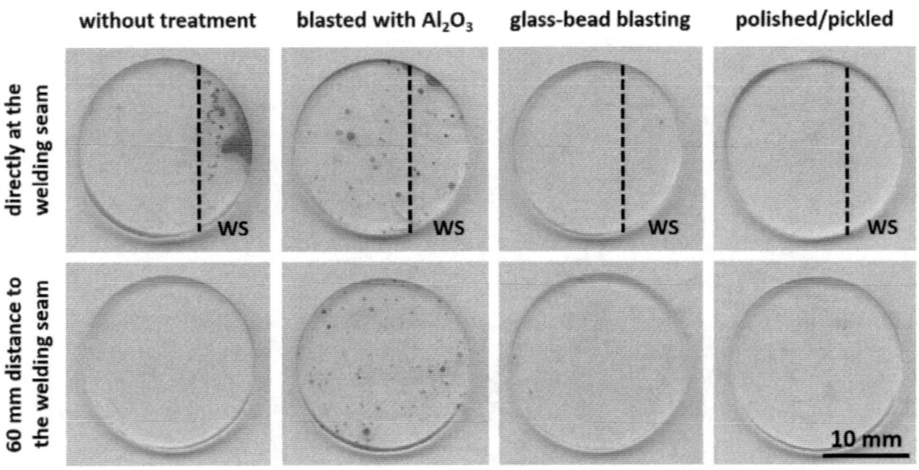

Figure 6. Results of the KorroPad test after various surface treatments at the welding seam (WS, top line) and at 60 mm distance to the welding seam (bottom line).

If the tempering colours are not removed after welding (untreated, directly at the welding seam), large areas with blue indications arise at the welding seam (WS) on the right side of the KorroPad. This indicates that iron ions are released at a high rate and a high susceptibility to pitting corrosion or a loss of passivity can occur. The untreated sheet material (60 mm distance to the welding seam) shows no KorroPad indications and therefore the high pitting corrosion resistance of the used duplex steel 1.4062. Surfaces blasted with Al_2O_3 (corundum) show a very high number of small (metastable) indications within the KorroPad at the welding seam and also with 60 mm distance. The indications are larger at the weld seam, which indicates a more critical condition.

Glass-bead blasting shows a significantly lower susceptibility to pitting corrosion, since the KorroPad has only isolated small indications. Polished and pickled surfaces are comparable to the untreated sheet and do not show any indications during the KorroPad test. The supplementary determination of critical pitting potentials resulted in significantly lower values for blasted surfaces compared to pickled or polished surfaces, which is presented in Figure 7 [15]. The blasted surfaces were examined by SEM in order to determine the cause of the lower pitting corrosion resistance, see Figure 8.

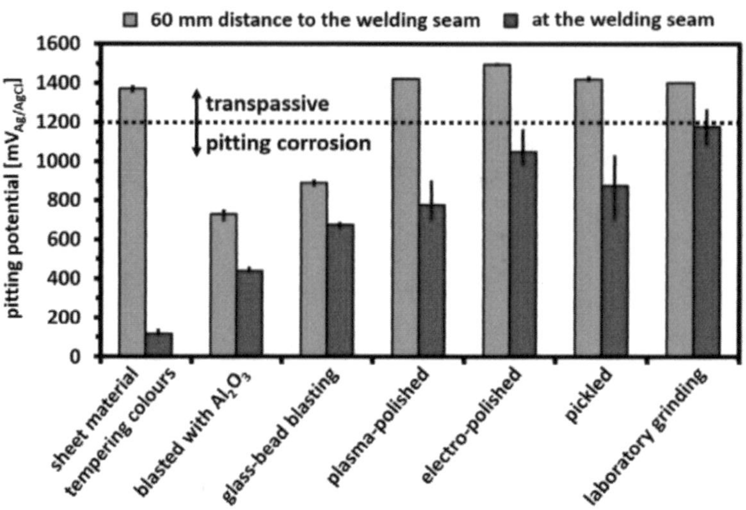

Figure 7. Pitting potentials after various surface treatments at 60 mm distance to the welding seam (blue) and directly at the welding seam (red), Electrolyte: 3000 ppm Cl^-, pH 4; dU/dt: 1 mV/s [15].

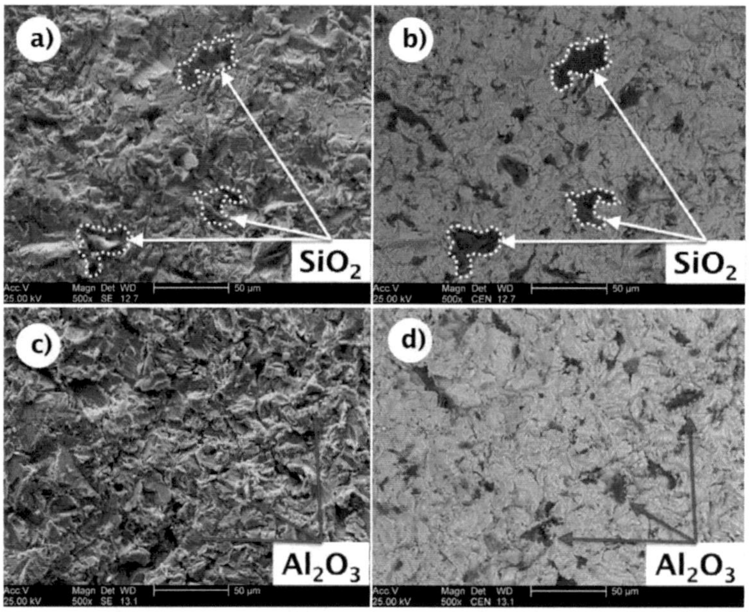

Figure 8. SEM images of surfaces: top - glass-bead blasted a) SE contrast, b) BSE contrast; bottom - blasted with Al_2O_3: c) SE contrast, d) BSE contrast [15].

The SEM images with SE contrast (topography) show rough surfaces for both blasting agents (corundum: $R_z = 26 \pm 2$ µm, $R_a = 3.8 \pm 0.2$ µm / glass-bead: $R_z = 29 \pm 8$ µm, $R_a = 4.1 \pm 0.5$ µm). If the same area of the surface is compared with element contrast (BSE), many dark areas can be seen, which indicate lighter elements. An analysis of these areas using EDX spots shows silicon and oxygen for glass-bead blasted surfaces, while aluminium and oxygen are present on surfaces blasted with corundum. At these points, when looking carefully at the SE contrast images, blasting particles can be recognized within the surface as result of the blasting process. These particles hinder the formation of a stable passive layer, which explains the increased susceptibility to pitting corrosion. Blasting (especially with Al_2O_3) may be suitable to remove tempering colours, but it generates surfaces that are more susceptible to pitting corrosion than other post-treatment processes for weldments made of 1.4062 duplex stainless steels.

4 Analysis and modification of the detection limit of the KorroPad test

The KorroPad was used in research projects on the corrosion resistance of martensitic stainless steels with 13-15% chromium in order to evaluate the influence of the heat treatment on the corrosion resistance and to detect suboptimal heat treatment conditions [11, 13, 14]. Non-critical surface conditions were produced by laboratory grinding (wet, SiC abrasive, single-layer grain, 180 grit) and passivated (24 h at > 95% relative air humidity) in order to exclude the influence of surface treatment and passivation. This preparation technique allows to study microstructure and alloy related factors influencing the pitting corrosion resistance with the KorroPad test. Based on the extensive studies on martensitic stainless steels, it was concluded that stable pitting corrosion occurs when the chromium content falls locally below 11 to 13 wt-% [13, 14]. This low chromium content was therefore initially adopted as the microstructure related detection limit of the KorroPad test. This statement should be viewed critically in terms of its general validity since martensitic stainless steels possess a high carbon content.

In order to determine the detection limit of the KorroPad more precisely, ferritic reference alloys with chromium contents between 7 and 18 wt.-% were manufactured and examined electrochemically as well as with the KorroPad. It was also examined how an increase in the KorroPad composition affects the detection limit. The aim of this adjustment was a higher detection limit in the range of 16 to 18 wt.-% chromium, in order to use the KorroPad test to detect sensitized states in ferritic, austenitic and duplex stainless steels.

For the preparation of the reference alloys, various chromium contents were specifically produced in an arc melting furnace (Compact Arc Melter MAM-1, Buehler GmbH). The melting chamber was previously evacuated three times and subsequently flushed with argon to minimize oxidation during melting. 20 g of each reference alloy were melted twice in the arc and the melt was cooled on a water-cooled copper plate. The chemical composition of all reference alloys is shown in Table 1.

Table 1. Chemical composition of the ferritic reference alloys in wt.-% [26].

Element	Cr	Ni	Mo	Mn	Si	N	C	P	S
7.3 %	7.32	0.28	0.01	0.78	0.28	0.005	0.019	0.011	0.003
9.3 %	9.34	0.36	0.01	0.99	0.37	0.009	0.025	0.013	0.002
11.2 %	11.19	0.43	0.01	1.17	0.44	0.007	0.029	0.015	0.002
13.6 %	13.56	0.33	0.02	0.90	0.46	0.019	0.025	0.016	0.002
15.8 %	15.80	0.25	0.03	0.68	0.49	0.015	0.029	0.020	0.002
17.4 %	17.36	0.18	0.03	0.48	0.49	0.024	0.022	0.018	0.002
18.0 %	17.96	0.16	0.03	0.42	0.51	0.023	0.025	0.019	0.002

The reference alloys were solution annealed at 1050 °C for 0.5 h, quenched in water and then annealed at 800 °C for 1 h to transform the martensitic microstructure at chromium contents of less than 10 wt.% chromium to ferrite. The alloys were then cold rolled to 50 % of their initial height and annealed again at 800 °C for 1 h to achieve a recrystallized microstructure.

The corrosion resistance of the ferritic reference alloys was characterized regarding the passivation behaviour in 0.1 M H_2SO_4 solution and in relation to the critical pitting potentials in 0.1 M NaCl solutions [26] to verify the increasing corrosion resistance with the Cr content. The samples were ground with 180 SiC paper in preparation for these electrochemical investigations and in case of critical pitting potentials also passivated for 24 h at more than 95 % relative humidity. This should enable the formation of a stable passive layer and minimize the influence of surface treatment and passivation on the pitting corrosion resistance. The results regarding critical pitting potentials and passivation current density are presented in Figure 9.

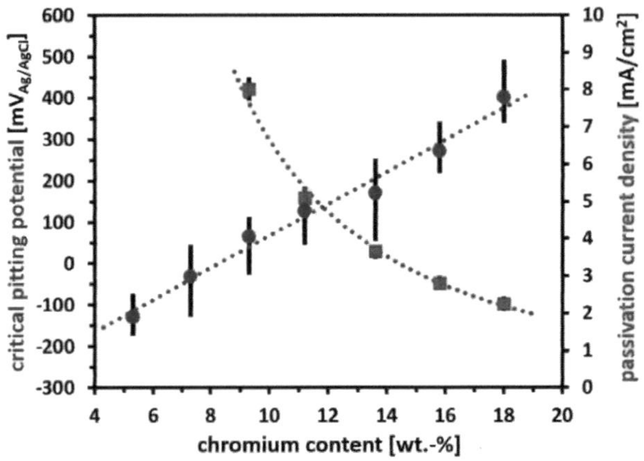

Figure 9. critical pitting potential and passivation current density of the ferritic reference alloys [26].

The critical pitting potential shows a linear increase with the chromium content and the passivation current density exhibit a logarithmic reduction with increasing chrome content. This is exactly the expected behaviour in relation to the chromium content and this allows the investigation of the detection limit of the KorroPad test. The reference alloys with increasing corrosion resistance were finally examined using the KorroPad test. In addition to the "standard" KorroPad with 0.1 M NaCl and 1 mM $K_3[Fe(CN)_6]$, modified KorroPads with 0.2 M NaCl and 2 mM $K_3[Fe(CN)_6]$, 0.5 M NaCl and 5 mM $K_3[Fe(CN)_6]$ and 1.0 M NaCl and 10 mM KHCF were used to determine the resulting detection limit. All KorroPads were applied for a test period of 15 minutes, scanned optically with 600 dpi and assessed according to their visual appearance (Figure 2). The results of the KorroPad tests with increasing corrosivity on alloys with different concentrations of chromium are presented in Figure 10.

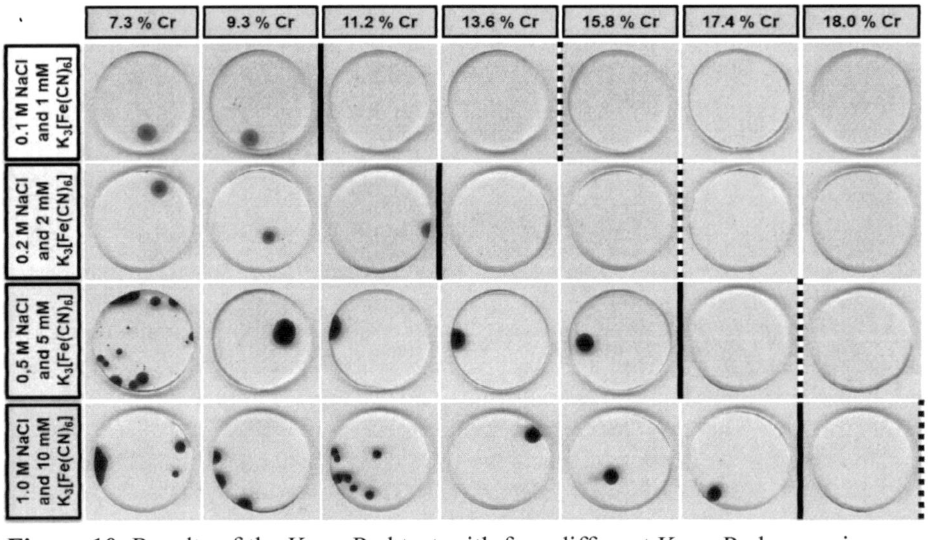

Figure 10. Results of the KorroPad test with four different KorroPad corrosiveness on ferritic reference alloys with different chromium content.

The standard KorroPads (0.1 M NaCl, 1 mM $K_3[Fe(CN)_6]$) shows a transition of stable pitting to metastable pitting is between 9.3 and 11.2 wt.-% Cr. With 13.6 wt.-% or more Cr, pitting corrosion no longer occurs and the passive layer remains stable (passive).

Increasing the corrosivity of the KorroPad by applying higher NaCl and $K_3[Fe(CN)_6]$ concentrations shifts the detection limits to higher chromium contents. In addition, less metastable pitting corrosion indications can be seen, since the higher redox potential favours the differentiation between stable pitting and the passive state. The KorroPad-results of all four KorroPad modifications allows an easy comparison of different alloy or microstructure related corrosion resistances. Furthermore, the KorroPads with 0.5 M NaCl, 5 mM $K_3[Fe(CN)_6]$ are more suitable for the detection of material related pitting susceptibility of stainless steels with 18 % Cr due to the higher detection limit. Figure 11 shows schematically how a change in the KorroPad concentration affects the open circuit potential and the critical pitting corrosion potential for a KorroPad test on the alloy with 15.8 wt.-% Cr.

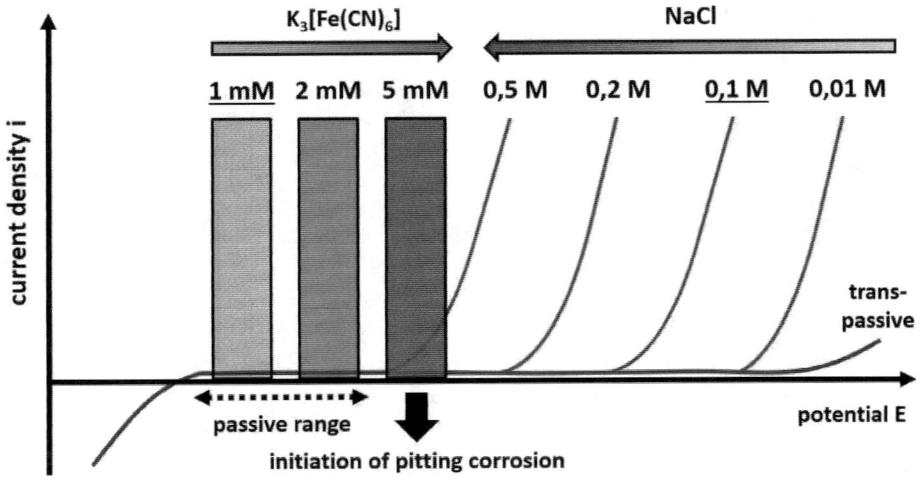

Figure 11. Change in the anodic polarization and the critical pitting potential when increasing the corrosivity of the KorroPad for a ferritic stainless steel with 15.8 wt.-% Cr.

The standard KorroPad (0.1 M NaCl, 1 mM $K_3[Fe(CN)_6]$) polarizes the surface to a potential within the passive range. The anodic polarization by 1 mM $K_3[Fe(CN)_6]$ is not sufficient to polarize the reference alloy with 15.8% chromium up to the pitting potential in a 0.1 M NaCl solution. Consequently, no stable pitting corrosion is initiated and no KorroPad indications occur.

The higher concentrations of $K_3[Fe(CN)_6]$ increase the anodic polarization and the distance to the pitting corrosion potential is reduced. At the same time, increasing the NaCl concentration lowers the critical pitting corrosion potential. The combination of 0.5 mM $K_3[Fe(CN)_6]$ and 0.5 M NaCl is sufficient to reach the critical pitting potential, initiate pitting corrosion and generate large indications on the reference alloy with 15.8 wt.-% Cr.

Since the critical pitting corrosion potentials are shifted to higher potentials with increasing chromium content, stable pitting corrosion does not yet occur in the reference alloys with 17.4 and 18 wt.-% Cr. However, local chromium depletion, which reduces the pitting corrosion resistance, can be detected on ferritic, austenitic and duplex stainless steels using KorroPads with higher corrosivity. This can be demonstrated on sensitized stainless steels which are susceptible to intergranular corrosion and pitting corrosion.

5 Sensitization behaviour of the lean duplex stainless steel X2CrNiN23-4

Secondary phases ($M_{23}C_6$, Cr_2N, σ, γ_2) can precipitate due to welding of the duplex stainless steel X2CrNiN23-4 (1.4062) within the heat affected zone. This can cause local chromium depletion at the ferrite-ferrite grain boundaries and reduce the corrosion resistance. The sensitization behaviour was characterized by aging studies at 600 °C with ageing times from 0.1 h to 10 h. The generated microstructures are presented in Figure 12. Significant precipitation is visible in the SEM images after aging for 0.2 h. Longer durations result in precipitation growth. Finally, precipitation starts at 10 h within the ferrite grains.

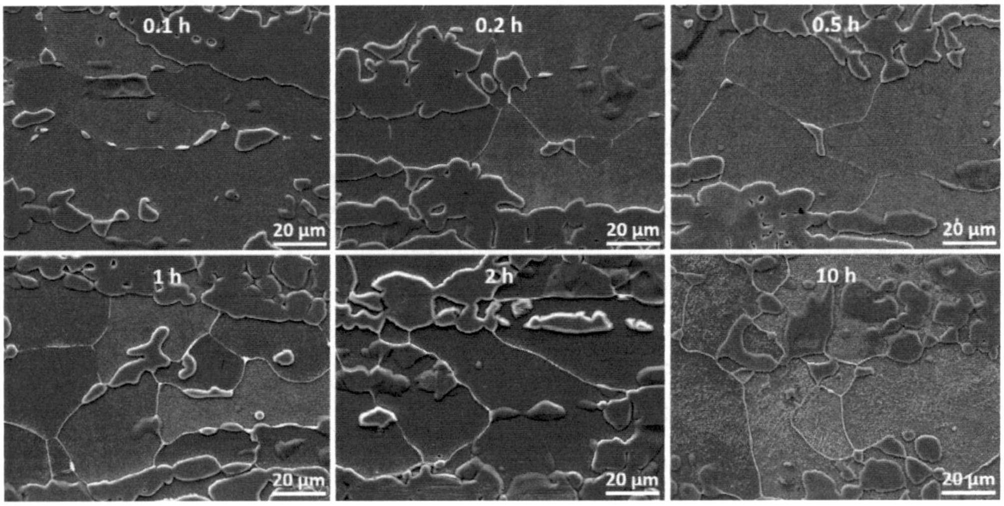

Figure 12. Microstructure of the duplex stainless steel X2CrNiN23-4 after aging at 600 °C [27].

The corrosion resistance of the aging conditions was examined by the EPR method, by critical pitting temperatures (CPT) according to ASTM G150 and with the modified KorroPads described in the previous chapter. Figure 13 shows the results regarding the degree of sensitization (DOS), critical pitting temperatures (CPT) and the results of the four KorroPad modifications.

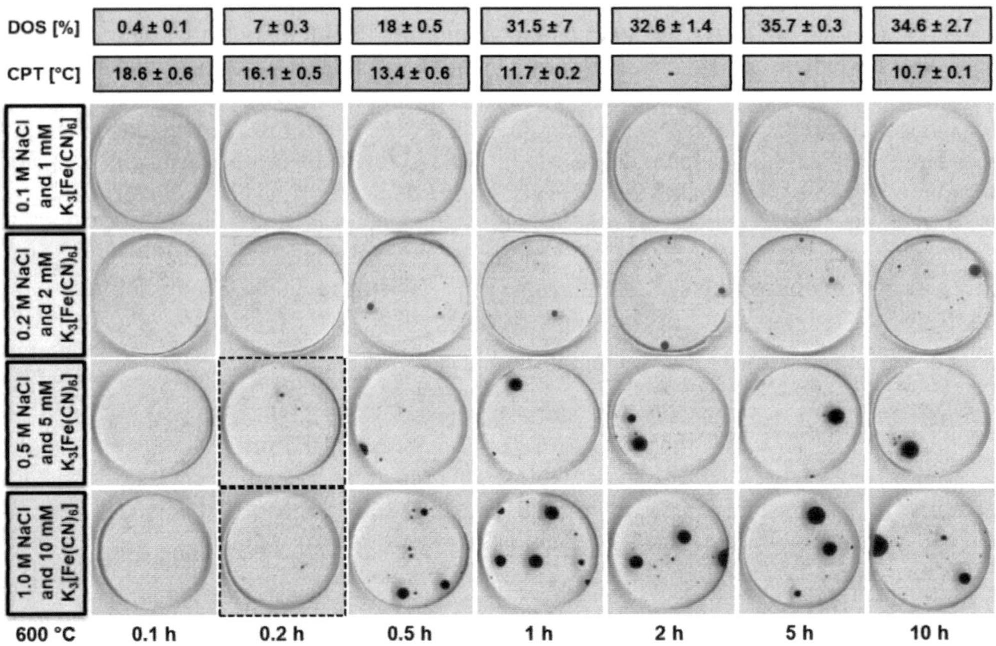

Figure 13. Results of the KorroPad test with various concentrations, the CPT values (°C) and the DOS [%] for the sensitization at 600 °C of the duplex stainless steel X2CrNiN23-4 [27]; dashed lines show the first reduction of pitting corrosion resistance.

KorroPads with 0.1 M NaCl and 1 mM $K_3[Fe(CN)_6]$) are not suitable to detect the reduction in corrosion resistance due to sensitization at 600 °C, while decreasing CPT values and increasing degree of sensitization (DOS) show the expected behaviour. This can be explained by the fact that the depletion zones formed during annealing do not fall below the detection limit of the standard KorroPad. Increasing the corrosivity of the KorroPads up to 0.5 M NaCl and 5 mM $K_3[Fe(CN)_6]$) detects the loss of pitting corrosion resistance with for aging times of 0.2 h or more. Therefore, sensitization can be detected with the modified KorroPad test and the EPR method with a comparable detection limit. The modified KorroPad test is therefore suitable to generate T-t-sensitization diagrams for different stainless steels. The critical pitting corrosion temperatures show the same trend. However, the modified KorroPad test can be carried out much more effectively using parallel tests. Electrochemical experiments (EPR method, ASTM G150) are much more time consuming and require suitable electrochemical devices.

6 Conclusions

The application examples of the KorroPad test from industry and research show that the pitting corrosion resistance of stainless steels can be assessed in a simple, quick and inexpensive way. Especially the visual indication of the pitting corrosion resistance is very advantageous, which is generated in a very short test duration (15 min) without a complex electrochemical test setup. The sensitivity of the KorroPad regarding the pitting corrosion resistance is comparable with other electrochemical methods and long-term experiments within artificial or natural atmospheres [1]. The KorroPad is therefore an interesting alternative to time-consuming long-term tests and complex electrochemical analysis methods for manufacturers, processors, users and researchers of stainless steels. The KorroPad is often used in current projects for screening tests and extensive parameter studies in order to limit the scope of systematic electrochemical investigations. Various questions and problems relating to stainless steels can be addressed, which are traced back to surface-related and material-related changes of the pitting corrosion resistance.

The current main application of the KorroPad test is the detection of surfaces prone to pitting corrosion, which can be correlated with an incomplete formation of the passive layer. The KorroPad can test whether the passive layer is formed properly or if there is a surface with critical defects. Furthermore, the KorroPad locates the most corrosion-prone surface areas, which was successfully illustrated by the example of ground surfaces. This location of critical defects makes it easy to clarify the causes of higher susceptibility to corrosion by further analytical examinations at these points.

If a defect-free surface is established by laboratory grinding and a stable passive layer is formed by suitable passivation, then the KorroPad can be used to examine the pitting corrosion resistance of alloys or microstructural states. KorroPad test on ferritic reference alloys with chromium contents from 7.3 to 18.0 wt.-% show, that chromium contents of less than 11.2 wt.-% result in stable pitting corrosion, even with non-critical surface and suitable passivation. This material related detection limit of the standard KorroPad can be shifted to higher chromium contents by increasing the concentrations of NaCl and $K_3[Fe(CN)_6]$. This modified KorroPads are suitable to detect material and microstructure related susceptibility to pitting corrosion for higher alloyed stainless steels like 1.4301, 1.4404 and 1.4062. This has been demonstrated using the example of sensitization of the duplex stainless steel X2CrNiN23-4. KorroPads with 5 times higher concentration (0.5 M NaCl, 5 mM $K_3[Fe(CN)_6]$) compared to standard KorroPad (0.1 M NaCl, 1 mM $K_3[Fe(CN)_6]$), characterised sensitization in perfect correlation to the well-established EPR method.

These results show that modified KorroPads can be used as a powerful tool to easily, quickly and effectively detect material-related causes of pitting corrosion and check the corrosion resistance of various stainless-steel batches. A vulnerability identified and localized with the KorroPad can then be investigated using electrochemical, metallographic and analytical methods to clarify the specific cause.

Acknowledgments
The authors thank the Department 7.6 Corrosion and Corrosion Protection of the Federal Institute for Materials Research and Testing for the provision of KorroPads for the presented investigations.

Dr. Heyn thanks the German Research Foundation (DFG) for funding the project "Agar-based gel electrolytes for corrosion diagnostics" (project number 330472124).

References
[1] Burkert A, Burkert A, Lehmann J, Müller T and Bohlmann T 2014 *final report of the research project 17136 N*
[2] Lehmann J, Burkert A and Steinhoff UM 2012 *german patent* DE 10 2010 037 775 A1
[3] Müller T, Lehmann J, Babutzka M and Burkert A 2015 *Fortschritte in der Werkstoffprüfung für Forschung und Praxis* 95
[4] Reinemann S 2015 *Untersuchung von Korrosionsvorgängen an nichtrostenden Stählen mittels Gelelektrolyten* (Magdeburg: Otto-von-Guericke-University Magdeburg)
[5] Rosemann P, Müller T, Babutzka M and Heyn A 2013 *HTM - Journal of Heat Treatment and Materials* **68** 224
[6] Lehmann J, Burkert A and Mietz J 2016 *Materials and Corrosion* **67** 84
[7] Burkert A, Klapper HS and Lehmann J, *Materials and Corrosion* **64** 675
[8] Gümpel P and Hörtnagel A, *Materials and Corrosion* **67** 607
[9] Rosemann P, Müller T, Babutzka M and Heyn A 2015 *Materials and Corrosion* **66** 45
[10] Müller T, Heyn A, Babutzka M and Rosemann P 2015 *Materials and Corrosion* **66** 656
[11] Rosemann P, Kauss N, Müller C and Halle T 2017 *Journal of Heat Treatment and Materials* **72** 87
[12] Rosemann P, Müller C, Kauss N and Halle T 2014, *Conference proceedings of the 17. Werkstofftechnisches Kolloquium, Chemnitz* 103
[13] Rosemann P, Kauss N, Müller C and Halle T 2015 *Materials and Corrosion* **66** 1068
[14] Rosemann P 2017 *Einfluss der Wärmebehandlung auf die Korrosionsbeständigkeit martensitischer nichtrostender Stähle* (Düren: Shaker Verlag)
[15] Rosemann P, Müller C, Baumann O, Modersohn W and Halle T 2017 *IOP Conf. Series Materials Science and Engineering* **181** 012019
[16] Reinemann S, Lehmann J, Rosemann P, Bohlmann T, Babutzka M and Burkert A 2019 *final report of the research project 18823 N*
[17] Reinemann S, Rosemann, P, Babutzka M, Lehmann J and Burkert A 2019 *Materials and Corrosion* **70**, 1776
[18] Gunn RN 1997 *Duplex stainless steels: Microstructure, properties and applications* (Cambridge: Abbington Publishing)
[19] Burkert A, Modersohn W, Müller T and Busch WB 2016 *GfKORR-Jahrestagung Duplex-Stähle - Nische oder Zukunft?* (Frankfurt: GfKORR - Gesellschaft für Korrosionsschutz e.V.) 82
[20] Eich G, Burkert A, Burkert A and Mietz J 2007 *Stahlbau* **76** 898
[21] Burkert A, Gümpel P, Lehmann J, Burkert A, Zietelmann C, Leu F and Hoffmann C 2011 *final report of the research project 16049 N*
[22] Schultze S 1999 *Selektive Korrosion von Duplexstahl* (Düren: Shaker Verlag)
[23] Vollmer T, Gümpel P, Blaise M and Racky W 1995 *Materials and Corrosion* **46** 92
[24] Wilke D 2016 *GfKORR-Jahrestagung Duplex-Stähle - Nische oder Zukunft?* (Frankfurt: GfKORR - Gesellschaft für Korrosionsschutz e.V.) 13
[25] Pohl M 2016 *GfKORR-Jahrestagung Duplex-Stähle - Nische oder Zukunft?* (Frankfurt: GfKORR - Gesellschaft für Korrosionsschutz e.V.) 106
[26] Rosemann P and Kauss N 2018 *WOMag - Kompetenz in Werkstoff und funktionelle Oberflächen* **7** 1
[27] Kauss N, Heyn A, Halle T and Rosemann P 2019 *Electrochemical Acta* **317** 17

IOP Conf. Series: Materials Science and Engineering **882** (2020) 012018 doi:10.1088/1757-899X/882/1/012018

Sensitization behaviour of the nitrogen alloyed austenitic stainless steel X8CrMnMoN18-19-2

P Rosemann, C Müller and T Halle

Otto-von-Guericke-University Magdeburg, Faculty of Mechanical Engineering, Institute of Materials and Joining Technology, Universitätsplatz 2, 39106 Magdeburg, Germany

paul.rosemann@ovgu.de

Abstract. Austenitic stainless steels have been used in many different industrial branches. Nitrogen alloyed austenite´s with up to 20 wt. % manganese and 0.8 wt. % nitrogen have been available on the market as nickel-free grade for some years. In these steels, the austenitic matrix is stabilized without the addition of nickel, while the corrosive resistance and mechanical properties are significantly improved by addition of Mn and N. As with all stainless steels, the chemical composition and heat treatment have a decisive influence on structure, mechanical properties and corrosion resistance. Solution annealing, quenching and work hardening are applied to combine excellent mechanical properties (R_m from 900 MPa to 2,000 MPa, $A_5 > 50$ %, $A_v > 350$ J) with high corrosion resistance. Solution annealing is applied to eliminate undesired precipitation phases (Cr_2N, $M_{23}C_6$ and sigma phase), distribute the alloy elements homogeneously in the austenite and assure the high corrosion resistance. If the distribution of the alloy elements (Cr, Mo and N) is impaired by suboptimal heat treatment, processing or high temperature conditions, the corrosion resistance can be reduced. Precise knowledge of the sensitization behaviour of these high nitrogen alloyed steels is necessary. Therefore, the sensitization behaviour of the nitrogen alloyed austenitic steel X8CrMnMoN18-19-2 (1.4456) was investigated by thermal ageing in the temperature rage of 500 °C to 900 °C for up to 100 h. Thermodynamic calculations were used to predict the formation of different phases. The microstructure of the various sensitization states was evaluated by scanning electron microscopy (SEM) and compared with the corrosion resistance characterized by the electrochemical potentiodynamic reactivation (EPR) test and the KorroPad method. The reduction of corrosion resistance is correlated with the occurrence of M_2N nitrides in the microstructure and two sensitization diagrams were created based on the experimental results.

1 Introduction

Besides the widely used chromium-nickel austenitic stainless steels, there are promising alloying concept austenite's stabilized with manganese and nitrogen. The substitution of Ni with Mn and N ensures the fcc (face-centered cubic) crystal structure, improves the corrosion resistance and the mechanical properties [1]. Nitrogen alloyed austenitic stainless steels with approx. 0.8 wt.% nitrogen offer an exceptional combination of corrosion resistance, strength and toughness [2-5]. In general, a homogeneous distribution of the alloying elements (especially Cr, Mo and N) in the austenitic matrix is of decisive importance for the formation of a stable passive layer and thus the corrosion resistance. The formation of precipitates (Cr_2N, $M_{23}C_6$ and sigma phase) by selective ageing at elevated temperatures can be used not only for work-hardening but also for optimizing strength, while processing errors can cause a significant reduction in corrosion resistance [1, 6-8]. When formed, chromium carbides and chromium nitrides locally reduce the chromium content of the surrounding matrix and decrease the corrosion resistance [6-10]. Despite the known theories on chromium depletion and precipitation formation of these steels, no systematic investigations have been carried out to date on different heat treatment conditions with regard to their specific corrosion resistance. Since the tendency to form chromium nitrides (Cr_2N) increases with the nitrogen content, see Figure 1, such characterization is essential for the safe use of these steels.

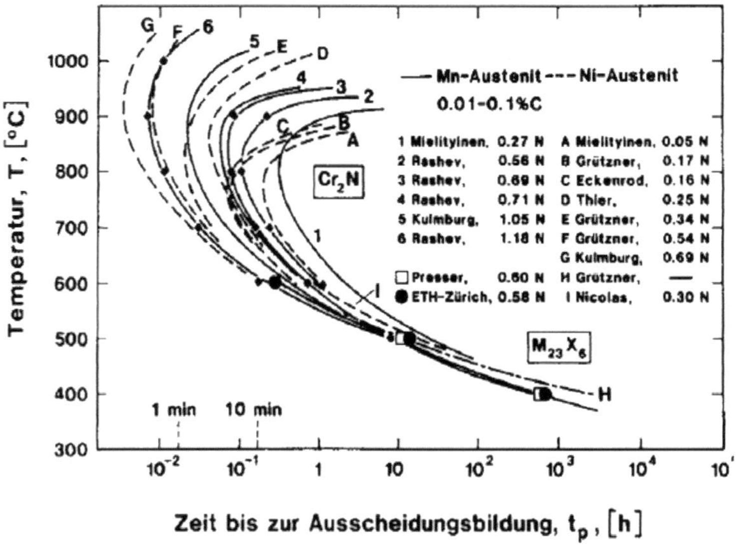

Figure 1. Time-temperature-precipitation-diagram (TTT) for various nitrogen-alloyed austenitic stainless steels [11].

The connection between sensitizing heat treatment, phase formation in the microstructure and corrosion resistance is investigated using the example of 1.4456 (X8CrMnMoN18-19-2) in order to enable future use in new technical areas and to initiate new impulses for their processing.

2 Material, heat treatment and methods

The nitrogen alloyed austenitic stainless steel 1.4456 (X8CrMnMoN18-19-2) was used for the investigations. The steel was produced by Energietechnik Essen GmbH using pressure electro-slag remelting. The chemical composition of the suited steel is presented in Table 1.

Table 1. Chemical composition of the nitrogen alloyed austenitic stainless steel 1.4456 (wt.%).

	C	N	Mn	Cr	Mo	Ni	Si	V	P	S
1.4456	**0.06**	**0.81**	**19.1**	**17.5**	**2.2**	**0.13**	**0.65**	**0.08**	**0.018**	**0.001**

The phases occurring in thermodynamic equilibrium were calculated based on the chemical composition (Table 1) using the commercial software *ThermoCalc* (database TCFe7). These calculations were used to determine the solution heat treatment temperature (1150 °C) being necessary to dissolve all soluble phases and to select the temperature range for artificial ageing. Temperatures < 800 °C are used to characterize the precipitation dynamics during the simultaneous formation of the three precipitation phases (Cr_2N, $M_{23}C_6$ and sigma phase). Artificial aging at 900 °C only causes the formation of Cr_2N, which means that the influence of $M_{23}C_6$ and sigma phase is suppressed. Table 2 summarizes the temperature range for artificial ageing of 1.4456 (X8CrMnMoN18-19-2).

Table 2. Artificial ageing conditions of 1.4456 after solution heat treatment at 1150 °C for 15 min.

900 °C	1 min	5 min	10 min	30 min	1 h	2 h	4 h	8 h
800 °C	1 min	5 min	10 min	20 min	40 min	1 h	2 h	4 h
700 °C	1 min	5 min	10 min	30 min	1 h	2 h	4 h	8 h
600 °C	5 min	10 min	30 min	1 h	2 h	4 h	8 h	16 h
500 °C	10 min	1 h	5 h	10 h	15 h	20 h	50 h	100 h

All samples were solution annealed at 1150 °C for 15 min and quenched in water prior to artificial ageing to ensure the dissolution of all precipitation phases and the homogeneous distribution of the alloying elements in the austenite. To characterise the microstructure, grinding, polishing and etching according to Beraha II (35 s) were performed and the microstructure was examined by scanning electron microscopy (SEM). The formation of precipitates as a result of the increasing artificial ageing time was thus detected microscopically and evaluated in relation to the results of the corrosion investigations.

The susceptibility to pitting corrosion of the various ageing conditions was characterized using the KorroPad method [8, 12-13]. The method was originally developed to detect an incompletely formed or instable passive layer on the austenitic stainless steels X5CrNi18-10 (1.4301). The KorroPad is a water-based agar gel, which contains 3 wt.% agar, 1 mM potassium hexacyanoferrate III ($K_3[Fe(CN)_6]$) and 0.1 M sodium chloride (NaCl). The agar forms a gel matrix, which holds and immobilizes the electrolyte, while the $K_3[Fe(CN)_6]$ causes the KorroPad to achieve a redox potential of up to +320 $mV_{Ag/AgCl}$ (measured on platinum) and is thus able to polarize the surface of a stainless steel up to 200 $mV_{Ag/AgCl}$. In addition to the anodic polarization, the NaCl dissolved in the KorroPad also ensures that pitting corrosion is initiated and eventually stabilized.

The KorroPad is used for the detection and evaluation of conditions susceptible to pitting corrosion in stainless steels [13]. Prior to the test, the specimens were wet ground with 180 grit SiC abrasive paper and passivated for 24 h at a relative humidity of more than 95 %. This enables the formation of a stable passive layer and ensures the comparability of the individual artificial ageing states. For the subsequent evaluation of the blue indication, the KorroPads were scanned, digitized and visually evaluated. Three typical appearances can be distinguished in the KorroPad test, which allow a comparative evaluation of the pitting corrosion resistance:

- no indication → stable passive layer → no pitting corrosion
- small indication → metastable pitting corrosion → low susceptibility to pitting corrosion
- large indication → stable pitting corrosion → high susceptibility to pitting corrosion

The sensitization behaviour was also investigated by electrochemical potentiodynamic reactivation (EPR test) to detect changes in the passivation and reactivation behaviour [7-8]. For this purpose, the samples were wet ground with 600 SiC abrasive paper to achieve an active state. A potentiostat was immediately anodically polarized wit 2 mV/s from -550 $mV_{Ag/AgCl}$ to +100 $mV_{Ag/AgCl}$ in 0.5 M sulfuric acid with 10 mM potassium thiocyanate as activator. The polarization direction was reversed at +100 $mV_{Ag/AgCl}$ and followed by reverse polarisation back to -550 $mV_{Ag/AgCl}$ with 2 mV/ to reactivate sensitized areas within the microstructure. With the EPR method, the passivation behaviour can be characterised by the maximum passivation current density and the passive layer stability by the maximum reactivation current density. These characteristic values provide information about the average alloy content of the matrix and the existence of chromium-depleted areas. A reactivation current density of more than 0.1 mA/cm^2 corresponds to sensitized states.

3 Results and discussion

Figure 2 shows the results of the thermodynamic equilibrium calculations with the *ThermoCalc* software. The calculations show that solution annealing at 1150 °C can dissolve all soluble phases (Cr_2N, sigma phase, $M_{23}C_6$ and M_6C) in the austenite and thus achieve a homogeneous initial state after water quenching. In the further course of the investigations, these calculations also serve as a basis for explaining changes in corrosion resistance during artificial ageing.

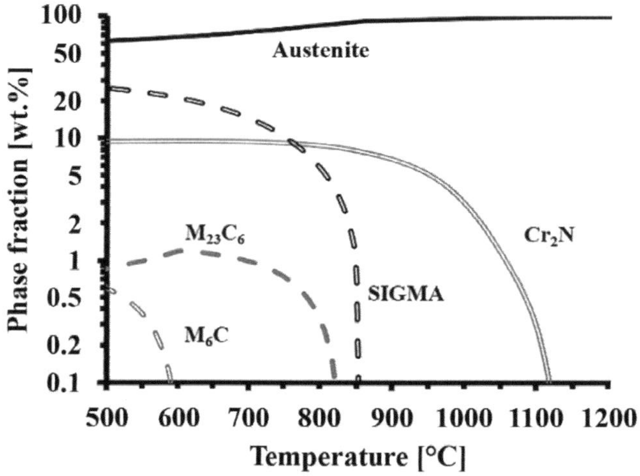

Figure 2. Results of thermodynamic calculations with *ThermoCalc* for the stainless steel 1.4456 (X8CrMnMoN18-19-2).

At artificial ageing temperatures < 800 °C all important precipitation phases (Cr_2N, $M_{23}C_6$ and sigma phase) can precipitate within the austenitic matrix. The calculations also show that at 900 °C only chromium nitrides are formed in the austenitic matrix. A comparison of the results at 700 °C and 900 °C makes it possible to evaluate the influence of Cr_2N on corrosion resistance. Figure 3 shows SEM images of three selected artificial ageing states at 700 °C. With increasing artificial ageing time, a clear occupation of the grain boundaries with small precipitates can be seen. Only after 5 min at 700 °C no precipitates are present in the austenitic matrix.

Figure 3. SEM images of the microstructure as a function of the artificial ageing time at 700 °C: a) 5 min, b) 2 h and d) 8 h.

Based on the thermodynamic calculations, the precipitates at the grain boundaries can be $M_{23}C_6$, Cr_2N or the sigma phase. The phases and phase fractions are too small for a local chemical analysis by energy dispersive X-ray radiation (EDX) or a phase analysis by X-ray diffraction. An exact identification would therefore only be possible by transmission electron microscopy (TEM). Whether the precipitates at the grain boundaries already have a detrimental effect on the pitting corrosion resistance was subsequently investigated using the KorroPad test. The corresponding appearances after the KorroPad test are shown in Figure 4.

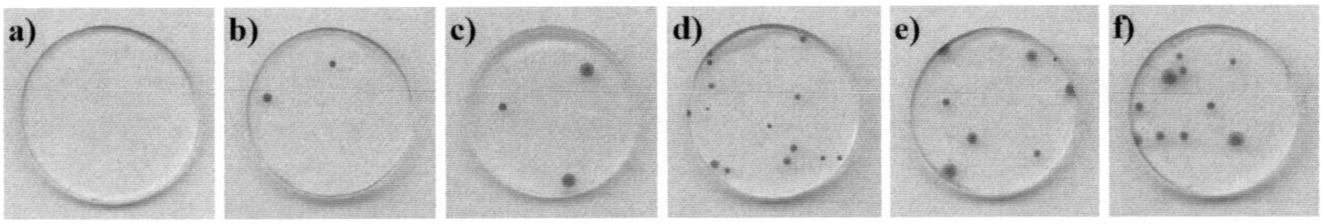

Figure 4. KorroPad test of the 1.4456 as a function of artificial ageing at 700 °C: a) 5 min, b) 10 min, c) 30 min, d) 1 h, e) 2 h, f) 4 h.

It can be clearly seen that even very short artificial ageing times of only 10 min at 700 °C cause increased susceptibility to pitting corrosion. The number of stable pitting corrosion indications tends to increase with the ageing period. The susceptibility to pitting corrosion correlates with the first occurrence of precipitation at 10 min. The further decrease in pitting corrosion resistance can be explained by the pronounced precipitation at the grain boundaries. The results of the EPR test are shown in Figure 5 using selected current density potential curves.

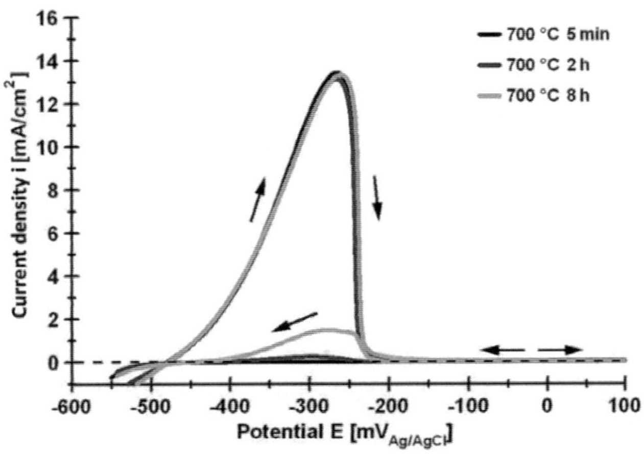

Figure 5. EPR curves of 1.4456 after solution heat treatment and artificial ageing at 700 °C.

An increasing reactivation loop is visible in the EPR curves with increasing ageing time at 700 °C. This proves that the precipitations at the grain boundaries lead to chromium depletion, which also explains the decreasing pitting corrosion resistance in the KorroPad test. The passivation current density remains constant because the average alloy content in the matrix hardly changes with increasing precipitation. The ability to form the passivation layer is therefore only affected locally at the grain boundaries. In correlation with the microstructural images in Figure 3, there is a direct correlation between the increasing phase content of the precipitates and the increasing chromium depletion. Artificial ageing of this nitrogen alloyed stainless steel destabilizes the passive layer.

Comparing the results of ageing at 700 °C and 900 °C is intended to clarify whether chromium nitrides are causing the reduction in corrosion resistance, since only chromium nitrides are formed at 900 °C, see Figure 2). The results of artificial ageing at 900 °C are presented in the following figures. Figure 6 shows the microstructure after artificial ageing at 900 °C. Cr_2N precipitation can already be observed after 30 minutes of artificial ageing at the grain boundaries. The phase content of these lamellar nitrides increases with the ageing time. In comparison to the ageing at 700 °C, large parts of the microstructure show perlite-like precipitation of Cr_2N within whole grains.

IOP Conf. Series: Materials Science and Engineering **882** (2020) 012018 doi:10.1088/1757-899X/882/1/012018

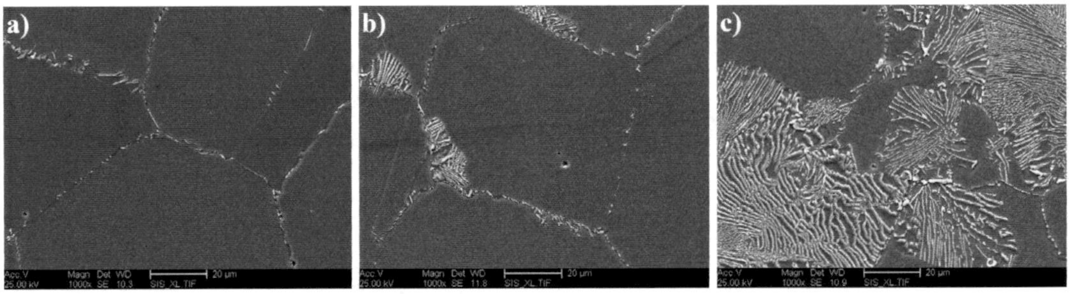

Figure 6. Microstructure of 1.4456 as a function of the artificial ageing time at 900 °C: a) 30 min, b) 1 h and c) 8 h.

The susceptibility to pitting corrosion was also investigated with the KorroPad test in dependence on the time of artificial ageing at 900 °C, see Figure 7. Despite the formation of Cr_2N precipitates at 900 °C, no reduction in pitting corrosion resistance can be detected by the KorroPad method. The strong decrease in pitting resistance at 700 °C is therefore not primarily due to the formation of Cr_2N. However, this does not mean that the chromium nitrides do not have a damaging effect on the corrosion resistance in more severe test conditions.

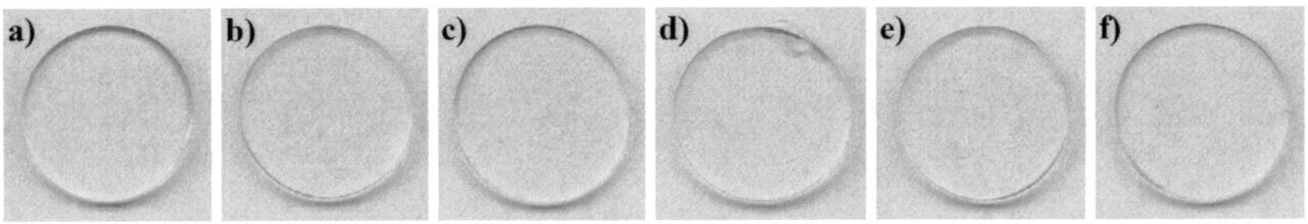

Figure 7. KorroPad test on 1.4456 as a function of artificial ageing at 900 °C: a) 5 min, b) 10 min, c) 30 min, d) 1 h, e) 2 h, f) 4 h.

The results of the EPR test regarding sensitization did not show any reactivation due to artificial ageing at 900 °C, see **Figure 8**. Despite the high content of chromium bond in the nitrides and their lamellar morphology, there is no detectable chromium depletion.

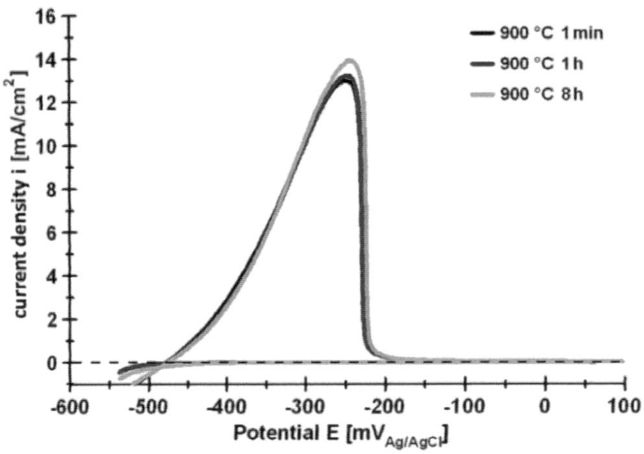

Figure 8. EPR curves of 1.4456 after solution heat treatment and artificial ageing at 900 °C.

At 8 h artificial ageing time, only a slight increase in the passivation current density is visible, which can be explained by the chromium bond in the Cr_2N phase. In the context of processing these steels and the hardening effect of the chromium nitrides, artificial ageing at 900 °C offers interesting application potential with regard to corrosion resistance.

Finally, the results of all artificial ageing conditions were used to specify the sensitization behaviour of the nitrogen alloyed steel 1.4456 on the basis of the KorroPad method and the EPR test, see Figure 9. It is noticeable that artificial ageing at 500 °C and 600 °C does not cause stable pitting corrosion. This can be attributed to too short ageing times or too low diffusion rates at these low temperatures. Artificial ageing at 700 °C and 800 °C proves detrimental to the corrosion resistance due to the fact that stable pitting corrosion sets in after only 10 min and 40 min, respectively. The EPR process also shows chromium depletion only in this temperature range. The KorroPad test is therefore much more sensitive in detecting silently corrosion-prone conditions. Artificial ageing at 900 °C, on the other hand, is not as critical as 700 °C and 800 °C since only metastable indications occur even after 8 hours of artificial ageing.

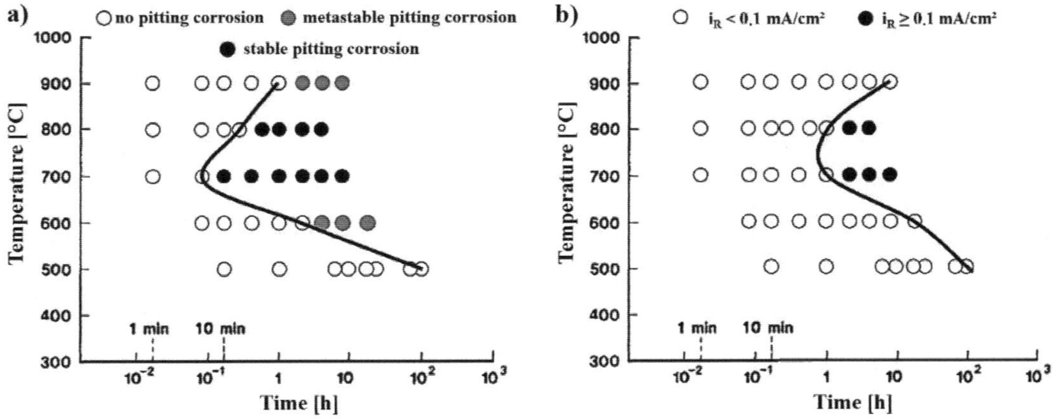

Figure 9. Sensitization diagrams of the nitrogen alloyed austenitic stainless steel 1.4456 (X8CrMnMoN18-19-2) based on the results of: **a)** the KorroPad method and **b)** the EPR test.

4 Conclusions

The sensitization behaviour of the nitrogen-alloyed austenitic, stainless steel X8CrMnMoN18-19-2 was investigated as a function of the artificial ageing treatment. The EPR (electrochemical potentiodynamic reactivation) test and the KorroPad method were used to demonstrate the effect of precipitation at different temperatures. The KorroPad method was more sensitive to the precipitation-related reduction of corrosion resistance compared to the EPR method, so that a susceptibility to corrosion was detectable at shorter artificial ageing times. Using the EPR test, significant chromium depletion ($i_R \geq 0.1$ mA/cm^2) could only be detected for five conditions in the temperature range of 700 °C and 800 °C. The general trend visible in the results of the KorroPad method was confirmed with the results of the EPR test. No sensitization was detected at 500 °C and 600 °C due to the low diffusion rate and the limited ageing times.

No increased susceptibility to corrosion could be detected at 900 °C as well. This is quite remarkable, since chromium nitrides precipitate at the grain boundaries and growth within the austenitic grains. However, this does not impair the corrosion resistance. This may lead to the assumption that the susceptibility at 700 °C and 800 °C was not caused by chromium nitrides and the formation of chromium carbides may be the decisive effect. The reason why the chromium-rich nitrides (approx. 80 wt.% chromium) do not cause chromium depletion is still to be clarified. One possibility is that the diffusion of chromium at 900 °C is faster than the consumption of chromium by the formation of chromium nitrides. Another possibility is that molybdenum is not incorporated into chromium nitrides. This means that molybdenum compensates the local depletion of chromium to such an extent that no detectable susceptibility to corrosion occurs.

Acknowledgments

The authors thank the Department 7.6 Corrosion and Corrosion Protection of the Federal Institute for Materials Research and Testing for the provision of KorroPads and the Energietechnik Essen GmbH for the provision of the studied stainless steel.

References

[1] Lo KH, Shek CH and Lai JKL 2009 *Materials Science and Engineering* R **65** 39
[2] Uggowitzer PJ, Magdowski R and Speidel MO 1996 *ISIJ International* **36** 901
[3] Uggowitzer PJ and Speidel MO 1991 *Proceedings of International Conference on Stainless Steels*
[4] Simmons JW 1996 *Materials Science and Engineering* A **207** 159
[5] Koch S, Büscher R, Tikhovski I, Brauer H, Runiewicz A, Dudzinski W and Fischer A 2002 *Material sciences and material technology* **33** 705
[6] Rechsteiner AA 1994 *Metallkundliche und metallurgische Grundlagen zur Entwicklung stickstoffreicher, zäher, hochfester austenitischer Stähle (*Zuerich: ETH Zuerich)
[7] Schilling K 2005 *Selektive Korrosion hochlegierter Stähle (*Aachen: Shaker Verlag)
[8] Rosemann P, Müller C, Kauss N and Halle T 2014, *Conference proceedings of the 17. Werkstofftechnisches Kolloquium, Chemnitz* 103
[9] Tostmann KH 2001 *Corrosion* (Weinheim: Wiley-VCH)
[10] Rosemann P, Kauss N, Müller C and Halle T 2015 *Materials and Corrosion* **66** 1068
[11] Harzenmoser MAE 1990 *Massiv aufgestickte austenitisch-rostfreie Stähle und Duplexstähle* (Zuerich: ETH Zuerich)
[12] Rosemann P, Müller T, Babutzka M and Heyn A 2013 *HTM - Journal of Heat Treatment and Materials* **68** 224
[13] Lehmann J, Burkert A and Steinhoff UM 2012 g*erman patent* DE 10 2010 037 775 A1

IOP Conf. Series: Materials Science and Engineering **882** (2020) 012019 doi:10.1088/1757-899X/882/1/012019

Einfluss des Umwandlungs- und Erstarrungsverhaltens nichtrostender Chrom-Nickel-Stähle auf die Schweißeignung beim Hochleistungsschweißen

J Schuster

Schweißtechnische Lehr- und Versuchsanstalt Halle GmbH, Köthener Str. 33a, 06118 Halle (Saale), Deutschland

Inhalt. Die Zielstellung der vorgestellten Untersuchungen bestand in der Klärung des Einflusses der gegenüber den herkömmlichen Schweißprozessen größeren Abkühlgeschwindigkeit des Laserstahlschweißens auf das Erstarrungs- und Umwandlungsverhalten hochlegierter Chrom-Nickel-(Molybdän)Stähle. Wie nachgewiesen werden konnte, treffen für Strahlschweißprozesse u. U. die Iso-Ferritlinien im WRC-92- und im SCHAEFFLER-Diagramm nicht mehr zu. Dabei zeigt das SCHAEFFLER-Diagramm Erstarrungswechsel besser an als das WRC-92-Diagramm. So verschiebt sich im WRC-92-Diagramm der Erstarrungswechsel zu höheren Werten des Ni-Äquivalents.

1 Einleitung

Die Schweißeignung hochlegierter Chrom-Nickel-(Molybdän-)Stähle wird im Wesentlichen durch die Anfälligkeit dieser Werkstoffe gegenüber der Bildung von Heißrissen der Typen Erstarrungs- und Wiederaufschmelzriss bestimmt. Untersuchungsergebnisse in der Literatur zeigen, dass in Abhängigkeit von der Erstarrungsgeschwindigkeit bei Legierungen zu beiden Seiten der Linie doppelter Sättigung im ternären System Eisen-Chrom-Nickel Veränderungen im Erstarrungs- und Umwandlungsablauf gegenüber dem Gleichgewicht auftreten können. Zu diesen Werkstoffen zählen insbesondere die klassischen metastabilen Cr-Ni-Stähle, also die am weitest verbreiteten rost- und säurebeständigen Stähle überhaupt. So sind diesbezügliche Unterschiede bei Vergleichen zwischen Blockerstarrung und der Kristallisation von Lichtbogenschweißbädern bekannt. Ein analoger Gradient hinsichtlich der Erstarrungsgeschwindigkeiten besteht ebenfalls zwischen dem Laserstrahl- und dem herkömmlichen Lichtbogenschweißen.

Die Problematik der Entstehung von Heißrissen der Typen Erstarrungs- und Wiederaufschmelzriss nimmt mit zunehmender Wandstärke bzw. Blechdicke zu. Somit wird deren Auftreten mit der zunehmenden wirtschaftlichen Verfügbarkeit von Laserschweißanlagen mit Leistungen über 10 kW zu einem Problem für die verarbeitenden Unternehmen.

Im Unterschied zur amorphen Erstarrung metallischer Schmelzen können beim Schweißen mit dem Laserstrahl jedoch noch ausreichend Diffusionsvorgänge ablaufen, so dass die hier stattfindenden Prozesse noch nicht in den Bereich der sogenannten „Kurzzeitmetallurgie" fallen.

Das Hauptziel der vorgestellten Untersuchungen bestand daher in der Klärung des Sachverhalts, welchen Einfluss die gegenüber den herkömmlichen Schweißprozessen größere Abkühlgeschwindigkeit des Laserstahlschweißens auf das Erstarrungs- und Umwandlungsverhalten nichtrostender Chrom-Nickel-(Molybdän-)Stähle ausübt.

IOP Conf. Series: Materials Science and Engineering **882** (2020) 012019 doi:10.1088/1757-899X/882/1/012019

2 Einfluss der Kristallisationsgeschwindigkeit auf den Ablauf der Primärkristallisation

Schwankungen der Erstarrungsgeschwindigkeit sind für die Kristallisation von Schweißbädern typisch [1]. Sie werden u. a. durch Leistungsschwankungen der Energiequelle, Änderungen beim Durchwirbeln des Schweißbades, Dickenschwankungen der diffusiven Erstarrungsfront, Zusammenwirken von Oberflächenspannung der Schmelze und Lichtbogendruck, aber auch durch Temperatur- und Konzentrationsunterschiede innerhalb des flüssigen Schweißbades hervorgerufen [1, 2].

Die Geschwindigkeit der Kristallisation wirkt sich auf den Erstarrungsablauf durch Beeinflussung von Liquidus- und Solidustemperatur, der Temperatur der peritektischen Reaktion im Intervall der Erstarrung, der Art der primär gebildeten Phasen einschließlich ihrer Mengenanteile und Morphologie aus und kann zu örtlichen Unterschieden in der Legierungszusammensetzung führen [1, 3, 4, 5]. Als Folge sind Wechsel in der Art der Primärerstarrung möglich.

Untersuchungen in der Literatur zeigten, dass mit steigender Abkühlgeschwindigkeit die Kristallisation von primärem Austenit begünstigt wird [2, 5, 6, 7, 8, 9]. FREDRIKSSON [10] konnte demgegenüber eine ferritstabilisierende Wirkung beobachten, die mit Zunahme der Kristallisationsgeschwindigkeit eintritt. SUUTALA [7] erklärt diesen Widerspruch damit, dass FREDRIKSSON [10] sehr geringe Wachstumsgeschwindigkeiten (0,01 bis 2 K/s) bei hohen thermischen Gradienten benutzte und somit möglicherweise Formen des nichtdendritischen Wachstums einstellte. Ein ähnliches Phänomen konnten SUUTALA und Mitarbeiter [1] auch in Schweißbädern feststellen, wo bei ebenem oder zellularem Wachstum der Erstarrungsfront in der Nähe der Schmelzlinie, also im Bereich mit großem thermischen Gradienten, die ferritische Primärkristallisation dominierte, wobei die chemische Zusammensetzung der Schmelze durchaus eine primäre Austenitausscheidung erlaubt hätte.

Entsprechend den Überlegungen von VITEK und Mitarbeitern [2] ist der Zusammenhang zwischen Abkühlgeschwindigkeit und begünstigter Primärphase abhängig von der Lage der Legierungen zur Linie doppelter Sättigung im ternären System Eisen-Chrom-Nickel, d. h. von ihrer chemischen Zusammensetzung. Mit steigender Geschwindigkeit der Erstarrung wird bei primär ferritisch kristallisierenden Chrom-Nickel-Stählen die Austenitausscheidung und bei primär austenitisch erstarrenden Eisen-Chrom-Nickel-Legierungen die Ferritausscheidung begünstigt.

Dieses Verhalten kann aus dem Verlauf der sogenannten „T_0-Linien restloser Erstarrung“ [2] abgeleitet werden. Diese Linien liegen im $(L + \gamma)$- bzw. $(L + \delta)$-Zweiphasenraum und überdecken ein breites Feld von Zusammensetzungen (vergleiche Bild 1). Durch sie werden die Temperaturen bezeichnet, unter denen entweder die einphasig austenitische oder die ferritische Erstarrung eine geringere freie Energie aufweist und somit begünstigt abläuft. Je weiter eine Legierung von der Liquidusschnittlinie entfernt ist und je tiefer damit T_0 liegt, desto stärker muss unterkühlt werden, um unter diese kritische Temperatur zu kommen, was mit steigendem Abstand von der Linie doppelter Sättigung immer schwieriger wird. Wird die Unterkühlung unter T_0 nicht erreicht, resultiert das in einem zweiphasigen Gefüge.

VITEK und Mitarbeiter [2] konnten in diesem Zusammenhang für eine primär ferritisch erstarrende Legierung vom AISI-Typ 308 (26 % Chrom, 21 % Nickel) bei Abkühlgeschwindigkeiten zwischen 101 und 103 K/s eine typische Duplexstruktur (Ferrit + Austenit) und für Abschreckgeschwindigkeiten zwischen 105 und 106 K/s eine primär austenitische Erstarrung nachweisen.

Eine weitere Wirkung steigender Abkühlgeschwindigkeiten äußert sich in der Vergrößerung des Erstarrungsintervalls durch Absinken der Solidustemperatur bei weitest gehender Konstanz der Liquidustemperatur [11, 12].

FREDRIKSSON [8] weist auf den Einfluss zunehmender Erstarrungsgeschwindigkeiten auf die peritektische Reaktion hin. Der Bereich der peritektischen Reaktion wird einerseits durch Absinken der Starttemperatur der peritektischen Reaktion und andererseits durch eine flächenmäßige Verbreiterung im Konzentrationsdreieck Eisen-Chrom-Nickel vergrößert.

Steigende Abkühlgeschwindigkeiten wirken sich ebenfalls auf die Verteilung der Legierungselemente infolge der Verkürzung der Diffusionszeit aus. Die Seigerung der Elemente Chrom und Nickel im Restferrit verringert sich mit wachsender Abkühlrate [2, 8, 13].

Entsprechend WEBER [14] lässt sich ein Zusammenhang zwischen der Abkühlgeschwindigkeit, dem δ-Ferritanteil bei Raumtemperatur und der Art der Primärkristallisation herstellen. Ein solcher zeigt sich auch bei Untersuchungen von LIPPOLD [27]. So bewirkt bei primärer Ferritkristallisation eine schnellere Abkühlgeschwindigkeit einen höheren Prozentsatz δ-Ferrit im Gefüge bei Raumtemperatur als bei normaler Abkühlung solcher Legierungen. Demgegenüber sollen bei primärer Austeniterstarrung höhere Abkühlgeschwindigkeiten einen niedrigeren δ-Ferritanteil im Gefüge zur Folge haben [14]. Diese Aussage erklärt sich dadurch, dass bei primärer Ferritausscheidung die Zeit der Umwandlung in den Austenit mit steigender Abkühlrate verkürzt wird und damit ein größerer Teil an Deltaferrit bis auf Raumtemperatur erhalten bleibt. Demgegenüber verringert sich bei austenitischer Primärerstarrung die Zeit zur Bildung des interdendritischen Ferrits, was sich wiederum in einer Absenkung des Deltaferritanteils bei Raumtemperatur mit steigender Abkühlgeschwindigkeit ausdrückt.

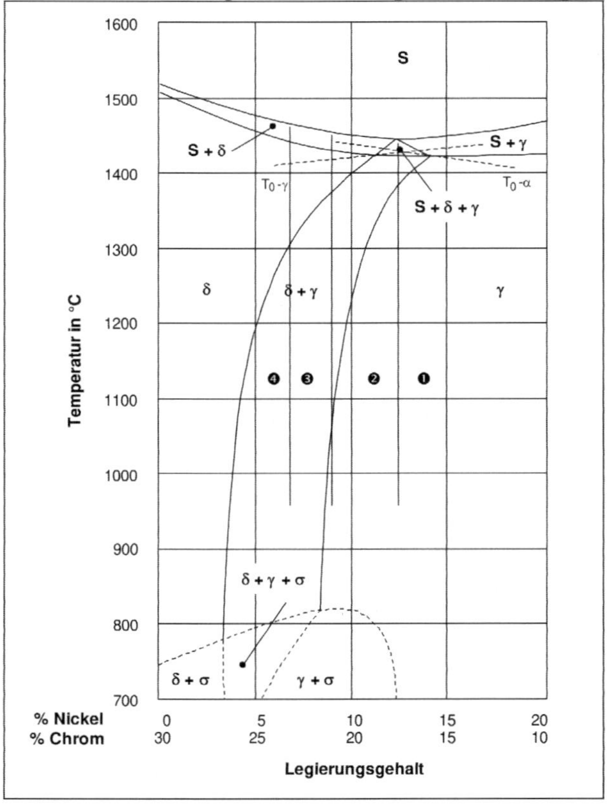

Bild 1. Konzentrationsschnitt durch das System Eisen-Chrom-Nickel bei 70 % Eisen mit Gebieten unterschiedlicher Deltaferritmorphologie nach LIPPOLD und SAVAGE [15] sowie „T_0-Linien restloser Erstarrung“.

Wird der Einfluss der Erstarrungsbedingungen mit der Wirkung der chemischen Zusammensetzung auf die Art der Primärerstarrung verglichen, lässt sich feststellen, dass dieser weitgehend von sekundärer Bedeutung ist [7, 16]. Nur in Legierungen mit simultaner Ferrit-Austenit-Kristallisation (Erstarrungstyp II/II bez. Gefügetyp BAC nach SIEGEL [3], vergleiche Tabelle 1) sollen sich geringe Änderungen der Erstarrungsbedingungen besonders deutlich in Wechseln der Primärkristallisation ausdrücken.

Die gegenwärtig einfachste Methode zur quantitativen Abschätzung der Art der Primärerstarrung ist die Funktion Φ nach HAMMAR und SVENSSON [17] (Gleichung 1). Der Wechsel im Erstarrungsmodus tritt hier entweder bei Φ-Werten um Null, Cr_0/Ni_0-Verhältnissen um 1,35 [17] oder Cr_1/Ni_1-Verhältnissen um 1,52 [18] ein (Tabelle 2).

Speziell bei solchen Fe-Cr-Ni-Werkstoffen, deren chemische Zusammensetzung im unmittelbaren Bereich der eutektischen Rinne lag, konnten SCHOBBERT und Mitarbeiter [19, 20] signifikante Einflüsse der Abkühlgeschwindigkeit auf den Ablauf der Primärerstarrung feststellen, die sich ebenfalls auf die Heißrissempfindlichkeit der untersuchten Werkstoffe auswirkten. Als Ursache dafür wird angenommen, dass sich bedingt durch die Mechanismen eines schnellen Zellwachstums eine metastabile austenitische Grenzfläche „fest-flüssig“ bereits bei höheren Temperaturen als beim Übergang zum stabilen Ferrit ausbilden kann. Die Anteile der primär austenitischen Erstarrung im Schweißgut erhöhten sich in diesem Zusammenhang mit zunehmender Erstarrungsgeschwindigkeit, d. h. allgemein auch mit zunehmender Schweißgeschwindigkeit [19]. Messungen des Deltaferritanteils in laserstrahlgeschweißten Verbindungen austenitischer Cr-Ni-Stähle vom Legierungstyp 18-10 durch KRASS und Mitarbeiter [21] sowie LIPPOLD [27] bestätigen diesen Sachverhalt.

Tabelle 1: Definition der Erstarrungs- und Gefügetypen nach SIEGEL [3].

Erstarrungstyp		Gefügetyp	Farbcodierung	Temperaturen			
				$> T_l$	T_l	T_2	T_s
III		C		L	$L_1 + \delta_1$	$L_2 + \delta_2$	δ
I		A		L	$L_1 + \gamma_1$	$L_2 + \gamma$	γ
II	II / III	BC		L	$L_1 + \delta_1$	$L_2 + \delta_2 + \gamma_1$	$\delta_3 + \gamma_2$
	II / I	AB		L	$L_1 + \gamma_1$	$L_2 + \gamma_2 + \delta_1$	$\gamma_3 + \delta_2$
	II / II	BAC; BCA		L	$L_1 + \delta_1 + \gamma_1$	$L_2 + \delta_2 + \gamma_2$	$\delta_3 + \gamma_3$

Anmerkungen: $T_l > T_l > T_2 > T_s$
T_l..... Liquidustemperatur *δ...... Deltamischkristall (δ-Ferrit)*
T_s..... Solidustemperatur *γ...... Gammamischkristall (Austenit)*
L...... Schmelze

Tabelle 2: Zusammenhang zwischen der chemischen Zusammensetzung und dem Erstarrungstyp charakterisiert durch den Φ-Wert [17, 18].

Φ-Wert	Cr_0/Ni_0	Cr_1/Ni_1	Art der Erstarrung	Erstarrungstyp	Gefügetyp
< 0	> 1,35	> 1,52	primär ferritisch	II/III; III	BC; C
= 0	= 1,35	= 1,52	δ-γ Simultanerstarrung	II/II	BAC; BCA
> 0	< 1,35	< 1,52	primär austenitisch	I	A

$$\Phi = Ni_0 - 0{,}75\ Cr_0 + 0{,}257 \quad (1)$$

$$Cr_0 = \%Cr + 1{,}37\ \%Mo \quad (2)$$

$$Ni_0 = \%Ni + 0{,}31\ \%Mn + 22\ \%C + 14{,}2\ \%N \quad (3)$$

$$Cr_1 = \%Cr + 1{,}37\ \%Mo + 1{,}5\ \%Si + 2\ \%Nb + 3\ \%Ti \quad (4)$$

$$Ni_1 = \%Ni + 0{,}31\ \%\ Mn + 22\ \%C + 14{,}2\ \%N + \%Cu \quad (5)$$

3 Experimentelle Untersuchungen

3.1 Versuchsschweißungen

Für die Untersuchung des Einflusses der Abkühlgeschwindigkeit des Laserstrahlschweißens wurden unter Anwendung eines 3,5 kW CO_2- und eines 4,5 kW Festkörperlasers (Ytterbium Faserlaser) im Vergleich zum WIG-Prozess ringförmige Umschmelzraupen auf Proben der Abmessungen 30 x 30 x 5 mm mit unterschiedlichen chemischen Zusammensetzungen (vergleiche Tabelle 2) aufgebracht. Diese Form und die Ausführung einer kreisförmigen Einschweißung wiesen im Vergleich zu einer rechteckigen Probengeometrie die nachfolgenden Vorteile auf:

- Reduzierung des benötigten Probenmaterials auf ein Minimum.
- Vermeidung des Aufmischens von Fremdmaterial bei Strichraupentechnik durch An- und Ablaufbleche.
- Unkomplizierte Präparation für die metallographischen und magnetinduktiven Untersuchungen zur Gefügebewertung und Bestimmung des Ferritanteils.

Zur Simulation einer größeren Blechdicke und der damit verbundenen größeren verfahrensabhängigen Wärmeableitungsgeschwindigkeiten, wurde die Prüfvorrichtung vollständig aus technisch reinem

IOP Conf. Series: Materials Science and Engineering **882** (2020) 012019 doi:10.1088/1757-899X/882/1/012019

Kupfer gefertigt (Bild 2 und Bild 3). Somit ergaben sich in Abhängigkeit vom jeweils untersuchten Schweißprozess und den damit verbundenen Schweißparametern die nachfolgenden mittleren $t_{12/8}$-Zeiten:

- WIG-Prozess: 750 ms
- CO_2-Laser: 350 ms
- Festkörperlaser: 250 ms

Die Messung der Abkühlzeiten erfolgten unter Verwendung einer speziellen Wärmebildkamera.

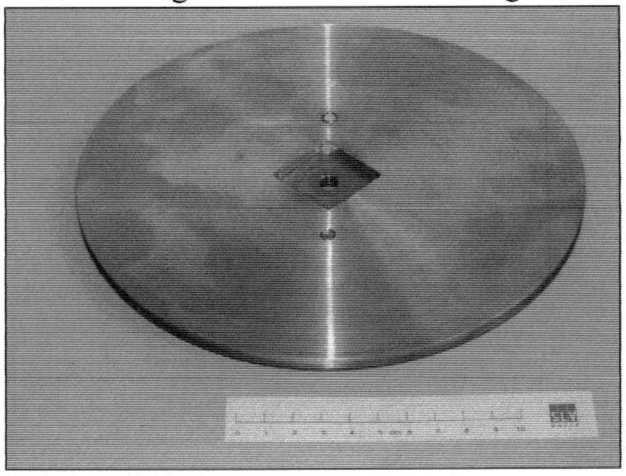

Bild 2. Prüfvorrichtung aus technisch reinem Kupfer mit Probenöffnung.

Bild 3. WIG-Schweißkopf in Position und eingelegte Versuchsprobe.

3.2 Probenmaterial

Die experimentellen Untersuchungen erfolgten an 39 kommerziellen und experimentellen Legierungen. Diese wurden so ausgewählt, dass sie hinsichtlich ihrer chemischen Zusammensetzung zu beiden Seiten der Trennlinie zwischen austenitischer und ferritischer Primärkristallisation (Linie doppelter Sättigung) lagen (vergleiche Bild 4). Einen Überblick über das untersuchte Legierungsfeld (minimale und maximale Legierungsgehalte) dieser Werkstoffe enthält Tabelle 3. Die berechneten Spannen der Legierungsäquivalente nach SCHAEFFLER [22], DeLONG [23] und WRC-92 [24] sowie des Erstarrungsparameters Φ (Gleichung 1) sind in Tabelle 4 zusammengestellt.

Tabelle 3. Übersicht über das Legierungsfeld des untersuchten Probenmaterials.

Legierungsfeld	Chemische Zusammensetzung in Masse-%														
	C	Si	Mn	P	S	Cr	Mo	Ni	Al	B	Cu	Nb	Ti	V	N
min.	0,025	0,12	0,539	0,010	0,001	15,18	0,16	8,45	0,006	0,000	0,123	0,019	0,012	0,033	0,001
max.	0,090	0,78	1,720	0,042	0,067	22,98	8,88	54,38	1,180	0,002	0,234	0,041	0,630	0,055	0,066

Tabelle 4. Feld der Legierungsäquivalente des untersuchten Probenmaterials.

Legierungsfeld g	Äquivalente							Abgeleiteter Erstarrungstyp
	Cr_{Sch}	Ni_{Sch}	Cr_{DeL}	Ni_{DeL}	Cr_{WRC}	Ni_{WRC}	Φ-Wert	
min.	15,99	10,32	15,99	11,06	15,36	11,00	-3,746	primär ferritisch
max.	32,08	54,38	32,08	54,38	31,84	54,38	28,638	primär austenitisch

4 Versuchsergebnisse

Durch alle Versuchsschweißungen wurden metallographische Querschliffe gelegt und unter Anwendung des Farbniederschlagätzverfahrens nach LICHTENEGGER und BLÖCH [25] präpariert. Dieses erlaubt eine sehr gute Bewertung der Art der Primärkristallisation und möglicher, im Verlauf des Abkühlens stattgefundener, Umwandlungen. Die dabei erzielten qualitativen Aussagen zum Erstarrungsablauf wurden zusammen mit den im Bereich der metallographisch bewerteten Umschmelzraupen magnetisch gemessenen Ferritanteilen (in FN, je fünf Parallelmessungen, Messfehler maximal 0,2 FN) in die Gefügediagramme nach SCHAEFFLER [22] und WRC [24] eingetragen. Die sich dabei ergebenden Abhängigkeiten sind für die Ergebnisse des Festkörperlasers in Bild 4 (WRC-Diagramm) und Bild 5 (Deltaferritanteil und Legierungsparameter) dargestellt. Nähert sich die chemische Zusammensetzung der Linie doppelter Sättigung können in Abhängigkeit von der Abkühlgeschwindigkeit spontane Wechsel in der Art der Primärerstarrung stattfinden.

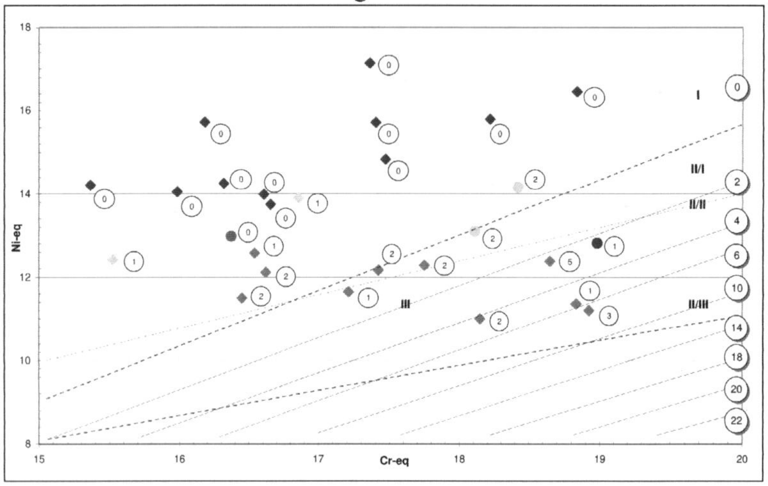

Bild 4. WRC-92-Diagramm (Ausschnitt) mit Lage der Versuchslegierungen (Festkörperlaser), die dargestellten Farben entsprechen den ermittelten Gefügetypen nach Tabelle 1.

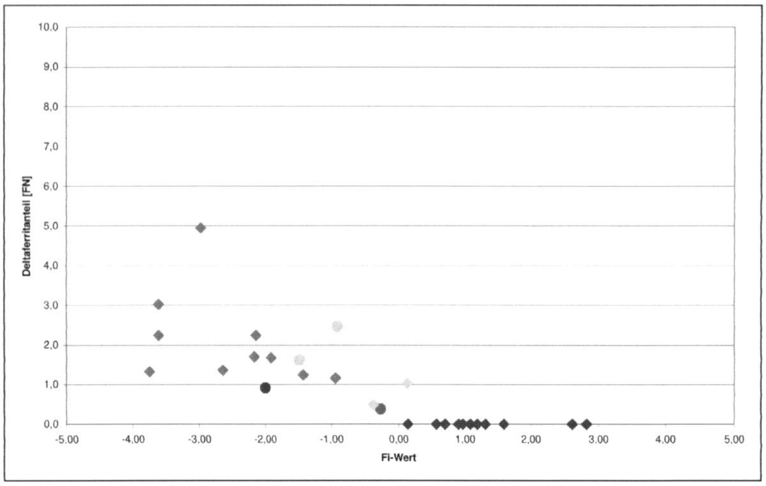

Bild 5. Deltaferritanteil und Legierungsparameter Φ (Festkörperlaser), die dargestellten Farben entsprechen den ermittelten Gefügetypen nach Tabelle 1.

5 Bewertung der Versuchsergebnisse

Wie die Untersuchungsergebnisse erkennen lassen, überschreitet die Lage der primär ferritisch erstarrenden Legierungen deutlich die im WRC-92-Diagramm eingetragene Trennlinie zwischen ferritischer und austenitischer Primärkristallisation (Linie doppelter Sättigung). Dabei war es unerheblich, welcher der drei untersuchten Schweißprozesse (und die damit verbundenen Abkühlgeschwindigkeiten) zur Anwendung kam. Diese Aussage trifft ebenfalls auf die Begrenzungslinie zum vollaustenitischen Phasenraum zu. Auch diese ist zu höheren Gehalten beim Nickeläquivalent verschoben. Es kann somit geschlussfolgert werden, dass diese Begrenzungslinien im WRC-92-Diagramm [24] nicht den tatsächlichen Umwandlungsablauf anzeigen. Eine Korrektur des Verlaufs dieser Linien erscheint somit erforderlich. So verschiebt sich die den Erstarrungswechsel charakterisierende Linie um 1 bis 2 % zu höheren Werten im Nickeläquivalent nach WRC (vergleiche Bild 6).

Legierungen, die gemäß den Angaben in diesem Diagramm eine austenitische Primärerstarrung annehmen lassen, können somit durchaus noch primär ferritisch kristallisieren, was sich signifikant verbessernd auf ihre Neigung zur Heißrissbildung auswirkt.

Demgegenüber charakterisiert die im SCHAEFFLER-Diagramm durch THIER [26] eingetragene Linie doppelter Sättigung sehr gut den Wechsel in der Art der Primärkristallisation. Auch hier wirkte sich das jeweils zur Anwendung gekommene Schweißverfahren nicht auf die Lage dieser Linie aus.

Wurden beide Diagramme ursprünglich für das Lichtbogenhandschweißen und die damit verbundenen Abkühlbedingungen aufgestellt, charakterisieren sie ebenfalls mit ausreichender Genauigkeit die beim Schweißen mit dem WIG-Prozess zu erwartenden Deltaferritanteile im Gefüge des reinen Schweißgutes.

Die mit dem CO_2- und dem Festkörperlaser ausgeführten Umschmelzraupen wiesen dagegen geringere Ferritanteile auf. Dabei zeigte sich, dass sich der Deltaferritanteil umgekehrt proportional zur Abkühlgeschwindigkeit verhält. D. h., mit zunehmender Abkühlgeschwindigkeit nimmt der Ferritanteil in der Regel ab. Somit charakterisieren die Iso-Ferritlinien in beiden Gefügeschaubildern beim Strahlschweißen nicht die tatsächlich nach dem Schweißen zu erwartenden Ferritanteile im Schweißgut (vergleiche Bild 4). So ist es mit beiden Diagrammen nicht mehr möglich, genaue Aussagen zum Deltaferritanteil zu treffen. In diesem Zusammenhang konnten ebenfalls Unterschiede im Deltaferritanteil zwischen den Schweißgütern des CO_2- und des Festkörperlasers festgestellt werden.

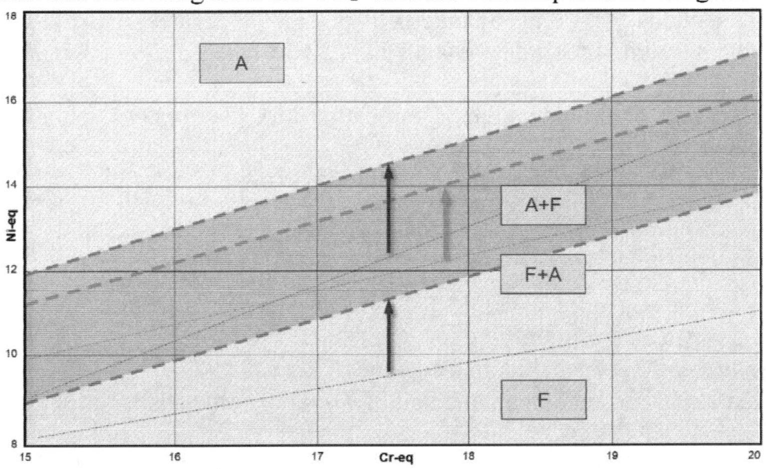

Bild 6. Verschiebung der originalen Umwandlungstrennlinien (dünn gepunktete Linien) zu höheren Ni-Äquivalenten (dicker gestrichelte Linien) im WRC-Diagramm (Ausschnitt), graphische Extrapolation.

6 Zusammenfassung

Die Untersuchungsergebnisse können wie nachfolgend beschrieben, zusammengefasst werden:

- Der Deltaferritanteil ist abhängig von der Wärmeeinbringung.

Für Strahlschweißprozesse treffen unter Umständen die Iso-Ferritlinien im WRC-92- und im SCHAEFFLER-Diagramm nicht mehr zu.

- Das SCHAEFFLER-Diagramm zeigt den Erstarrungswechsel besser an als das WRC-92-Diagramm.

Im WRC-92-Diagramm verschiebt sich der Erstarrungswechsel zu höheren Werten des Ni-Äquivalents nach WRC (um ca. 1 bis 2 %).

- Spontane Wechsel in der Art der Primärerstarrung konnten festgestellt werden.

Legierungen im unmittelbaren Bereich des Erstarrungswechsels können in der Art ihrer Primärerstarrung variieren. Damit werden Untersuchungsergebnisse von LIPPOLD [27] bestätigt. So verschiebt sich mit Zunahme der Abkühlgeschwindigkeit die primär austenitische Erstarrung in Richtung des primär ferritischen Kristallisationsgebietes.

7 Hinweis

Die in dieser Veröffentlichung vorgestellten Untersuchungsergebnisse wurden im Rahmen eines Forschungsvorhabens aus Mitteln des Bundesministeriums für Wirtschaft und Arbeit innerhalb des Förderprogramms „Förderung von Forschung, Entwicklung und Innovation in kleinen und mittleren Unternehmen und externen Industrieforschungseinrichtungen in den neuen Bundesländern" unter dem Kennzeichen 364/03 gefördert.

8 Schrifttum

[1] Suutala N, Sivonen S und Moisio T Einfluß der Erstarrungsgeschwindigkeit auf die Gefügeausbildung des Schweißguts aus austenitischem Chrom-Nickel-Molybdän-Stahl *Schweißen und Schneiden,* Düsseldorf 30 (1978) 5, S. 170/173

[2] Vitek J M, Dasgupta A and David S A Microstructural Modification of Austenitic Stainless Steel by Rapid Solidification. *Metallurgical Transactions* 14A (1983) 9, S. 1833/1841

[3] Siegel U Beitrag zur Klärung des Mikroseigerungsverhaltens von austenitischen Chrom-Nickel-Stählen Dissertation Bergakademie Freiberg 1984

[4] David S A Ferrite Morphologie and Variations in Ferrite-Content in Austenitic Stainless Steel *Welds Welding Journal* 60 (1981) 4 S. 63-s/71-s

[5] Kou S and Le Y The Effect of Quenching on the Solidification Structure and Transformation Behavior of Stainless Steel Welds Metallurgical Transactions 13A (1982) 7 S. 1141/1152

[6] Héritier J and Lévy J Fusion et solidification des alliages Fe-Cr-Ni *Mem. Sci. Rev. Métallurg.* 73 (1976) 9, S. 523/535

[7] Suutala N Effect of Solidification Conditions on the Solidification Mode in Austenitic Stainless Steels *Metallurgical Transactions* 14A (1983) 2 S. 191/197

[8] Fredriksson H The mechanism of the peritectic reaction in iron-base alloys Met. Sci. 10 (1976) 3 S. 77/86

[9] Stjerndahl J and Däcker C A Sulphide Formation in Stainless Steel *Scand. J. Metallurgy* 9 (1980) S. 217/224

[10] Fredriksson H The Solidification Sequence in an 18-8 Stainless Steel Investigated by Directional Solidification *Metallurgical Transactions* 3A (1972) 11 S. 2989/2997

[11] Anonym A Guide to the Solidification of Steels Jernkontoret Stockholm, 1977

[12] Leffler B and Malm S Volume changes accompanying solidification of some austenitic stainless steels *Met. Technolog.* 4 (1977) 2 S. 81/90

[13] Hein K und Buhrig E Kristallisation aus Schmelzen Leipzig: Deutscher Verlag für Grundstoffindustrie, 1983

[14] Weber H Einfluß der Abkühlgeschwindigkeit auf die δ-Ferritausscheidung im austenitischen Schweißgut *Z. Werkstofftechn.* 10 (1979) S. 221/229

[15] Lippold J C and Savage W F Solidification of Austenitic Stainless Steel Weldments: Part II - The Effect of Alloy Composition on the Ferrite Morphology *Welding Journal* 59 (1980) 2 S. 48-s/58-s

[16] Suutala N Solidification Studies on Austenitic Stainless Steels in: „Acta Universitatis Ouluensis" Serie C Nr. 23 University of Oulu 1982

[17] Hammar Ö and Svensson U Influence of Steel Composition on Segregation and Microstruction during Solidification of austenitic stainless Steels in: „Solidification and Cast Metals." Proc. Int. Conf. Solidification Sheffield 1977 S. 401-410 London: 1979

[18] Suutala N Effect of Manganese and Nitrogen on the Solidification Mode in Austenitic Stainless Steel Welds *Metallurgical Transactions* 13A (1982) 12 S. 2121/2130

[19] Schobbert H, Noack V, Engert J u. a. Zusammenhänge zwischen Erstarrungsmorphologien und Heißrissentstehung beim Laserstrahlschweißen von vornehmlich austenitischen Werkstoffen DVS-Berichte Band 205 S. 18-24 Düsseldorf: Deutscher Verlag für Schweißtechnik DVS Verlag GmbH 1999

[20] Schobbert H, Böllinghaus T und Schöler A Heißrisssicherheit hochlegierter austenitischer Stähle beim Laser- und Hybridschweißen VDI-Jahrestagung Schadensanalyse Würzburg 18.-19. Oktober 2000 in: VDI-Berichte Düsseldorf 1556 (2000) S. 225/240

[21] Krass M, Exner H, Bujara, S and Müller, L.Structure of austenites after laser welding in: „LANE 97, Laser Assisted Net Shape Engineering 2" Proc. of the 30th Internat. CIRP Seminar on Manufacturing Systems Erlangen 23. -24. September 1997 Band 30 (1997) S. 251/256

[22] Schaeffler A L Constitution Diagram for Stainless Steel Weldments Metal Prog. Cleveland /Ohio 56 (1949) 11 S. 680/684

[23] DeLong W T Ferrite in Austenitic Stainless Steel Weld Metal *Welding Journal* 53 (1974) 7 S. 273s-286s

[24] Kotecki D J Ferrite Measurement and Control in Duplex Stainless Steel Welds IIW-Dok. 1116-91 (ehem. Dok. II-1154-91) *Welding in the World* 30 (1992) 1/2 S. 4-13

[25] Lichtenegger P und Blöch R Beitrag zur Farbätzung hochlegierter Stähle *Praktische Metallographie*, 12 (1975) 11 S. 67-73

[26] Thier H Deltaferrit und Heißrisse beim Schweißen chemisch beständiger austenitischer Stähle DVS-Berichte Band 41 S. 100-104 Düsseldorf: Deutscher Verlag für Schweißtechnik DVS Verlag GmbH 1976

[27] Lippold J C Solidification Behavior and Cracking Susceptibility of Pulsed-Laser Welds in Austenitic Stainless Steels *Welding Journal* 73 (1994) 6 S. 129s-139s

IOP Conf. Series: Materials Science and Engineering **882** (2020) 012020 doi:10.1088/1757-899X/882/1/012020

Measuring technology and quality assurance in packaging steel production

J Schwarzmann

Development Department for Gauging Technology, IMS Messsysteme GmbH, 42579 Heiligenhaus, Germany

Abstract. Tinplate, being used primarily for packaging purposes, went through a considerable development over the last decades. Its quality was continuously improved enabling significant savings in material consumption and production economy and ecology. A premise for that progress is the development of advanced measurement and inspection equipment. It utilizes different physical principles to allow an online monitoring of important material parameters for regulation feedback or quality control. One of the latest developments is an online inspection system for non-metallic inclusions. It is based on magnetic flux leakage which is contactless detected by arrays of differential GMR sensors. The sensors are mounted to compact sensor modules containing all required peripheral equipment. Multiple sensor modules are connected to hubs which convert their data stream to the GigE camera standard. The resulting device resembles a magnetic line camera covering the whole surface of the inspected strip. The recorded images are processed by a software to detect and classify defects from the material noise.

1 Tinplate

The use of rolled iron or steel for packaging purposes ranges back to the early 19th century. Since, a typical form of packaging steel is tinplate: thin, rolled iron or steel plate with tin plating for the purpose of corrosion protection. Zinc plating has better anticorrosive properties and protects the substrate even in case of surface damage, unlike tin, by cathodic protection. However, zinc plating is not food safe and does not provide a decorative surface quality as tin plating. Hence tinplate has become an important material for food as well as high-grade decorative packaging applications.

Initially, tinplate was produced by hot rolling of bundles containing multiple plates. The tin layer was applied by putting the plates into baths containing molten tin (fire tinning). By the middle of the 20th century, the bundle rolling method for obtaining thin steel or iron plate was replaced by cold rolling, first on reversing and later on tandem rolling mills. Now cast steel slabs are hot rolled into plate with a thickness of 1,5 mm - 4 mm. After the removal of scale and rust by electrolytic pickling, the plate is cold-rolled to the final thickness, which is 100 µm - 500 µm. The rolling is followed by annealing and temper rolling to define the plates mechanical properties and surface texture. At the same time, the fire tinning process was replaced by electrolytic tin plating. This plating method produces a matte surface finish. The surface structure and thus the materials appearance can be modified by inductively heating the tin plated strip, liquefying the tin coating and thus creating a smooth, shiny surface. Depending on the application, it may be followed by a film or lacquer coating. All these production steps have gradually developed to continuous processes. The latest progress is the online laser-welding of the ends of individual strips enabling truly continuous production. 97% of the produced tinplate is processed for packaging applications. Typical products are decorative packaging, food, beverage and technical cans as well as lids and bottle caps. In that function, tinplate offers a number of advantages over other packaging materials. Steel packaging can be easily separated due to its ferromagnetic behavior and, due to its high strength, used at a very small thickness. It also can be recycled infinitely with nearly no loss of quality. The resulting recycling rate in Germany is over 90 %. The reduction of material consumption per packing is another approach to enhance the economy and ecology of steel as a packaging material. It is made possible by the usage of specialized materials with precisely maintained properties and allowed an average reduction of material usage in packaging application of approximately 20 % over the last three decades. For deep drawn beverage cans, the reduction by the course of 20 years even was 32 %: While 1985 the weight of a 0,33 l cans body was 32 g, it could be reduced to 21,5 g in 2005.

The permanent improvement of tinplate quality as well as the ability to supply a high number of material grades with different properties, among others, became possible by the development of advanced measuring and inspection equipment.

2 Measuring and inspection equipment in tinplate production

A wide range of different measuring and inspection equipment using different physical principles is required for the production of tinplate meeting today´s quality demands.

2.1 Thickness measurement

The production of packaging with the smallest possible material input requires very tight tolerances in material thickness. Hot as well as cold rolling of steel plate to precise dimensions requires a fast and accurate online measurement of the materials thickness. Because of the high material velocity during rolling as well as the high temperatures during hot rolling and the delicate surface after cold rolling, a contactless thickness measurement is necessary. A possible solution is a measurement by laser triangulation: A laser projects a dot or a line onto the surface. The scattered and reflected light is observed by a detector at a defined angle. A change in the position of the surface causes a change in the position of the reflected light on the detector. The usage of two suchlike devices, placed on both sides of the strip at the same spot, with an accurate calibration allows a precise thickness measurement. While the utilised measuring principle is simple, it leads to a number of difficulties. The accuracy of the measurement depends on the mechanical stability of the measuring gauge. As typical strip widths are over 1000 mm, the construction of a gauge of such dimensions and a long-term stability in the micrometer region is challenging. The measurement requires some amount of light scattering on the materials surface. Thus, very shiny surfaces, as often found on rolled and especially plated metal, can cause insufficient or excessive illumination of the detector. The same applies for extraneous light, radiation of the hot strip during hot rolling and obstacles in the optical path, like steam, mist and heat haze.

As with any other measurement method, the measurement of material at an angle will lead to excessive thickness values. With the measuring principle described above, any misalignment of the upper and the lower measuring points will dramatically intensify that effect.

A different, technically very sophisticated, approach, is the radiometric thickness measurement. A highly stable radiation source and an also stable and sensitive detector are placed on different sides of the strip. The radiation is partially absorbed by the strip, dependent mainly on its thickness but also on its temperature and alloy. Mechanical inaccuracy of the gauge and slight obstacles in the beam path do not have a high impact on radiometric thickness measurement.

Initially, only encapsulated radionuclides were used as a radiation source. They offer the advantage of a very constant activity with a narrow energy band which simplifies the calculation of the materials thickness from the measured intensity. Their decay energy is chosen to be convenient for the measured materials thickness and density. The drawback of radionuclides as a radiation source is their limited service life (half-life usually in the years or decades region) and their permanent radiation. To get the ability to turn off the radiation, the capsules containing the radionuclides are mounted into tungsten blocks which can be mechanically shut.

With the availability of highly stable and durable x-ray sources as well as more performant computers, x-ray tubes replaced radionuclides as a radiation source in many radiometric thickness measurement applications. While offering the advantage of being easily turned off and having adjustable operating values, they have the drawback of emitting bremsstrahlung with a complex energy spectrum instead of nearly monochromatic radiation. That complicates the calculation of the materials thickness from the measured intensity as the absorption coefficient is energy- and alloy-dependent. However, the required calculations are easily handled by modern computers. The current measuring accuracy of radiometric gauges for thin steel plate is in the micrometer range.

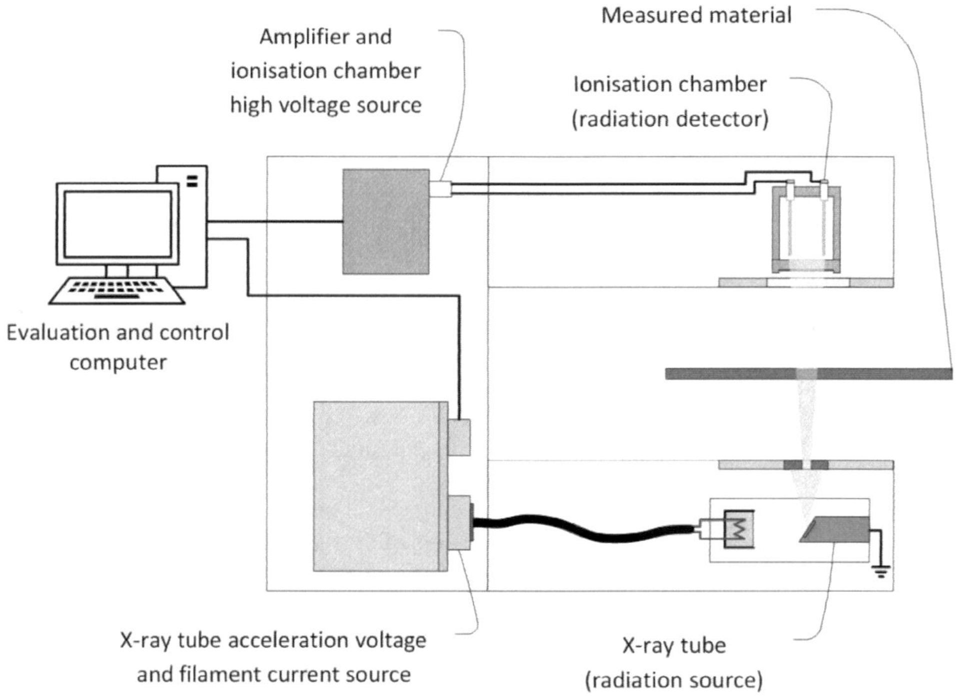

Figure 1. Sketch of an x-ray thickness gauge.

Both types of radiometric gauges require regular calibration to compensate radiation source and detector drifts and obstacles in the beam path. For that purpose, they contain a set of accurate metal plates which are, for a calibration procedure, automatically pushed into the beam path in different combinations. Radiometric gauges with one or three stationary or traversing measuring points as well as profile gauges with a high number of detectors arranged in a line are available.

Optical as well as radiometric thickness gauges require a temperature compensation, especially for hot rolling: The materials outer dimensions increase with rising temperature. On hot material optical thickness gauges will measure a higher thickness compared to cold material as it would also be the case for a simple mechanical measurement. On the contrary, the materials density decreases with temperature leading to lower thickness values measured by radiometric gauges.

2.2 Width measurement

The production or distribution of material with excessive width makes necessary edging and thus causes scrapping of unused material. Therefore, the production of strip with an accurate width leads to immediate material saving.

Width measurement is accomplished optically by measuring the shading of a light source or the projection of a laser line on the material. The first principle uses a fluorescent or LED light source providing an uniform lighting, which is placed under the measured material. For very thin material with a stable position, a single camera is placed above the strip, gauging both edges. The width of the material can be calculated from the position of the materials shadow on the camera. For thicker material, the position of the shadow also depends on the materials thickness. In this case a stereoscopic measurement, thus two shifted cameras, are necessary for an independent width measurement. Another option are traversing cameras which are automatically moved over the material edge, resulting in a true vertical measurement and thus eliminating the need for any thickness or position compensation. For complex edge geometries, which are often found on hot rolled material, or a wide range of material widths, a stereoscopic measurement of both edges (a total of 4 cameras) is applied.

The second principle uses a high number of small cameras aligned in a line over the material. Each camera only views a small section of the strip. A laser line is projected onto the material. Outside of the materials edges, there is no reflection of the line to be detected by the cameras. The length of the detected line allows a calculation of the material width. This measuring principle can only be applied on thin material.

2.3 Coating measurement

The enhancement of the quality of the tin plating and its passivation allowed a gradual reduction of the applied coating weights, which requires an examination of the latter.

Coating weights of metal plating can be measured radiometrically utilising the effect of x-ray fluorescence. A x-ray tube exposes the materials surface to radiation. Some of the photons interact with the substrate or the coating by ejecting electrons from the k-shell of their atoms. The missing k-shell electrons are replaced l-shell electrons, which release their excessive energy by the emission of a photon. The energy of an incident photon must be higher than the binding energy of an k-shell electron to cause k-shell ionization. At the same time, the closer the energy of the incident photon is to the k-shell electrons binding energy, the higher the probability for the described interaction.

The acceleration voltage of the x-ray tube is adjusted for the radiation to be well-absorbed by the iron substrate but to pass the tin plating, which has a significantly higher k-shell binding energy.

The iron substrate fluoresces at an energy of approximately 6,4 keV, which is well absorbed by the tin coating. The backscattered fluorescence radiation of the iron substrate passing the tin coating is measured by multiple, very sensitive ionization chamber detectors placed around the x-ray tube. The reduction of the detected radiation intensity compared to an iron sample without a tin layer is used to calculate the tin plating thickness based on a calibration curve.

2.4 Flatness measurement

A strip is termed unflat if different longitudinal sections have different lengths. This causes the strip to bulge at regions with excessive length, resulting in an uneven appearance when laid on a flat table. The amount of excessive length can be approximately calculated from the height of the appearing waves and their density. The origin of unflatness are inaccuracies during the rolling process. It is undesirable as it can cause problems in automatic processing of the plate or with the quality of the finished products.

For strip with low or no tension, the flatness can be measured optically: Two laser lines are projected close to each other across the strip. They are monitored by a line of cameras, calculating their height by triangulation. A periodical height difference indicates bumps caused by unflatness.

That principle is not applicable for tensioned strip, as possible bumps caused by unflatness will be smoothed by elastic elongation. In this case, the strip is guided over a roll consisting of a thin shell over a number of pressure-sensitive elements arranged axially along the roll's axis. Unflat material will cause an uneven force distribution along the rolls surface which is detected by those elements.

2.5 Pinhole, edgecrack detection and surface inspection

Imperfections in the rolling process can result in small holes or in cracks at the edge of the material. While at least resulting in defective products during final processing, they can even cause a rupture of the strip under tension in a processing line due to the caused notch effect. Surface defects like scratches, dents or staining can impair the finished products.

IOP Conf. Series: Materials Science and Engineering **882** (2020) 012020 doi:10.1088/1757-899X/882/1/012020

Holes and cracks are detected by a line of cameras picking up light passing through the holes and cracks, provided by a strong light source placed under the material. Surface inspection is also performed by fast cameras observing the materials surface, identifying irregularities by a specialized software.

2.6 Oil film measurement
Tinplate is typically coated by a very thin oil film for lubrication and corrosion protection purposes during the subsequent processing. The application of additional layers like painting without further cleaning requires the oil film not to exceed a critical thickness. At the same time, a too thin or incomplete oil film can cause the plate to stick to itself or to processing tools.

A recently developed optical measuring system enables the online measurement of thin oil films with up to 30 nm thickness by ellipsometry: The material is illuminated at the Brewster angle of the metal substrate by linear polarized light from a laser source. The oil film as well as other translucent surface layers transfer the initially linear polarization of the light to elliptical polarization with its ellipticity being dependent on the layers thickness. By empirical calibration with known surface conditions and layer thicknesses a monitoring of the applied layer thickness is possible.

3 Internal defect detection
Tinplate can contain internal defects like pores and non-metallic inclusions (NMIs) which initially remained in the slab during casting at the very beginning of the plates production process. At the same time, a progressive saving on tinplate consumption for packaging purposes requires a high purity. The production of packing often involves processes subjecting the tinplate to high deformation ratios. In that applications, even small pores or inclusions of non-metallic materials inside the strip can cause holes in the finished product or material rupture during production. As that kind of defects are mostly hidden inside of the material, they cannot be detected by optical inspection. As an online measuring system would have to inspect material at speeds of up to 1000 m/min, a radiometric inspection, as widely applied in general NDT, is not feasible.

A common offline inspection method for NMIs in ferromagnetic material is magnetic powder testing, which is based on the effect of magnetic flux leakage: The inspected material is magnetized by an external magnetic field and sprayed by ferromagnetic particles. Non-ferromagnetic NMIs cause a local change in the material's magnetic cross-section, inducing an inhomogeneity in its magnetic reluctance. This results in a variation of the magnetic field strength on the materials surface leading to an aggregation of that particles indicating a defect. The refraction of the magnetic field upon passing out of the materials surface, caused by the large permeability difference between the material and the surrounding air, enlarges the leakage field extension, which benefits the measurement [1].

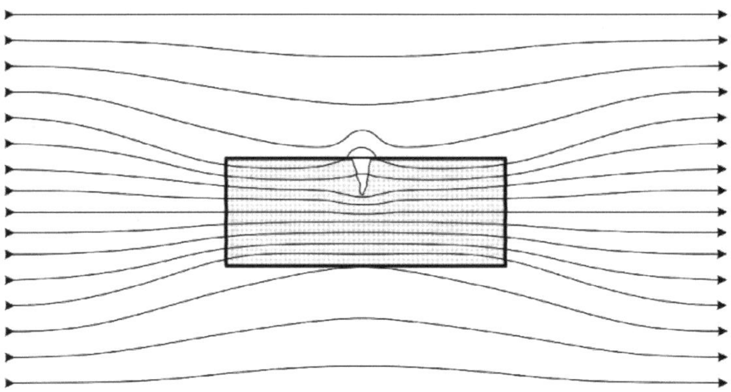

Figure 2. Sketch of the magnetic flux leakage of a surface crack in ferromagnetic material.

While magnetic powder testing is a protracted process with limited sample size when conducted automatically, magnetic flux leakage can also be used for online NMI detection. The leakage field is detected by recently developed, compact sensor modules, containing an array of GMR sensors, amplification, signal filtering and digitalization as well as an electromagnet for magnetization and power supplies for the listed components.

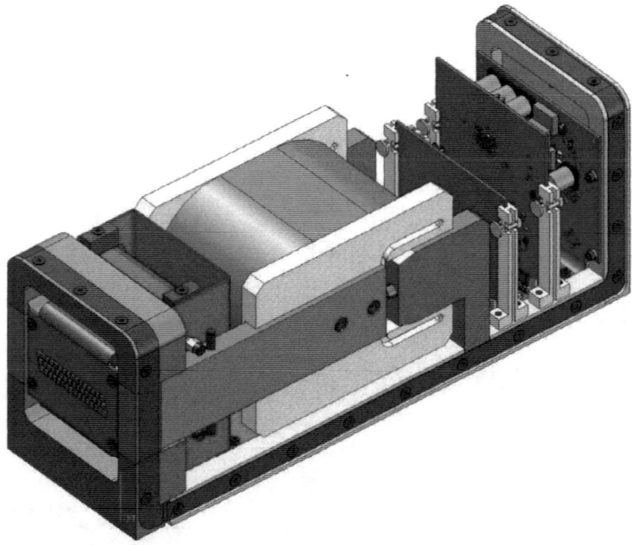

Figure 3. CAD image of a sensor module without its outer housing.

The measurements on moving material required for the design of the sensor module were conducted on a test stand. It uses non-magnetic discs on which samples can be glued, containing natural or artificial defects. That defects can be reproducibly passed under a sensor, testing different sensor types and orientations, liftoffs and magnetization configurations. While the steel plate used for samples may have structural anisotropies caused by the rolling process, there was no observable effect on the material's magnetic behavior by the change of the material's moving direction to its original rolling direction.

The used sensors are gradiometers with a sensitive area distance of 1 mm. They are arranged in three rows, resulting in a spatial resolution of 1 mm. A higher resolution would not be useful as the required smaller sensors are less sensitive due to their smaller active areas. At the same time the smallest defects detectable by amplitude induce a leakage flux with dimensions of around 1 mm.

Different magnetization directions were tested. A magnetization parallel to the material moving direction has several advantages for the construction of a measuring system. That configuration also gains a higher SNR for compact defects like round holes. A magnetization across the materials moving direction results in a lower SNR for compact and a higher SNR for elongated defects [2]. A 45 ° angle between the magnetization and the materials moving direction was found to reduce the amplitude of compact defect signals while providing no increase to the signals of elongated defects. The sensor modules were designed for a crosswise magnetization despite the constructional drawbacks, as most defects occurring in the measured material have elongated shapes due to rolling [3]. The electromagnet's current can be regulated by a buck converter to adapt its magnetomotive force to different measured materials.

The leakage field of small defects, although being extended by refraction, still has only small dimensions. It is thus desirable, for a high sensitivity, to reduce the distance between the active sensor areas and the material. As the material is moving at high velocities, very small distances result in technical difficulties and reduce the reliability of the measuring system. A 0,5 mm liftoff between the sensor housing and the material was found to be a good compromise between sensitivity and the risk of material contact, which would result in sensor damage. Liftoffs as low as 0,25 mm are found on resembling measuring systems. Laboratory tests have shown that an increase in liftoff from 0,25 mm to 0,5 mm results in a decrease in SNR of approximately 1,2 dB while significantly increasing reliability.

With crosswise magnetization on moving material, the width of the magnet is limited: The exceeding of a critical value leads to additional signal noise and a reduction of defect signal amplitudes [4]. For material thicknesses up to 500 μm and speeds up to 1000 m/min, a useful magnet width was found to be 69 mm, allocating a 48 mm wide space in its middle with the magnetic field homogenous enough for measurement. As the defined spatial resolution is 1 mm, each sensor module contains 48 GMR sensors. The limited magnet width leads to pole pieces interrupting the sensor line. Thus, the complete surface of the inspected material cannot be covered by one line of sensor modules. As the sensor line width of one sensor module is limited to 48 mm, the modules were designed not to exceed a total width of 95 mm. This allows the whole material surface to be covered by two lines of sensor modules including a tolerance margin of 1 mm per module.

Digitalization is performed at a variable rate of up to 200 kSa/s. With a lengthwise resolution of 0,1 mm per sample, the sampling rate is sufficient for material speeds of up to 1200 m/min. Figure 3 shows a defect recorded by an online test setup containing two sensor modules operating with a reduced sampling rate. Recorded defects are categorized and archived by a long-term storage system.

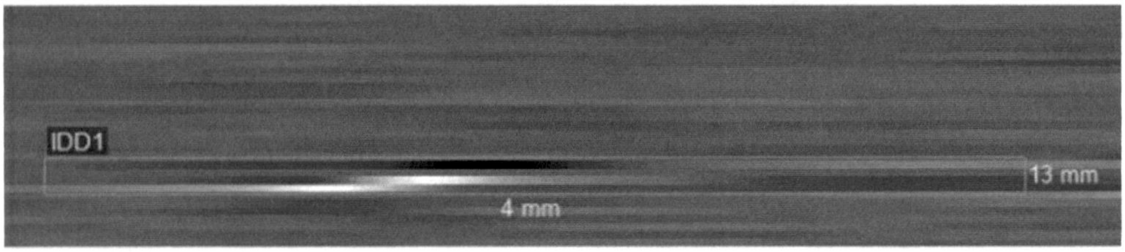

Figure 4. Image of defect recorded by an online test setup.

For the prevention of contamination and damage, the sensor arrays are molded into aluminum frames using an epoxy casting compound. Their front is machined to achieve a smooth surface. The machining also reduces the front housing of the individual GMR sensors by 50 μm for an additional reduction of the distance between the measured material and the active sensor areas. That sensor blocks are additionally machined to a precise thickness to be easily interchangeable in case of damage.

IOP Conf. Series: Materials Science and Engineering **882** (2020) 012020 doi:10.1088/1757-899X/882/1/012020

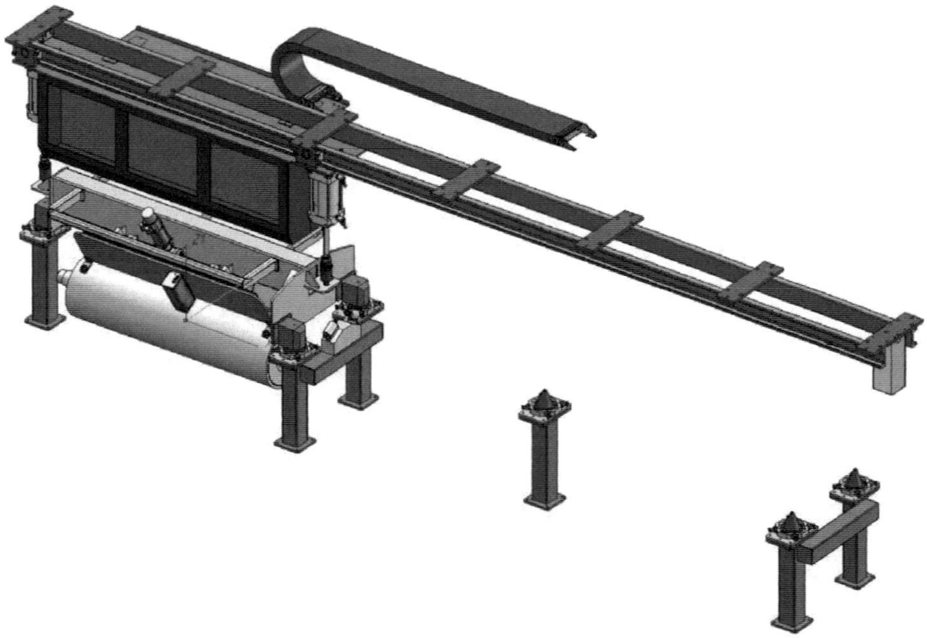

Figure 5. CAD image of an online test system containing 48 sensor modules.

Recently (2020), the first full-size test system was installed in a production line. It contains 28 sensor modules, thus covering a maximum strip width of 1344 mm. The modules are arranged in two lines at an angle of 60 ° to measure on the surface of a single roll. The roll is made of non-magnetic austenitic steel to offer a stability high enough for a precise guiding of the material over its whole width, while not disturbing the measurement. The individual sensor module lines are mounted to water-tempered plates for cooling and to prevent thermal warping. The plates are driven by servo drives and ball screws for an accurate adaptation to different strip thicknesses. Capacitive distance sensors monitor the position of the plates regarding the roll and material surface and trigger a retraction of the sensor module lines if the distance falls below a critical value. The whole system can be pneumatically lifted for the protection of the sensor modules in case of disturbances in the production flow and for maintenance purposes. Additionally, the system can be driven out of line to a maintenance position. It contains a calibration setup for the distance as well as GMR sensors. The distance sensors are calibrated by being pushed against pockets with a defined depth. At the same time all GMR sensors are pushed against circuit boards containing a meander shaped trace. That trace passes every single sensor and is energized by a calibration signal.

References

[1] Heptner H, Stroppe H, Magnetische und magnetinduktive Werkstoffprüfung, Leipzig
[2] Deutsch V, Morgner W, Magnetpulver-Rißprüfung, Düsseldorf
[3] Angani C, Kim J, Le M, Magnetic leakage testing using linearly integrated hall and GMR sensor arrays to inspect inclusions in cold-rolled strips, Chosun University
[4] Nestleroth J B, Davis R J, The effects of magnetizer velocity on magnetic flux leakage signals, Boston

IOP Conf. Series: Materials Science and Engineering **882** (2020) 012021 doi:10.1088/1757-899X/882/1/012021

Ceramic oxidation protection coatings for refractory alloys from filler-loaded preceramic polymers: The role of particle size and volume fraction of particulate fillers

I Smokovych, H Eley, M Bosiaha and M Scheffler

Otto-von-Guericke University Magdeburg, Faculty of Mechanical Engineering, Institute for Materials and Joining Technology, Universitätsplatz 2, 39106 Magdeburg, Germany

iryna.smokovych@ovgu.de

Abstract. Oxidation protection coatings on the base of perhydridopolysilazane (PHPS) polymer and Si/B filler materials were developed to prevent Mo-basis alloys against oxidation at 800 °C. Within this work the influence of the particle size and the volume fraction of particulate fillers on the oxidation protective effect of the coatings was investigated. Double and triple coatings were applied in order to determine the effect of layer thickness on the oxidation resistance properties.

1 Introduction

In stationary gas turbines and aircraft engines, nickel-based superalloys are used as turbine blades at high temperatures. However, these superalloys (in monocrystalline and coated forms) have reached the point of maximum application temperatures at 1150 °C, which results in a homologues temperature of around 0.8-0.9·Tm [1]. Potential materials for the high-pressure turbine application are molybdenum alloys, which due to their high melting temperature of over 2000 °C and high temperature mechanical properties represent an excellent alternative to substitute nickel-based alloys [2, 3]. Nevertheless, the problem of massive oxidation arises for fine-grained molybdenum alloys in the temperature range from 750 °C to 900 °C [4]. For this reason, it is necessary to apply a protective barrier coating against oxidation on the surface of most Mo-based materials.

Polymer-derived ceramics provide a cost-effective and perspective method of coating applications. For an effective Mo-alloy protection a coating system based on pre-ceramic polymer filled with ceramic and/or metal particles is considered. A polymer precursor forms a thermoset during a crosslinking process, which in a subsequent temperature treatment step – the pyrolysis – is converted into an amorphous ceramic caused by a polymer-to-ceramic transformation process. [5]. The fillers serve three main functions: they increase a coating thickness compared to unfilled coating systems; they reduce the shrinkage of the coating during polymer-to-ceramic transformation and they form new phases through reactions with the pre-ceramic polymer and the surrounding atmosphere as working medium that intend to compensate a possible volume change due to oxidative processes on the coated surfaces [6].

In this work coatings with a perhydridopolysilazane (PHPS) polymer based on Si and N in their molecular arrangement were produced as protective systems for Mo-based alloys. Powders of Si and B were used as filler materials, which, due to their strong chemical affinity to oxygen, may form SiO_2-B_2O_3 glass phases with a self-healing effect and have an average or a large increase in volume [7, 8].

Determining corresponding oxidation kinetics, the specific surface area of the particles plays a decisive role that can be controlled by the particle size and the volume fraction of fillers. Thus, in the present work coatings with fillers of various sizes were developed, and the effect of the particle size and their volume fraction on the oxidation kinetics of coating systems was investigated.

2 Experimental procedure

For the coating system a commercially available perhydridopolysilazane PHPS NN 120-20 (Clariant Advanced Materials GmbH, Sulzbach, Germany) was used. Boron amorphous powder with a particle size of 1-5 µm (Chempur, Feinchemikalien und Forschungsbedarf GmbH, Karlsruhe, Germany) and silicon crystalline powders of 1-5 µm and 5-17 µm in particle size (Thermo Fisher GmbH, Kandel, Germany) were received from commercial supplier in high purity quality. A silicon powder with the specific particle size of 4-12 µm was received by a powder metallurgical processing route: -100 mesh silicon was planetary ball milled for 0.5 h at 200 rpm. The particle size distribution of the modified and unmodified particles was determined using a laser diffraction analyser (Mastersizer 2000, Malvern Instruments), which is suitable for the non-destructive analysis of wet and dry dispersed particles with a size ranging from 0.02-2000 µm. The particle size distribution of filler materials is presented in figure 1. The main important properties of the filler materials and their schematic sample codes are summarized in table 1.

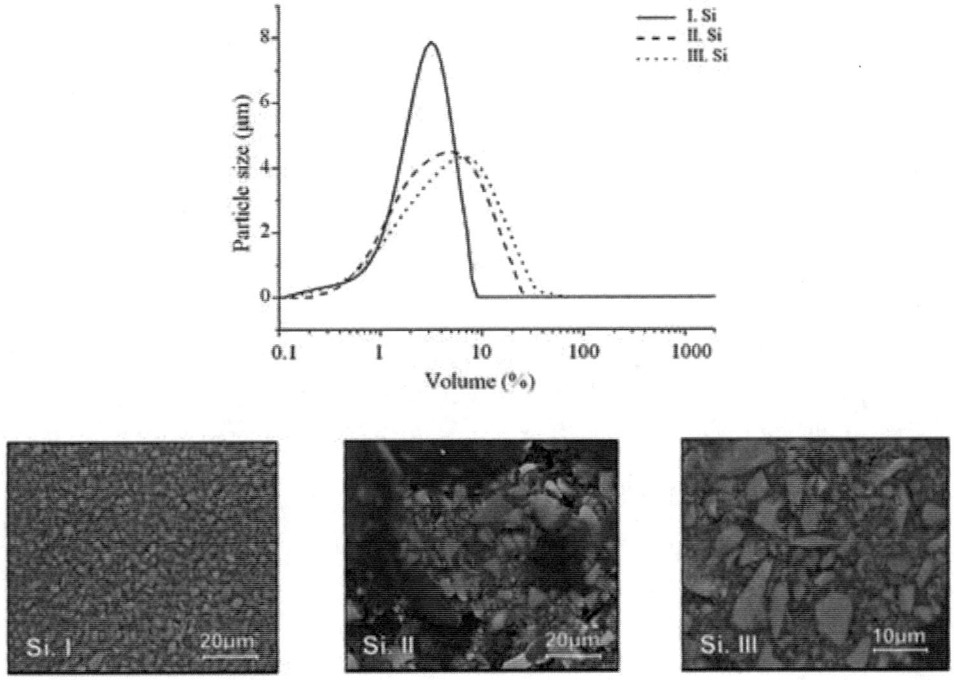

Figure 1. Silicon particle size distribution.

Preceramic slurries for the coating systems were mixed in a specific precursor-to-filler volume ratio: 72 PHPS + 17.5 Si + 10.5 B and 65 PHPS + 22 Si + 13 B, all numbers in vol. %. Powders were introduced into the precursor solution (20 wt. % of PHPS in dibutyl ether) by vigorous stirring under air. The coating systems were developed on the base of our previous work and the preparation rout is described in detail in [7, 8]. Also, the process of the PHPS polymer/(Si+B) filled precursor-to-ceramic conversion, the coatings′ phase composition and the microstructure are reviewed in the references above. The coating consists of Si and B particles surrounded by the polymer derived (PDC) matrix phase, SiON, along the filler grains and in the vicinity of a SiO_2 glassy phase [7]. The aim of this work is to investigate an influence of the particle size and volume fraction of filler materials on the oxidation kinetics of the coatings.

Table 1. Filler properties and samples codes.

Filler	Producer	Purity, %	Particle size, (µm)	Density, (g/cm³)
I. Si	Silicon Powder, crystalline, APS 1-5 µm, Thermo Fisher GmbH	99.9	2.9-5.6	2.3
II. Si	Silicon Powder, crystalline, -100 mesh, Thermo Fisher GmbH	99.9	4.0-12.4	
III. Si	Silicon Powder, crystalline, -325 mesh, Thermo Fisher GmbH	99.5	5.0-16.3	
B	Chempur GmbH, amorphous	98.0	2.3-8.2	2.1

For applying the coatings a dip coating process was carried out with a withdrawal speed of 3 mm/s and a sample holding time in the coating system of 10 s. The coatings were applied on the previously prepared samples of a plain Mo alloy (E. WAGENER GmbH, Heimsheim, Germany, 99.95 % purity) and arc melted Mo-14Hf-23B and Mo-14.8Zr-26B (in at. %) alloys. Processing of Mo-14Hf-23B and Mo-14.8Zr-26B is carefully described elsewhere [8]. It was found that the coating process must be carried out on the surface of Mo-basis alloys repeatedly, since all Mo is vaporized through the areas where the substrate was fixed on the holder for dip coating [8]. Thus, in this work double and triple coating was applied to Mo-basis alloys. The coated samples were heated to 110 °C with a slope of 2 K/min in air for 1 – 1.5 h to allow crosslinking. Pyrolysis was conducted in nitrogen atmosphere in a tube furnace (HTRRH 70-600/18, Fa. Gero Hochtemperaturöfen GmbH & Co. KG, Neuhausen) for 1 h at 1000 °C; heating and cooling rates were 3 K/min.

To produce thick coating layers with the dip coating method an important issue for filler loaded precursor systems is a uniform dispersion of solid phases in the fluid medium, aimed to avoid sedimentation [9]. Another issue is, that the filler loaded systems should have a low viscosity under high withdrawal speed at the beginning of the dip coating process, and a high viscosity after the discontinuation of the withdrawal stresses. In this study a viscosity of the coating slurries was measured using a rotational viscometer MCR 301 (Anton Paar Germany GmbH, Ostfildern-Scharnhausen/Germany) in a plate-to-plate configuration with a gap size of 1 mm. The measurement was carried out at room temperature a variation of the shear rate from 10^{-1} s^{-1} to 10^{-3} s^{-1}. In accordance to the plate-to-plate measuring/sensor system the share rate values correspond to the withdrawal speed during dip coating [10]. The correlation between the shear rate and the withdrawal speed is represented by equation (1):

$$\dot{\gamma}_a = \frac{2}{3} mm^{-1} \cdot U, \qquad (1)$$

where $\dot{\gamma}_a$ is the average shear rate und U is the withdrawal speed.

In accordance with the Landau-Levich theory the thickness of the coating, h_0, could be estimated by equation (2) [11]:

$$h_0 = c_1 \sqrt{\frac{\eta\, U}{\rho\, g}}, \qquad (2)$$

where c_1 is a proportionality constant, η is the viscosity of the suspension, ρ is the density of the suspension and g is the gravitational acceleration.

The coating thickness is proportional to the viscosity of the coating system and the withdrawal speed and inversely proportional to the density. Thus, changing the withdrawal speed and the viscosity of the coating slurry the thickness of the coating can be adjusted. These relationships relate to an idealized continuous process, in which a sample is hoisting up at a constant speed. It means that in the case of discontinuous operation, the coefficient in the formula must be adapted to the respective system.

Moreover, the Landau-Levich theory requires a purely Newtonian behaviour, i.e., no shear thinning or thickening of the slurry. The particle-filled coating system cannot be directly regarded to this. Nevertheless, M. Günthner et al. [12] use this theory to explain the thickness of precursor/BN composite coatings. In our work, we have considered the possibility of Landau-Levich theory application for previous calculations of the coating thicknesses regarding developed coating systems. The value of the layer thickness relates to the layer after being withdrawn immediately from the slurry. The reason for this is that the suspension continuously flows from the substrate until it is completely dry, and the resulting coating shrinks during pyrolysis due to the polymer component.

The microstructure characterization and phase analysis of compact samples with the coating composition, prepared for comparison, were carried out with an X-ray diffractometer (X'PERT PRO PANalytical, Malvern Panalytical GmbH) using Co-K-α radiation and scanning electron microscopy (HR-FESEM, Zeiss Merlin equipped with an EDS detector, Carl Zeiss Microscopy GmbH, Jena, Germany; and FEI DualBeam Scios, FEI Company, Hillsboro, USA).

Cyclic oxidation tests of coated samples were performed in a convection furnace (Model L9/S17, Nabertherm, Lilienthal) at 800 °C in air. Interval withdrawal of samples for characterization was carried out in 20 h steps up to 100 h. The specific mass change of the samples was measured using an analytical balance with an accuracy of 0.0001 g.

3 Results and discussion

3.1 Viscosity measurements and determination of the coating thickness.

Important for the coating process is the rheological behavior of the pre-ceramic slurries. In the first stage of our investigation viscosity measurements of the 72 PHPS + 17.5 Si + 10.5 B, vol. %, coating system with different particle size of silicon were carried out, figure 2. The measurements were conducted in air soon after preparing the coating slurries, because they tend to crosslink even at room temperature and their viscosity may increase.

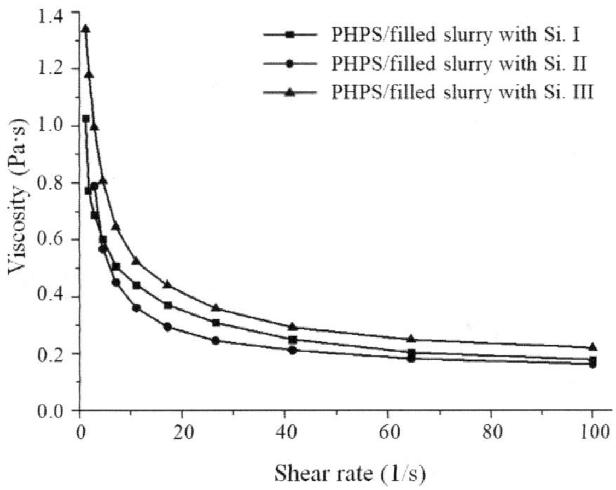

Figure 2. Viscosity versus shear rate of the filler loaded pre-ceramic coating slurry with Si fillers of different particle sizes.

All coating slurries independent from their silicon particle size demonstrate a sheer thinning behaviour, owning a low viscosity under high shear rates, and a high viscosity after the discontinuation of the shear stress. Such kind of behaviour provides an opportunity of dip coating application on all type of substrate materials: a low viscosity of the slurry while dipping the sample, and a high viscosity after withdrawing the sample for coating deposition.

Retaining a constant volume fraction of particles in the slurry while reducing the silicon particle size leads to an increase of the particle number in the coating system. A higher number of smaller particles results in more particle-particle interaction processes and an increased resistance to flow. Therefore, the coating slurry with Si I (1-5 μm) particles demonstrates higher viscosity compared to slurry with Si II (4-12 μm) at low shear rates. An exception is the coating slurry based on Si III (5-16 μm) particles, which has a higher viscosity than Si I and Si II. This can be explained by a broader particle size distribution of the silicon powder and a partial obstruction to shear by some larger particles. This may be evidenced by the higher viscosity of this coating slurry compared to the other two at both, low and high shear rates.

The particle-particle interactions are relatively weak and broke down with an increase of the share rate. Thus, the particles size effect became less visible at high shear rates for all coating slurries.

Using respective viscosity values at definite shear rates, the expectable coating thickness was calculated according to equation (2), considering a correlation between the shear rate and the withdrawal speed represented in equation (1). The calculated coating thickness for the three coating systems at a given withdrawal speed of 3 mm/s is summarized in table 2.

Table 2. Calculated values of coatings thickness for different Si particle sizes.

Coating thickness in μm		
System: 72PHPS+17.5Si+10.5B, vol. %		
I. Si	**II. Si**	**III. Si**
32.4 μm	29.5 μm	35.8 μm

In accordance with calculations, the theoretical coating thickness should be in the range from 30-35 μm. As it was mentioned above, these values respond to the coating layer direct after withdrawing the sample from the slurry and excluding the shrinkage of the coating caused by crosslinking and pyrolysis. Thus, the resulting coating thickness is expected to be lower.

The coating deposition on the surface of plain Mo and Mo-14Hf-23B and Mo-14.8Zr-26B alloys was carried out according to the distinguished procedure described in section 2. Before deposition, the alloy samples were ground with SiC paper FEPA P1200 (17.3-19.3 μm) and cleaned in an acetone (80 vol. %) / ethanol (20 vol. %) mixture by ultrasonic treatment, and subsequently dried. The double and triple coating layers were achieved by one and two times overcasting the first layer followed by crosslinking and pyrolysis after each step. The microstructures of coated Mo-alloy samples are presented in the figure 3.

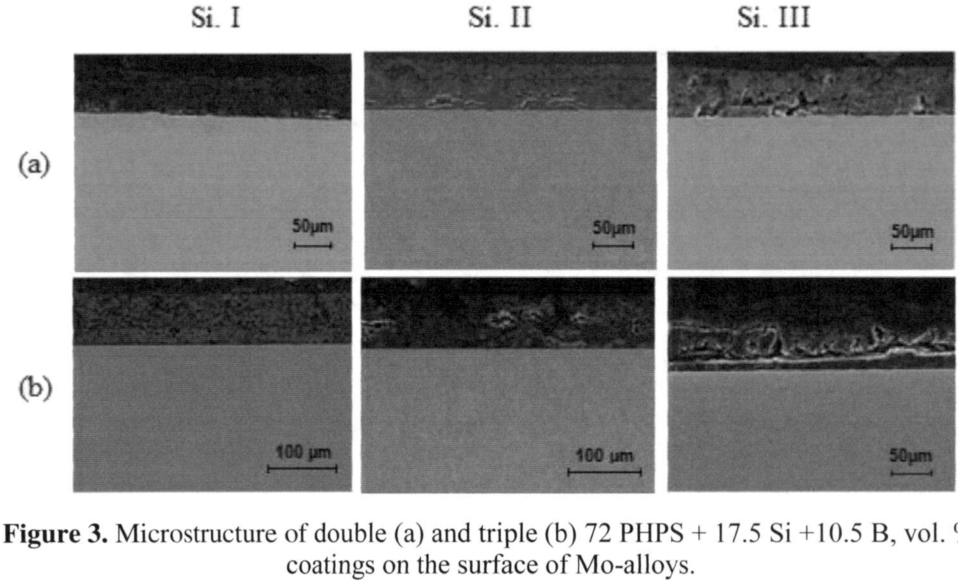

Figure 3. Microstructure of double (a) and triple (b) 72 PHPS + 17.5 Si +10.5 B, vol. %, coatings on the surface of Mo-alloys.

The resulting coating composition and the phase analysis are detail described elsewhere [7, 8]. According to macro structural studies, the coating layers were homogeneous across the entire surface of the samples and did not have any phase segregation. However, some pores in the inner layers of the coatings were obtained using Si II and Si III fillers as imaged by SEM. This may be caused by in the broader particle size distribution of the silicon powder in these types of fillers that might interferes the package density of the particles. Nevertheless, no through pores or pores entering the outer protective layer were formed. In turn, the coatings with Si I filler was homogeneous across the sample's cross section and any pores were detected between coating layers. It must be mentioned, that the overcasting of the first coating layer leads to an increase in thickness of the second and the third layer compared to the first layer. This may be caused by a purity change of the sample's surface and an increase of the surface roughness.

The thickness of the coatings ranged from 50 μm to 54 μm for double layers and from 70 μm to 79 μm for triple layers. In conformity with theoretically calculated coating thicknesses with the Landau-Levich theory, these values are somewhat lower in compared to 30-35 μm for a single layer; this was expected due to shrinkage processes. Considering the lack of prediction of coating shrinkage during pyrolysis, the Landau-Levich theory may be used to estimate the coatings thickness.

3.2 Oxidation resistance of the PDC coatings.

Aim of this work is the oxidation resistance study of coatings with respect to the role of particle size and volume fraction of particulate. As the first step the oxidation protection properties of the double and triple coatings were examined on Mo-alloys. The oxidation investigations were conducted at 800 °C, the harshest temperature for Mo-based alloys because of the pesting phenomenon, which means the formation of a volatile MoO_3 phase [13]. The oxidation curves of double and triple coated Mo-samples are presented in figure 4 (a).

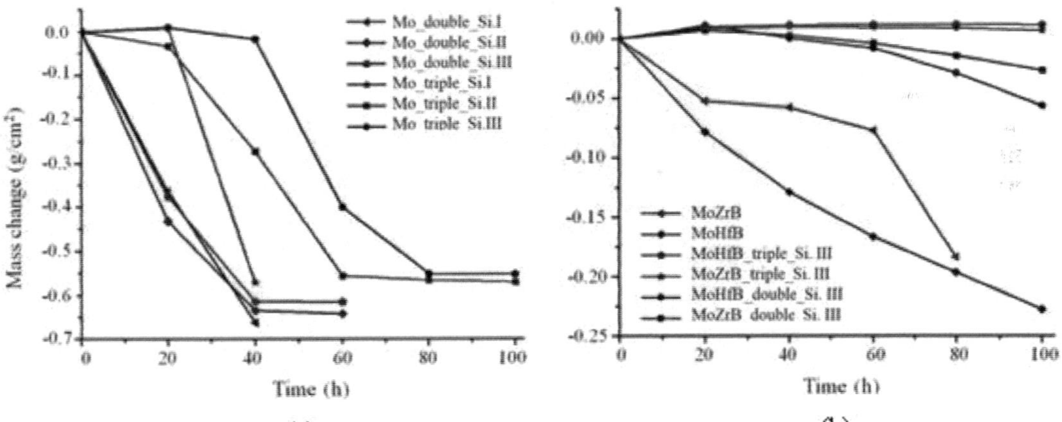

Figure 4. Oxidation curves at 800 °C in air (up to 100 h) of double and triple coated Mo- (a) and Mo-14Hf-23B and Mo-14.8Zr-26B (b) samples with coatings of the composition 72 PHPS + 17.5 Si + 10.5 B; numbers in vol. %.

Double coated samples show a significant mass loss already after 20 h of high temperature (800 °C) exposure to air with the plain Mo samples. Up to 40 h of oxidation the samples lost their mass in a catastrophic course. The destruction of the coating layers is observed starting at 20 h and the further mass loss is caused by evaporation of MoO_3 through damaged coating areas.

The triple coatings demonstrate a more stable behavior up to 20 h, but it cannot be considered as a protective coating. For the coatings with Si I and Si III for the first 20 h a slight increase in mass was observed. This might be assumed as a formation of the SiO_2-B_2O_3 passivation layer as a result of B and

Si oxidation with oxygen from air. However, after 20 h of oxidation the mass loss indicates the beginning of MoO_3 evaporation. The highest mass loss is identified for the samples coated with Si I in comparison to samples with Si II and Si III. It is initially surprising, because smaller particulate fillers should provide a higher specific surface area and as a result a higher reactivity of Si with oxygen to form a protective glassy layer. Evidently, the viscosity of SiO_2-B_2O_3 coating layer at the initial stage of oxidation is still too high to prevent from oxygen penetration into the metal substrate [14]. The formation and evaporation of MoO_3 destroys the glassy layer and provokes a further penetration of O_2 into the matrix that repeats the process.

For the coatings on Mo-14Hf-23B and Mo-14.8Zr-26B surfaces the system with Si III was applied. The cyclic oxidation curves of double and triple coated and uncoated samples are presented in figure 4 (b). As can be seen in the graph, uncoated Mo-14Hf-23B and Mo-14.8Zr-26B alloys show dramatic mass losses when the peak temperature (800 °C) is reached. The more detailed description of the cyclic oxidation behaviour of Mo-14Hf-23B and Mo-14.8Zr-26B at 800 °C is described elsewhere [8]. The oxidation proceeds due to the formation of MoO_2 and MoO_3 oxides and the evaporation of the last one. As a result of MoO_x evaporation $Hf(MoO_4)_2$ and $Zr(MoO_4)_2$ ternary oxides are detected by XRD analysis on the surface of the samples. A boron-containing oxide was not found by XRD.

Nevertheless, the triple coated Mo-14Hf-23B and Mo-14.8Zr-26B alloys did not show any mass change after 40 h of exposure up to 100 h. Up to 20 h a slight mass increase in these samples corresponds to the formation of a protective glassy layer. This oxidation behaviour is obviously related to the possibility of Mo-14Hf-23B and Mo-14.8Zr-26B to form B_2O_3, compared to plain Mo, which may control the viscosity of the glassy layer on the interface between coating and substrate [13]. The glassy nature of the coating was proven by EDS analysis of the triple coated Mo-14.8Zr-26B after 100 h of cyclic oxidation at 800 °C, see figure 5. It must be mentioned, that the total coating thickness was increased from 73 μm to 106 μm as a result of oxidation that, in turn, confirms the formation and growth of the SiO_2-B_2O_3 layer.

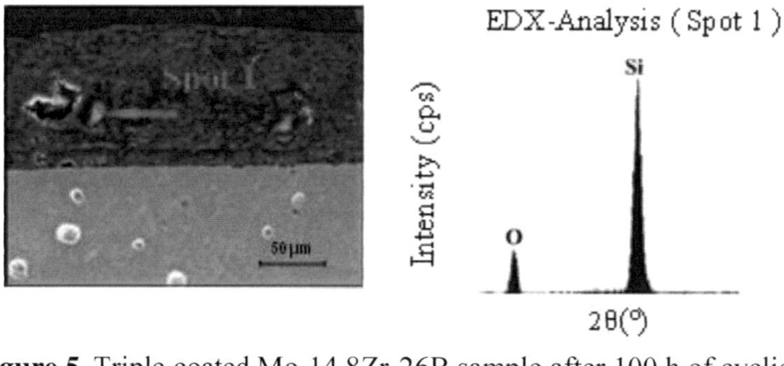

Figure 5. Triple coated Mo-14.8Zr-26B sample after 100 h of cyclic oxidation in air at 800 °C.

The thickness of the double coatings on Mo-14.8Zr-26B alloy was 50 μm in the initial state. Evidently, this thickness cannot protect the sample from oxidation, see figure 4 (b).

To investigate the influence of the volume fraction of particulate fillers on the oxidation behavior of the coatings slurry with an increased amount of silicon and boron was developed: 65 PHPS + 22 Si + 13 B, vol. %. To obtain a 70 μm to 90 μm thick coating that was found to be effective against oxidation using the previous slurry, a double coating was applied to simplify the procedure. The calculations of the expected coating thickness were conducted by equation (2) considering the modified density and viscosity of the modified slurry. The triple coatings were as well applied to the surface of plain Mo and Mo-14.8Zr-26B alloys; there thickness reached 150-155 μm.

The mass change curves of the double and triple coating with 65 PHPS + 22 Si + 13 B, vol. % slurry samples are presented in figure 6. It is evident that an increase of the volume fraction of fillers in the slurry and an enhancement of the coating thickness do not have a profitable effect on the improvement of oxidation resistive properties of plain Mo: a significant mass loss is observed for all samples, see figure 6 (a). The performance of the triple layered coating, however, is higher than that of the double layered coating, see figure 6 (b).

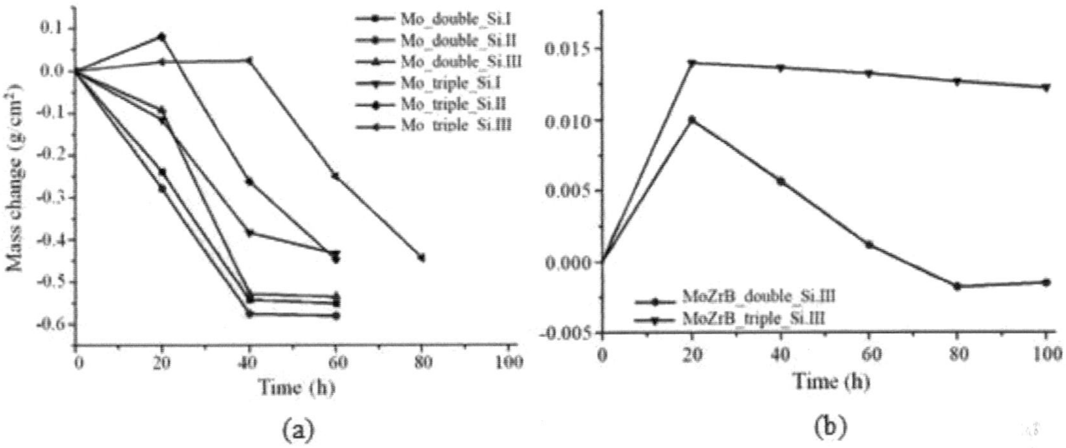

Figure 6. Oxidation curves of double and triple coated Mo- (a) and Mo-14Hf-23B and Mo-14.8Zr-26B (b) samples with 65 PHPS + 22 Si + 13 B, vol. %.

The triple coating with the thickness of 153 μm shows a minor mass loss after 20 h of oxidation. However, this mass loss is not caused by the MoO_3 evaporation, but by some detachment of the coating layer, as found after oxidation treatment and sample revision. Obviously, the thickness of the coating and the number of coating layers play a significant role for the protective function. Nevertheless, increasing the coating thickness above 90 μm does not improve the oxidation resistance of the alloys; the effect of coating spalling is currently under investigation.

4 Conclusions

Oxidation protection coating systems developed in this work were proven to be appropriate for improving the oxidation resistance of Mo-14Hf-23B and Mo-14.8Zr-26B alloys at 800 °C. These alloys, in contrast to plain Mo alloy, can generate a B_2O_3 phase during initial oxidation, while the oxygen-trapping coating systems slow down the oxygen intake into the alloy. This may provide time for the formation of a low-viscosity glassy layer on the interface between coating and substrate, which, in turn, provides a protection against oxidation in the initial stage of high temperature exposure.

An increase of the volume fraction of particulate Si and B fillers in the slurry lead to an increase of the coating thickness; this showed a negative influence on the oxidation resistance on coatings because of undesired spalling effects.

Acknowledgement

The authors gratefully acknowledge the financial support from Deutsche Forschungsgemeinschaft DFG (SCHE 628/20-1). Our special thanks are dedicated to Dr. Michael Schwidder (Institute for Chemistry, Otto-von-Guericke-University Magdeburg) for rheological characterization and to Dr. Volodymyr Bolbut (Institute of Materials and Joining Technology, Otto-von-Guericke-University Magdeburg) for providing us Mo-14Hf-23B and Mo-14.8Zr-26B alloys.

References

[1] Sinha N K 2006 *Mater. Sci. Eng. A* **432** 129
[2] Bolbut V, Bogomol I, Bauer C and Krüger M 2017 *Mat.-wiss. u. Werkstofftech.* **48** 1113
[3] Bolbut V, Bogomol I, Loboda P and Krüger M 2018 *J. Alloys Compd.* **735** 2324

[4] Smokovych I, Hasemann G, Krüger M and Scheffler M 2017 *J. Eur. Ceram. Soc.* **37** 4559
[5] Sōmiya S 2013 *Handbook of advanced ceramics: Materials, applications, processing, and properties* (Amsterdam: Academic Press) p 1229
[6] Torrey J D, Bordia R K 2010 *Polymer derived ceramics: From nano-structure to applications* ed Colombo P *et al* (Lancaster: DEStech Publications) chapter 5.2 pp 329–339
[7] Smokovych I, Krüger M and Scheffler M 2019 *J. Eur. Ceram. Soc.* **39** 3634
[8] Smokovych I, Bolbut V, Krüger M and Scheffler M *Materials* 2019 **12** 2215
[9] Brinker C J 2014 *Chemical Solution Deposition of Functional Oxide Thin Films* vol 18 ed Schneller T *et al* (New York: Springer) chapter 10 p 233
[10] Malcherek A 2010 *Die Küste 77* (Heide, Holstein: Boyens) pp 135-178
[11] Brinker C J and Scherer G W 1990 *Sol-gel science: The physics and chemistry of sol-gel processing* (Boston: Academic Press)
[12] Günthner M, Kraus T, Krenkel W, Motz G, Dierdorf A and Decker D 2009 *Int. J. Appl. Ceram. Technol.* **6** 373
[13] Parthasarathy T A, Mendiratta M G and Dimiduk D M 2002 *Acta Mater.* **50** 1857
[14] Perepezko J H, Rioult F and Sakidja R 2008 *Mater. Sci. Forum* **595-598** 1065

IOP Conf. Series: Materials Science and Engineering **882** (2020) 012022 doi:10.1088/1757-899X/882/1/012022

Manufacturing of Open-Cell Metal Foams by the Sponge Replication Technique

A Sutygina, U Betke, G Hasemann and M Scheffler

Otto-von-Guericke-University Magdeburg, Faculty of Mechanical Engineering, Institute of Materials and Joining Technology, Universitätsplatz 2, 39106 Magdeburg, Germany

alina.sutygina@ovgu.de

Abstract. Different manufacturing techniques for open-cell metal foams, such as investment casting and space holder casting, are compared with the sponge replication technique that is originally used for the preparation of ceramic foams. Processing of aluminum foams by the sponge replication technique has been reviewed. Based on the results obtained for aluminum open-cell foams, the feasibility of copper foam manufacturing with the sponge replica technique using an aqueous copper dispersion was demonstrated.

1 Introduction

Wood or bones are one of the common cellular materials in nature. Highly porous materials including metal and ceramic foams possess a combination of properties such as high stiffness-to-weight ratio, low specific weight, high energy and acoustic absorption [1–3]. Due to these combinations of properties porous materials have found application in nearly every sector of technology as aerospace industry (noise reduction), architecture (light-weight constructions and thermal/sound insulation materials), automotive industry (engine brackets, gear-wheel with an anti-vibration layer of aluminium foam), filters, heat exchangers, biomedical prostheses (bone implant material) and vibration dampers [1–6].

There are different types of metal foams: open and closed cells metal foams. These metal foams need different manufacturing processes and are characterized by their different properties. Closed-cell metal foams are used predominantly as constructional materials (automotive industry, architecture, and more). At the same time, open-cell metal foams have received a great deal of attention as functional materials such as heat exchanges for adsorption-driven heating/cooling applications [7], as catalyst support materials [5], pressure regulators, fuel cell components [8], battery components [9, 10] and filters, due to combination of properties such as thermal/electric conductivity, high specific surface area and liquid/gas permeability [11, 12]. These types of metal foams are used as a template for deposition or coating with active materials to enable chemical reactions, adsorption properties, mechanical stability or electrical conductivity [4, 5, 11]. Open cell metal foams for the above mentioned application are made from aluminium [9, 11, 13], nickel [8, 14], copper [10, 15] and titanium [16] or alloys such as stainless steel [17], Ni-Fe [18], Ni-Cr and Ni-Fe-Cr [19].

The most common conventional methods to manufacture open-cell metal foams are investment casting, casting around hollow spheres, and space holder or infiltration casting [20–22]. There are other methods such as layer by layer electro deposition on a polymer pre-form [19] or direct foaming [13], etc to obtained open-cell metal foams.

2 Review of manufacturing techniques for open cell metal foams

2.1 Manufacturing of open-cell metal foams by investment casting

The processing of open-cell metal foams consist of four main manufacturing steps, which are schematically illustrated in figure.1. On the first step, a reticulated polyurethane foam (PU) template is filled under pressure with a mould slurry of heat resistant material such as ceramic or gypsum (plaster) suspension. After drying the mould material, the PU foam is removed (burned out) and molten metal is filled into these open voids. The last steps are removal of the mould material with high pressure water or dissolving it with acids or leaching. The obtained metal foam is a replica of the shape of the original PU foam.

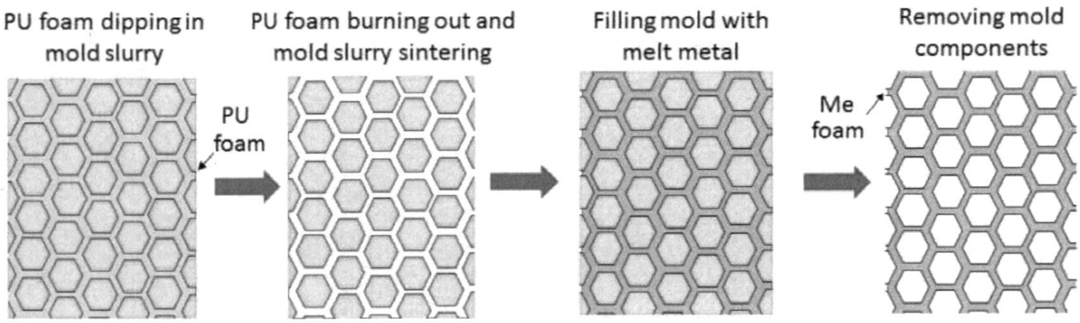

Figure 1. Schematic illustration of the manufacturing of open-cell metal foams by investment casting.

The advantage of the investment casting is the manufacturing of open-cell foams from different metals and a wide spectrum of desired cell sizes, relative density and porosity, which are determined by the polymer foam (porosity may range from 80 % to 97 %). However the problem of this method is no uniform filling of the mold with molten metal; therefore, the filling of the mold is carried out under high pressure in a gravity field and preheating of the mold [23–25]. In ref. [23], it was shown that changing the gravity coefficient from 300 to 900 enhanced the structural integrity of pores, surface morphology, and thickness of the cell struts of aluminum foams. Also, authors successfully manufactured open-cell metal foams with different cell sizes (from 4.5 mm to 0.1 mm) in a super-gravity field with the gravity coefficient of 500. However, it was found that removing the plaster materials was not effective for the foams with the cell sizes of 0.1 mm [23].

As a result, the manufacturing of large open-cell metal foams is problematic due to the difficulty of controlling the uniform heating of the mold form and its filling with liquid metals that has considerable influence on the macrostructure and phase distribution of metal alloys [25]. Also removing the heat resistant mold is not only complex but also can damage the foam structure. In addition, foams manufactured this way are characterized by low yield and have high prices [26].

2.2 Manufacturing of open-cell metal foams by casting with space holders

Manufacturing of open-cell metal foams by the space holder casting is a technique in which, instead of PU template, different space holder particles such as sodium chloride, sugar, carbonates or sand are used. The manufacturing route is shown in figure 2. The packing of these space holders are infiltrated with a liquid metal. After solidification space holders are removed by acids or leaching in solvents [27].

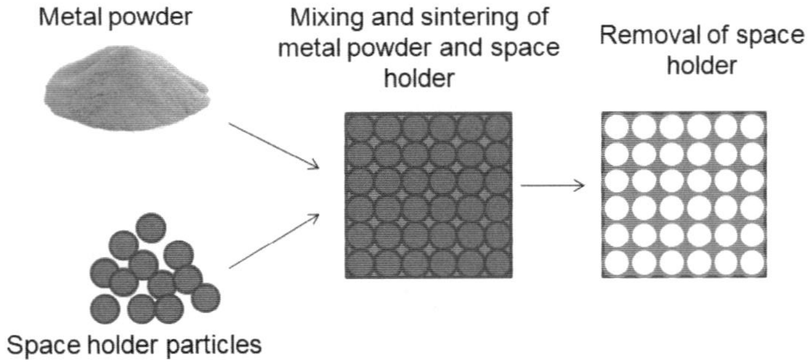

Figure 2. Schematic illustration of manufacturing of open-cell metal foams by space holder casting.

A high surface tension of liquid metals can be problematic during wetting of the space holders, which leads to an incomplete filling of the voids between the space holder particles. Therefore, for the uniform casting of space-holders with molten metal high pressure melt infiltration, vacuum and pre-heating of space holders are required to avoid solidification of the molten metal until complete filling the pre-form with liquid metal. It is possible to manufacture aluminium, magnesium, zinc, lead, tin foams, etc by space holder casting [27]. The main advantages of this manufacturing way are fabricating foams with a desired geometry and controllable pore size which depends on the size of space holders [29, 30]. At the same time, the maximum value of porosity is limited by the size of space holders; it is laborious to achieved foams with porosity >90%. The shape of cells in these foams is not uniform and repeats the shape of space holders.

Numerous research is focused on optimizing the manufacturing parameters to obtain stable foams with the desired functional properties. In ref. [30], the influence of vacuum, space holder size and compacting pressure of space holders on the cell structure and compressive strength of foams was investigated. Also, a super-gravity field was used to fill a pre-form by molten metal uniformly (figure 3) [29]. It is seen that the metal had completely filled the pre-form at higher centrifugal pressure.

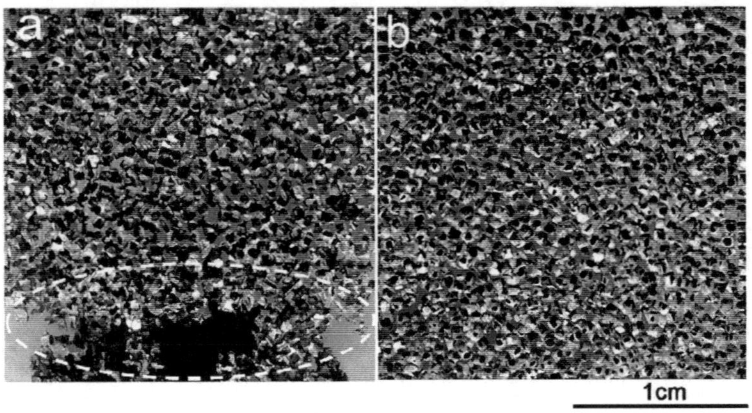

Figure 3. Typical cross-section images of aluminum foams with a pore size of 600 μm fabricated with different centrifugal pressures: (a) 24 kPa and (b) 32 kPa [29].

2.3 Manufacturing of open-cell foams by the sponge replication technique

The sponge replication technique was initially used for the manufacturing of ceramic foams [31], which consists of three manufacturing steps (figure 4). The first step is the coating of a PU template with an aqueous suspension of metal or ceramic powder mixed with a binder. The second stage is burning out the PU template and the binder. The last step is sintering or heat treatment of the foams.

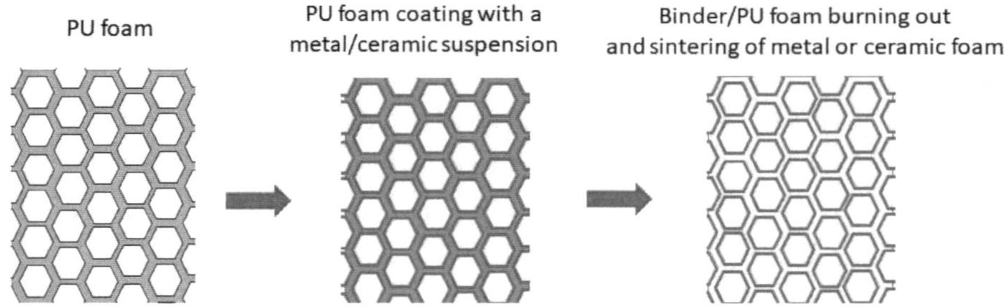

Figure 4. Schematic illustration of manufacturing of open-cell ceramic or metal foams by sponge replication technique.

The foams prepared by the sponge replication technique have controllable pore size (90%) which depends on the pore size of the initial PU template. The foams are characterized by a high surface-area-to-volume ratio; therefore, this manufacturing way is attractive for preparing metal foams used in chemical engineering/industrial chemistry (catalytic reactions, supports for different coatings, filters, etc) [20]. One of the limitations of the sponge replication technique for metal foam manufacturing is the presence of thin oxide layers on the metal powders used as starting materials; the oxide layers are often difficult to disrupt (depending in the metal), and this leading to an inhomogeneous microstructure of the struts due to incomplete powder particle sintering. As a result, metal foams may have a low compression strength and low thermal/electric conductivity [32, 33]. As an example, a natural thin alumina film and high oxidation rates of aluminum particles are still problematic for sintering aluminium metal foams [34]. A possible solution to this problem is varying the sintering temperature and atmosphere [35], as well as various additives used to dissolve the metal oxide layer [36].

2.3.1 Current manufacturing of open-cell metal foams by sponge replication technique. The manufacturing of open-cell metal foams by sponge replication technique has been successfully applied for titanium [37–39], Ti6Al4V [40] and austenitic steel [41]. Open-cell aluminum foams made by sponge replication technique are attractive for catalytic application and for heat exchange systems. In ref. [42], the feasibility of the aluminum foam preparation by the sponge replication technique in air at 620 °C for 4 h and 7 h was demonstrated. The aluminum foams were stable and had a porosity between 94.4 % and 95.5 %. This work was more focused on determining the optimal composition of the metal suspension and demonstrating applicability of this manufacturing method for aluminium starting powder.

Investigations carried out by our research group [32] have shown that the shape and size of aluminum powders used for the manufacturing of foams influence the thermal consolidation process of the foam resulting in a variation of structure features. It was found that the foams obtained from spheroidal aluminium particles have a lower strut porosity, a lower amount of residual alumina, and a higher compressive strength in comparison with foams manufactured from flaky, irregularly shaped aluminum powder. Figure 5 shows cross sections of struts of the foams made from the flaky, irregularly shaped Aldrich powder and spheroidal-shaped Ecka powder after heat treatment at 750 C for 3 h [32].

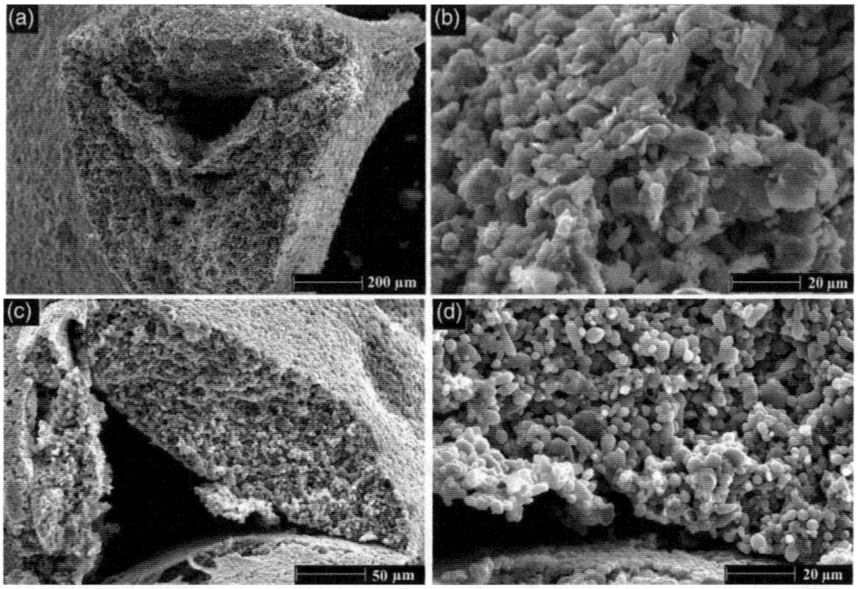

Figure 5. SEM images of cut struts of the foams made from the (a,b) flaky, irregularly shaped Al powder and (c,d) spheroidal-shaped Al powder after heat treatment at 750 C for 3 h [32].

The manufacturing of the open-cell aluminum foams from the spheroidal aluminum powder by a sponge replication technique was performed in ref. [33]. The thermal processing of aluminum foams at temperatures up to 800 °C in Ar reduced the strut porosity in comparison with the strut porosity of foams thermally processed in air. In addition, after thermal processing in Ar the Al metal foams were characterized by a lower amount of aluminum oxide, a higher thermal conductivity, and a higher compression strength due to a comparatively lower amount of aluminum oxide after thermal processing. In figure 6 Al foams are shown after thermal processing at different temperatures [33]. However, despite of the successful consolidation, the foams have porous struts resulting from incomplete thermal processing. This may be explained by a thin oxide shell around the starting aluminum particles and a steady disruption of the alumina shells during thermal processing. Therefore, for the further research it is necessary to focus on reduction of the alumina/oxygen content in the aluminum starting powders.

In general, the reticulated aluminum foams are stable, had high porosity and satisfying strength. Consequently, processing open-cell aluminum metal foams is possible to produce highly porous foams with a relatively large surface area by the sponge replication technique.

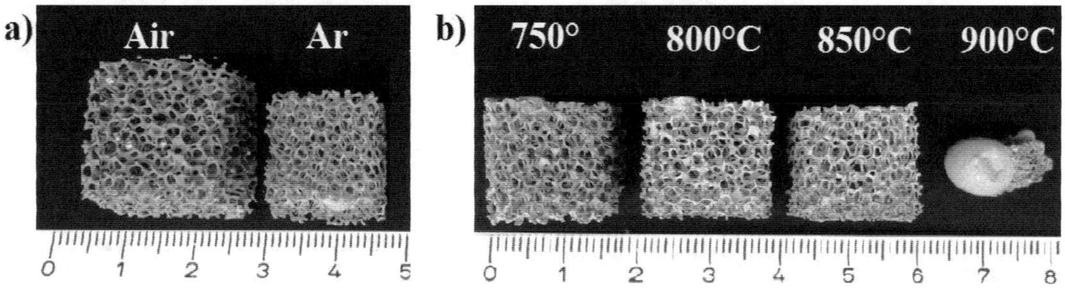

Figure 6. Al foams after thermal processing for 3 h in (a) air and Ar at 750 °C and (b) Ar at 750–900 °C [33].

Open-cell copper foams are attractive for advanced electronics, energy-related devices such as rechargeable batteries, electrochemical catalysts, heat exchangers, heat pumps, fuel cells, and super capacitors because of the combination of high thermal/electric conductivity and high surface area [10, 43, 44]. Thus, based on the results with aluminum open-cell foams, manufacturing open-cell metal foams by sponge replication technique may be applicable for processing copper foams. According to the above, this work aims to demonstrate the feasibility of copper foam manufacturing with the sponge replica technique with an aqueous copper-containing slurry.

3 Manufacturing of open cell copper foams

3.1 Specimen preparation

For the manufacturing of copper foams, copper powder was used, supplied by Ecka Granules (ECKA Kupfer CH UF 10, Ranshofen, Austria) with a purity of 99.81 wt.%. The average particle size was <10 μm. For the preparation of a coating slurry the copper powder was mixed with a 10.7 wt.% solution of a polyvinyl alcohol binder (1.2 wt%, Optapix PA 4G, Zschimmer and Schwarz Chemie GmbH, Lahnstein, Germany) in distilled water. The solid content of copper in the slurry was 82.8 wt.%. A planetary centrifugal mixer (THINKY Mixer ARE-250, THINKY Corp. Tokyo, Japan) was used for mixing the copper slurry at 2000 rpm for 6 min. As a PU template, an open-cell PU foam with a linear cell count of 20 ppi with a geometric size of 15 mm × 15 mm × 20 mm (Koepp Schaum GmbH, Oestrich Winkel, Germany) was used. A schematic illustration of the open-cell copper foam preparation by the sponge replication technique is shown in figure 7.

After dipping the PU template foam into the slurry the specimens were dried for 24 h at room temperature in air. The binder and PU burning out were carried out in air at 250 °C for 3 h and at 500 °C for 3 h in a circulating air furnace (KU 40/ 04/A, THERMCONCEPT Dr. Fischer GmbH, Bremen, Germany). The foams were thermally processed for 6 h in Ar atmosphere with 2% H_2 (99,999% purity) at 900 °C in a conventional tube furnace (alumina tube, HTRH 70-600/1800, Carbolite-Gero GmbH and Co. KG, Neuhausen, Germany).

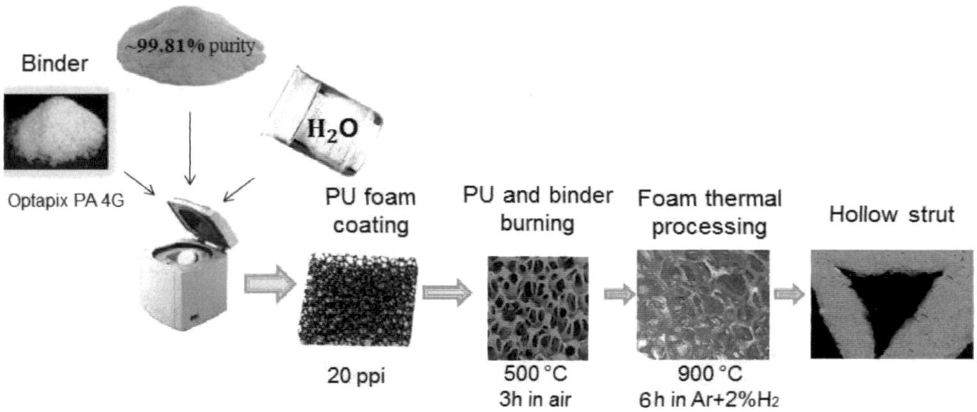

Figure 7. Schematic illustration of the open-cell copper foams preparation by the sponge replication technique.

3.2 Characterization

The microstructure imaging of the thermally processed foams was carried out by a scanning electron microscope (SEM; FEI ESEM XL30 FEG, Hillsboro, OR/USA). For the determination of the phase composition X-ray diffraction analysis (XRD) conducted with an X'Pert Pro diffractometer (PANalytical GmbH, Kassel, Germany, Co Kα1/2 radiation, 2θ, 40–85°) with Bragg–Brentano geometry was applied. The total porosity was calculated from the geometric foam density and the skeletal density of the strut material (V_{pores}/V_{foam}) [45], considering that the density of the pure bulk copper ρ_{Cu} equals to 8.96 g·cm^{-3}. The quantification of the strut porosity ($V_{strut\ pores}/V_{struts}$) of copper foams was conducted by the Archimedes method with water according to the DIN EN 623-2:1993-11 standard procedure [46]. For the measurements of the compression strength a TIRAtest 2825 testing machine was used (TIRA GmbH, Schalkau, Germany) with a loading plate of 150 mm in diameter and the applied load 2 mm·min^{-1}.

3.3 Results

After thermal processing in Ar with 2% H_2 the copper foams were characterized by a volumetric shrinkage of ~56%. The copper foams possess a total porosity of ~93%, which includes material pores, cavities (hollow strut pores) and cell pores. The total strut porosity consisting of porosity of hollow struts and material pores amounts to ~44 %.

Figure 8 shows SEM images of cross-sections of copper foams thermally processed at 900 °C. It can be seen, that their struts are hollow and reperesent the shape of the PU template foam, despite the high degree of shrinkage. No molten copper beads outside and inside the foam struts were detected, as it was the case for the of Al foams. The microstructure of the foam struts is characterized by homogeneous (figure 8, top) and inhomogeneous regions with pores in the strut walls, which may result from incomplete thermal processing (figure 8, bottom).

The results of the XRD phase analysis (figure 9) indicated that there is only a pure copper phase present after thermal processing for 6 h at 900 °C in Ar with 2% H_2. No copper oxide and/or carbide phases were found.

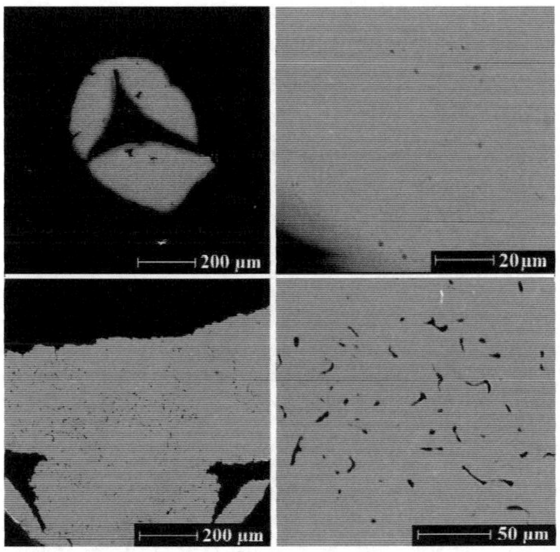

Figure 8. SEM images of copper foams thermally processed for 6 h at 900°C.

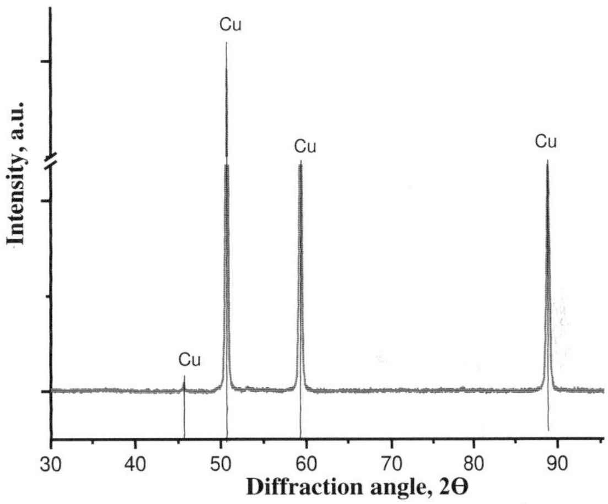

Figure 9. X-ray diffraction patterns of copper foams thermally processed for 6 h at 900°C in Ar with 2% H_2.

Figure 10 shows a typical stress-strain curve of the copper foams after thermal processing. The stress-strain curves on the initial stage had an elastic deformation region with a linear increase of the stress from the strain; in a point where the stress-strain changed its behaviour, the yield strength σ^* exceeded the strength of the cell struts, which is the onset of plasticity. Further increase of the stress is characterized by a long plateau region (plastic cell collapse) with a significant strain increasing. When the cells of the foams are compacted, the stress-strain curve increased drastically (densification region). The average yield strength value σ^* of 10 stress-strain measurements was σ^*=0.38±0.06 MPa.

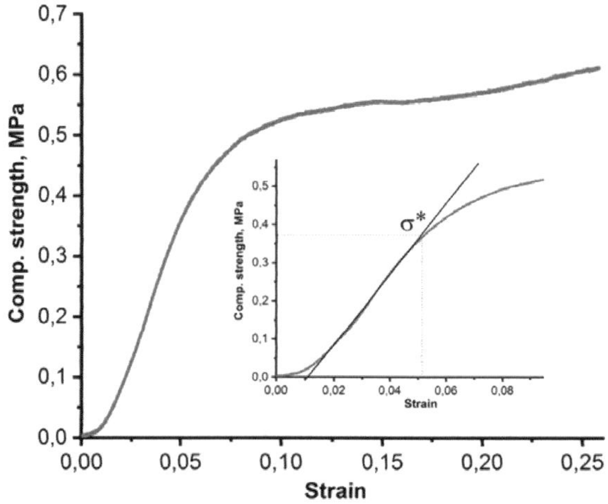

Figure 10. Typical stress–strain curves of copper foams after compressive strength test of thermally processed for 6 h at 900 °C in Ar with 2% H_2.

4 Conclusion

In this work, investment casting and space holder techniques for the manufacturing of open-cell metal foams were reviewed and compared to the sponge replication technique. As material examples, aluminum and copper foams were prepared by using polyurethane as foam templates. It was shown, that, despite of oxide impurities in the case of aluminum, or a melting point of copper at 1085 °C, manufacturing of intact foam structures of these metals is possible. Both replica-manufactured foams possess a high porosity, a high specific surface area and hollow struts. While more research is necessary to reduce the oxide share of the aluminum foams, copper foam-processing in an Ar/H_2 gas mixture led to a plain/elemental copper material. In comparison with the aforementioned processes a number of processing steps can be economized, and the hollow struts may be used for further functionalization of these type of cellular materials.

Funding

Financial support from the International Graduate School for Medical Engineering and Engineering Materials (MEMoRIAL) funded by the European Structural and Investment Funds (ESF) under the program “Sachsen-Anhalt Wissenschaft Internationalisierung” is gratefully acknowledged.

Acknowledgments

We are thankful to Holger Voß for the XRD measurements (OvGU-IWF) and Andreas Schlinkert (OvGU-MVT) for particle size measurements.

References

[1] Singh S and Bhatnagar N 2018 A survey of fabrication and application of metallic foams (1925–2017) *J. Porous Mater.* **25** 537–54
[2] Banhart J 2001 Manufacture, characterisation and application of cellular metals and metal foams *Prog. Mater. Sci.* **46** 559–632
[3] Kim S and Lee C-W 2014 A Review on Manufacturing and Application of Open-cell Metal Foam *Procedia Mater. Sci.* **4** 305–9
[4] Bortolozzi J P, Banús E D, Milt V G, Gutierrez L B and Ulla M A 2010 The significance of passivation treatments on AISI 314 foam pieces to be used as substrates for catalytic applications *Appl. Surf. Sci.* **257** 495–502
[5] Gancarczyk A, Sindera K, Iwaniszyn M, Piatek M, Macek W, Jodłowski P J, Wroński S, Sitarz M, Łojewska J and Kołodziej A 2019 Metal foams as novel catalyst support in environmental processes *Catalysts* **9**

[6] García-Moreno F 2016 Commercial applications of metal foams: Their properties and production *Materials (Basel).* **9** 20–4
[7] Bonaccorsi L, Freni A, Proverbio E, Restuccia G and Russo F 2006 Zeolite coated copper foams for heat pumping applications *Microporous Mesoporous Mater.* **91** 7–14
[8] Chen J, Zhao C X, Zhi M M, Wang K, Deng L and Xu G 2012 Alkaline direct oxidation glucose fuel cell system using silver/nickel foams as electrodes *Electrochim. Acta* **66** 133–8
[9] Li Q and Bjerrum N J 2002 Aluminum as anode for energy storage and conversion: A review *J. Power Sources* **110** 1–10
[10] Um J H, Park H, Cho Y H, Glazer M P B, Dunand D C, Choe H and Sung Y E 2014 3D interconnected SnO2-coated Cu foam as a high-performance anode for lithium-ion battery applications *RSC Adv.* **4** 58059–63
[11] Sanz O, Javier Echave F, Sánchez M, Monzón A and Montes M 2008 Aluminium foams as structured supports for volatile organic compounds (VOCs) oxidation *Appl. Catal. A Gen.* **340** 125–32
[12] Groppi G, Giani L and Tronconi E 2007 Generalized correlation for gas/solid mass-transfer coefficients in metallic and ceramic foams *Ind. Eng. Chem. Res.* **46** 3955–8
[13] Barg S, Soltmann C, Schwab A, Koch D, Schwieger W and Grathwohl G 2011 Novel open cell aluminum foams and their use as reactive support for zeolite crystallization *J. Porous Mater.* **18** 89–98
[14] Bidault F, Brett D J L, Middleton P H, Abson N and Brandon N P 2009 A new application for nickel foam in alkaline fuel cells *Int. J. Hydrogen Energy* **34** 6799–808
[15] Etiemble A, Adrien J, Maire E, Idrissi H, Reyter D and Roué L 2014 3D morphological analysis of copper foams as current collectors for Li-ion batteries by means of X-ray tomography *Mater. Sci. Eng. B Solid-State Mater. Adv. Technol.* **187** 1–8
[16] Manonukul A, Tange M, Srikudvien P, Denmud N and Wattanapornphan P 2014 Rheological properties of commercially pure titanium slurry for metallic foam production using replica impregnation method *Powder Technol.* **266** 129–34
[17] Kaya A C and Fleck C 2014 Materials Science & Engineering A Deformation behavior of open-cell stainless steel foams *Mater. Sci. Eng. A* **615** 447–56
[18] Huang X L, Wu G H, Lv Z, Zhang Q, Kang P C and Leng J F 2009 Electrical conductivity of open-cell Fe-Ni alloy foams *J. Alloys Compd.* **479** 898–901
[19] Pang Q, Wu G H, Xiu Z Y, Chen G Q and Sun D L 2012 Synthesis and mechanical properties of open-cell Ni-Fe-Cr foams *Mater. Sci. Eng. A* **534** 699–706
[20] Quadbeck P, Kümmel K, Hauser R, Standke G, Adler J, Stephani G and Kieback B 2011 Structural and material design of open-cell powder metallurgical foams *Adv. Eng. Mater.* **13** 1024–30
[21] Banhart J 2003 Aluminum foams: On the road to real applications *MRS Bull.* **28** 290–5
[22] Yang X, Hu Q, Du J, Song H, Zou T, Sha J, He C and Zhao N 2019 Compression fatigue properties of open-cell aluminum foams fabricated by space-holder method *Int. J. Fatigue* **121** 272–80
[23] Chang K, Gao J T, Wang Z and Guo Z C 2018 Manufacturing 3-D open-cell aluminum foam via infiltration casting in a super-gravity field *J. Mater. Process. Technol.* **252** 705–10
[24] Fischer S F, Groten V F, Brachmann J, Fix C, Vossel T and Bührig-Polaczek A 2016 Influence of die and casting temperatures and titanium and strontium contents on the technological properties of die-cast A356 in the as-cast and T6 condition *Miner. Met. Mater. Ser.* 1031–6
[25] Schüler P, Fischer S F, Bührig-Polaczek A and Fleck C 2013 Deformation and failure behaviour of open cell Al foams under quasistatic and impact loading *Mater. Sci. Eng. A* **587** 250–61
[26] Kulshreshtha A and Dhakad S K 2020 Preparation of metal foam by different methods: A review *Mater. Today Proc.*
[27] Banhart J 2001 Manufacture, characterisation and application of cellular metals and metal

foams *Prog. Mater. Sci.* **46** 559–632
[28] DeFouw J D and Dunand D C 2012 Processing and compressive creep of cast replicated IN792 Ni-base superalloy foams *Mater. Sci. Eng. A* **558** 129–33
[29] Wang Z, Gao J, Chang K, Meng L, Zhang N and Guo Z 2018 Manufacturing of open-cell aluminum foams: Via infiltration casting in super-gravity fields and mechanical properties *RSC Adv.* **8** 15933–9
[30] Jinnapat A and Kennedy A 2012 Characterisation and mechanical testing of open cell al foams manufactured by molten metal infiltration of porous salt bead preforms: Effect of bead size *Metals (Basel).* **2** 122–35
[31] Jamaludin A R, Kasim S R, Ismail A K, Abdullah M Z and Ahmad Z A 2015 The effect of sago as binder in the fabrication of alumina foam through the polymeric sponge replication technique *J. Eur. Ceram. Soc.* **35** 1905–14
[32] Sutygina A, Betke U and Scheffler M 2020 Open-Cell Aluminum Foams by the Sponge Replication Technique: A Starting Powder Particle Study *Adv. Eng. Mater.* **1901194**
[33] Sutygina A, Betke U and Scheffler M 2019 Open-cell aluminum foams by the sponge replication technique *Materials (Basel).* **12**
[34] Kwon H, Park D H, Park Y, Silvain J F, Kawasaki A and Park Y 2010 Spark plasma sintering behavior of pure aluminum depending on various sintering temperatures *Met. Mater. Int.* **16** 71–5
[35] Pieczonka T, Schubert T, Baunack S and Kieback B 2008 Dimensional behaviour of aluminium sintered in different atmospheres *Mater. Sci. Eng. A* **478** 251–6
[36] Xie G, Ohashi O, Sato T, Yamaguchi N, Song M, Mitsuishi K and Furuya K 2004 Effect of Mg on the sintering of Al-Mg alloy powders by pulse electric-current sintering process *Mater. Trans.* **45** 904–9
[37] Manonukul A, Srikudvien P, Tange M and Puncreobutr C 2016 Geometry anisotropy and mechanical property isotropy in titanium foam fabricated by replica impregnation method *Mater. Sci. Eng. A* **655** 388–95
[38] Li J P, Li S H, Van Blitterswijk C A and De Groot K 2005 A novel porous Ti6A14V: Characterization and cell attachment *J. Biomed. Mater. Res. - Part A* **73** 223–33
[39] Wang C, Chen H, Zhu X, Xiao Z, Zhang K and Zhang X 2017 An improved polymeric sponge replication method for biomedical porous titanium scaffolds *Mater. Sci. Eng. C* **70** 1192–9
[40] Li J P, Li S H, Van Blitterswijk C A and De Groot K 2005 A novel porous Ti6A14V: Characterization and cell attachment *J. Biomed. Mater. Res. - Part A* **73** 223–33
[41] Bakan H I and Korkmaz K 2015 Synthesis and properties of metal matrix composite foams based on austenitic stainless steels -titanium carbonitrides *Mater. Des.* **83** 154–8
[42] Zaman E and Kele O 2014 Open cell aluminum foams produced by polymer impregnation method *Acta Phys. Pol. A* **125** 445–8
[43] Park H, Choi M, Choe H and Dunand D C 2017 Microstructure and compressive behavior of ice-templated copper foams with directional, lamellar pores *Mater. Sci. Eng. A* **679** 435–45
[44] Yu Y, Wang H, Zhang H, Tan Y, Wang Y, Song K, Yang B, Yuan L, Shen X and Hu X 2020 Blanket-like Co(OH)2/CoOOH/Co3O4/Cu(OH)2 composites on Cu foam for hybrid supercapacitor *Electrochim. Acta* 135559
[45] Henon J, Alzina A, Absi J, Smith D S and Rossignol S 2013 Potassium geopolymer foams made with silica fume pore forming agent for thermal insulation *J. Porous Mater.* **20** 37–46
[46] Hochleistungskeramik; MonolithischeKeramik; Allgemeine und strukturelleEigenschaften; Teil 2: Bestimmung von Dichte und Porosität; DIN EN 623–2:1993–11; BeuthVerlag: Berlin, Germany, 1993. (In German)

IOP Conf. Series: Materials Science and Engineering **882** (2020) 012023 doi:10.1088/1757-899X/882/1/012023

Tungsten inert gas bead-on-plate weld chemical composition analysis by laser-induced breakdown spectroscopy

U A Taparli[1], T Kannengiesser[1,2] and A Griesche[1]

[1] Federal Institute for Materials Research and Testing (BAM), Unter den Eichen 87, 12205 Berlin, Germany
[2] Otto-von-Guericke University Magdeburg, Faculty of Mechanical Engineering, Institute of Materials and Joining Technology, Universitätsplatz 2, 39106 Magdeburg, Germany

ugur-alp.taparli@bam.de

Abstract. Chemical compositions of a weld can be varying locally as a result of the welding process. These local variations can be due to the vaporization of individual alloying elements. In this work, tungsten inert gas (TIG) bead-on-plate stainless steel welds of EN grade 1.4404 and 1.4435 were investigated using laser-induced breakdown spectroscopy (LIBS) on the completed welds. This study aims to reveal the welding parameters' influence on the resulting local chemical compositions of the stainless steel welds. We demonstrated Mn vaporize before Cr due to its lower latent enthalpy of vaporization. Hence, Mn accumulates on the heat-affected zone (HAZ) both sides across the weld bead by being swept away through the circulation flow of the welding plasma. Additionally, increasing the heat input tends to enhance the accumulated Mn content on the HAZ as well as increasing the shielding gas flow rate. The results are in good agreement with the literature and proved that LIBS is an effective method to inspect completed welds.

1 Introduction

Tungsten inert gas (TIG) welding process uses a free burning electric arc in a surrounding shielding gas between a non-consumable tungsten electrode and a workpiece. Consequently, the strong heat flux from the welding arc plasma partially melts the workpiece and forms a weld pool. Metal vapor can be released from the weld pool surface due to the vaporization of the molten material[1]. If the material's loss results in a critical change of the chemical composition, the mechanical properties of the weld may be impaired. Moreover, the chemical composition of the metal vapor and the arc plasma alters the arc stability and fume formation[2].

The alloying elements' vaporization from the weld pool is a complex process and includes several stages. Moreover, the rate of vaporization depends on the relative contributions of each following step. First, the alloying elements are transported to the surface of the weld pool either by convection or diffusion. Second, as the alloying element reaches the surface, several thermodynamic factors determine the rate of vaporization e.g. the local concentration of the alloying element, the extend of surface coverage by the surface active elements, the nature of the weld pool surface interface and the weld pool surface turbulence. Finally, the nature of the boundary layer and the diffusivities of the vaporized species influence the rate of vaporization[3]. The principal factors governing the rate of vaporization including various interdependent physical phenomena are e.g. the temperature distribution at the surface of the weld pool, partial pressure of the gas species above the weld pool, the chemical composition of the melt, individual elements' molecular weight and their diffusion coefficients[4].

Various experimental and numerical simulation studies have been conducted on the topic of alloying element vaporization to determine the metal vapor behaviour during welding. Yamamoto *et al.* have used a numerical model for TIG welding of stainless steel and concluded the concentration of the manganese in the arc plasma is much higher than the amount of manganese in the stainless steel, especially above the area with the lower weld pool temperature (farther locations from the welding arc) due to the higher partial pressure of manganese at lower temperatures[5]. Tanaka et al. have detected most of the chromium and manganese vapor released from the stainless steel weld pool surface are carried into a circulation flow in the welding plasma (cathode jet) and then swept away towards the surroundings of the welding arc. They have also concluded that if the concentration gradient in the arc plasma is large enough, chromium could be transported through the cathode jet by diffusion and could be carried towards the welding electrode[6]. Tashiro et al. have investigated the fume formation mechanism and the individual particle behaviour in stationary TIG-welding of pure iron. Hence, they observed the metal vapor escaping from the plasma arc to the atmosphere, instantaneously cooling down and condensing as fume[7]. Tanaka et al. have used imaging spectroscopy to observe stationary TIG welding processes and metal vaporization. They have detected that the atomic and ionic chromium are the dominant species in the welding arc plasma compared to iron due to the lower ionisation energy of Cr[8]. Taparli et al. have detected post-weld manganese enrichment at the heat affected zone (HAZ) in stainless steels using laser-induced breakdown spectroscopy (LIBS)[9]. In situ monitoring of the stationary weld pool using LIBS has also indirectly revealed the significant amount of manganese in the metal vapor above the weld pool and instantaneous condensation on the weld metal as soon as the welding process was terminated[10].

Laser-induced breakdown spectroscopy is one of the various methods in atomic emission spectroscopy. It is based on measuring the element specific photon emission. A pulsed laser beam is focused onto the surface of the substance to be analysed. Radiation energy is locally coupled into the material. Consequently, the material starts to evaporate. In this material vapor and the surrounding atmosphere, a plasma is generated, leading to the excitation of the material constituents and their spontaneous emission (Bremsstrahlung). As the plasma cools down and decays with time, it starts to emit element-specific radiation. This emission can be resolved and detected by a spectrometer[11].

This work presents post weld chemical composition analysis results of TIG bead-on-plate stainless steel welds using LIBS. Individual welds were investigated (after the welding process is completed) by perpendicular (to the welding direction) LIBS line-scans to determine the final chemical compositions in the fusion zone (FZ), heat affected zone and the base material. Furthermore, the influence of the welding heat (energy) input and the shielding gas (Ar) flow rate on the final chemical compositions of the welds were investigated.

2 Materials and methods

2.1 Materials

EN grade 1.4404 and 1.4435 (standardized in DIN EN 10088-1) austenitic stainless steels were chosen as sample materials for the experimental series. Chemical compositions of the main alloying elements are shown in Table 1. The values were measured by conventional mobile spark optical emission spectroscopy instrument Spektrotest (SPECTRO Analytical Instruments GmbH).

Table 1. Selected sample materials and their alloying element concentrations. (Source: Spark-OES)

Grade / Batch	Thickness (t)	Chemical composition (wt%)						
(EN 10088-1)	(mm)	C	Si	Mn	Cr	Ni	Mo	Fe
1.4404 / 31785	3	0.013	0.33	0.75	17.02	11.11	2.18	bal.
1.4404 / 327005	6	0.021	0.33	1.39	16.84	10.00	2.00	bal.
1.4435 / 315821	4	0.023	0.49	1.49	17.92	12.55	2.47	bal.

Sample plates with the dimensions 320 mm x 35 mm x t manufactured by Columbus Stainless Ltd.(1.4435), Outokumpu (1.4404 / 31785) and U&A (1.4404 / 327005) were used for the experiments. The respective surface of the plates was in as-manufactured condition without further mechanical treatment. Prior to welding, they had been cleaned with acetone to remove any resultant surface contaminants.

2.2 Experimental set-up

The LIBS system consisted of a 1064 nm Nd:YAG laser at 10 Hz pulse repetition rate with output energy of 200 mJ (Quantel CFR 200). The laser source was synchronized by a LIBS controller (Lasertechnik Berlin (LTB), Berlin Germany) with the Echelle spectrometer Aryelle 200 (LTB) equipped with intensified charged-coupled device (ICCD) detector (ANDOR iStar, Oxford Instruments). ICCD-detectors have a better time resolution to detect the very short lifetime of the laser-induced plasma (typically lies between 0.5-50µs[11]) and a higher gate-width flexibility compared to single CCD-type detectors[12].

Welding experiments were conducted using a tungsten inert gas (TIG) arc welding system by Castolin CastoTIG 1611DC power supply and Castolin TIG torch using WC-20 tungsten electrode (2% Ceriated) with 1.6 mm diameter. The contact distance from the base material to the electrode tip was set to approx. 3 mm. The electrode tip extended 3 mm from the ceramic gas nozzle (diameter 6 mm). Welding processes were conducted using a motorized linear stage. Sample plates were mounted on the stage and moved, whereby the welding torch and the LIBS system remained fixed. Relative movement of the sample plate resulted in weld beads of 80 mm in length. Welding parameters (welding current and speed) were arranged in order to realise various heat input values. Argon 5.0 (purity ≥ 99.999 %) was used as the shielding gas.

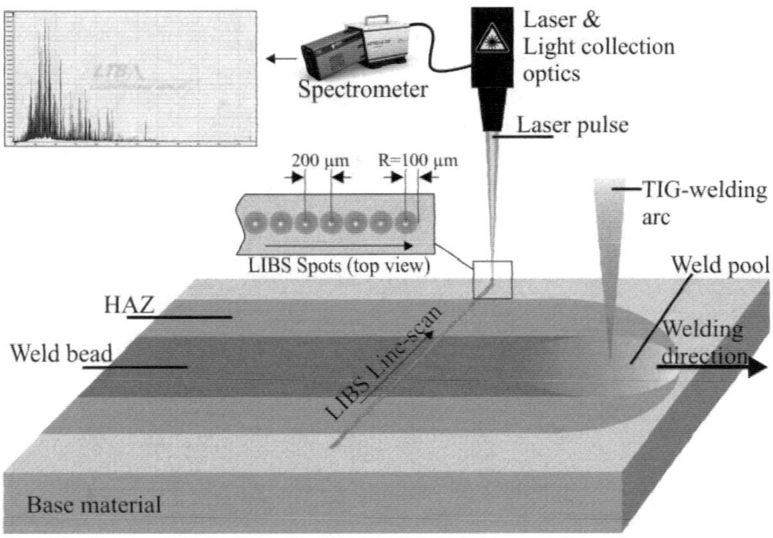

Figure 1. Experimental set-up.

IOP Conf. Series: Materials Science and Engineering **882** (2020) 012023 doi:10.1088/1757-899X/882/1/012023

The laser head was positioned normal to the sample surface. The laser beam was focused on the sample surface using a custom-built focusing and light collecting optics system (Figure 1). The emitted light was transmitted into the spectrometer through an optical fibre. A single measurement accumulated the emitted light from 10 laser shots. The delay time between the laser pulse and the beginning of exposure was set to 1.0 μs and ICCD gate width (how long the ICCD gate is open for each LIBS measurement) was set to 20 μs. LIBS line-scans were conducted perpendicular to the welding direction and at the midpoint (at approx. 40 mm from the start) of each weld seam, using the motorized linear stage with a step size of 200 μm. Individual LIBS measurement spots have radii of approx. r = 100 μm.

3 Results and discussion

Each LIBS measurement provides an individual spectrum for that specific measurement spot. Data processing was conducted for each spectrum in the following sequence; the spectral background was removed and individual peaks were fitted using the Gaussian function. Integrating the selected peaks results in the relative intensity values of the selected element. Using a selection of certified reference materials (CRM) and measuring their relative intensity values, one can establish a quantification model for the measured unknown substances and predict their chemical compositions. In this study, we used the basic linear regression method using CRMs to establish the quantified chemical composition values. Further reading on the topic can be found elsewhere[13, 14].

3.1 Space-resolved elemental distributions of Cr and Mn

Processing each spectrum of the line scan results in a chemical composition vs. position profile for each weld seam. Figure 2 demonstrates three processed spectra between approx. 402.8 nm to 403.8 nm (the welding parameters for the sample are indicated in the caption of Figure 2.). These spectra show the typical Mn I triplet peaks at 403 nm, which were also used for quantification of the manganese element in this study. These peaks were processed as it was explained in the previous section i.e. the spectral background was removed; the data were fitted by the Gaussian function and the area under the peak was calculated by definite integral of the curve. In Figure 2, one can see three spectra, however two of them overlap; namely the black curve (measured spectrum from the weld bead) and the blue curve (measured spectrum from the base material) are hard to distinguish. The reason for this can be seen in Figure 3, i.e. the Mn content at the base material and the weld bead are largely similar. However, Mn content for the HAZ is significantly greater compared to the weld bead and the base material. This result indicates that a higher Mn content was observed at the border of the FZ. Moreover, Mn accumulates on the HAZ as a result of the welding process. The spectrum in Figure 2 (red curve), respectively the Mn I peaks recorded from the HAZ, supports this argument with much higher relative intensity values compared to the other two curves.

Figure 3 reveals a slightly different trend for Cr. The chemical compositions present the lowest values for the weld bead or the FZ, and slightly higher values towards the base material (farther away from the FZ). This phenomenon indicates a possible Cr depletion in the solidified weld metal. This can be explained by a possible vaporization of the alloying element in this region. However, contrary to Mn, a significant local accumulation of Cr at any specific location does not occur.

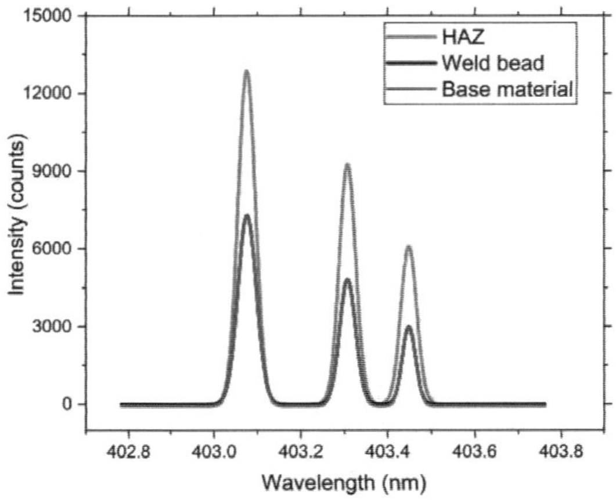

Figure 2. Fitted Mn I triplet peaks around 403 nm. Curves for the "weld bead" and "base material" overlap due to the similar chemical composition values. (EN 1.4404 / 327005 – Welding parameters: welding current = 65A, welding speed = 10 cm/min (Heat input= 491 J/mm), shielding gas flow rate = 18 l/min).

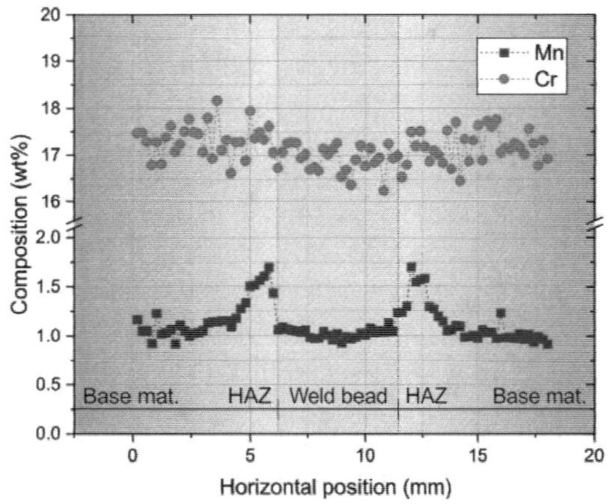

Figure 3. Space-resolved distribution of Cr and Mn elements on the LIBS line-scan perpendicular to the welding direction.

These results for both Mn and Cr are in good agreement with the literature[7-9]. Therefore, we extended our study to investigate the influence of the welding parameters on the local chemical composition changes.

3.2 Relation between welding heat input and Mn accumulation at the HAZ

The previous section has revealed a possible Mn accumulation phenomenon at the HAZ. Hence, the Cr compositions at the FZ differs slightly from the rest of the welds. Figure 4 presents the Mn (left) and Cr (right) compositions as a function of welding heat input for the material EN 1.4404 / 31785. Both diagrams demonstrate the chemical composition of the aforementioned elements measured in three pre-defined locations on the welds (base material, HAZ and weld bead (FZ)).

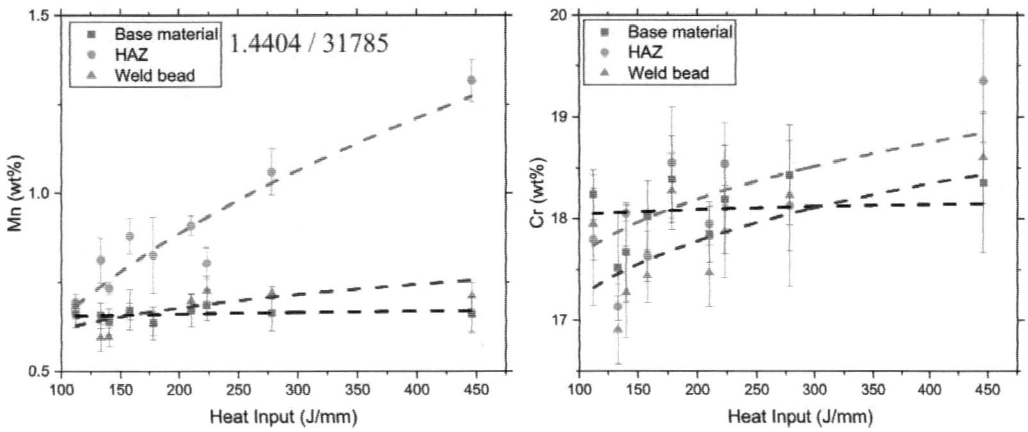

Figure 4. Mn (left) and Cr (right) chemical compositions at the base material, HAZ and the weld bead displayed as a function of welding heat input. Dashed lines are guide-to-the-eye.

Figure 4 shows that increasing the heat input from 110 J/mm up to 450 J/mm results in approx. 0.5 wt% increase of the resulting chemical compositions for both Mn and Cr at the HAZ. Similarly, at the weld bead (FZ) an increasing trend for Mn and Cr compositions with the increasing heat input is observed. However, this latter increasing trend for Mn at the weld bead is relatively slower than for Cr (blue dashed lines). Measurements conducted at the base material have showed only negligible composition variations as function of the heat input (black dashed lines).

Khan et al. reported that increasing the power input during laser welding of stainless steels increases the vaporization rate of individual elements [4]. Their results are in good agreement with the Figure 4. Hence, the Mn composition at the HAZ increases with the higher welding heat input values. Consequently, the amount of the condensed material on the weld surface after welding is also increased. Same principal applies for Cr but in a different location on the weld surface. Tanaka et al. have demonstrated that Cr is dominantly present in the welding arc plasma due to its lower ionisation energy. [8]. It has also been reported that the vaporized Cr could be carried towards the TIG welding electrode[6], which supports the finding of higher Cr composition directly above the weld bead or on the FZ.

3.3 Shielding gas flow rate and alloying element vaporization

This last section presents the chemical composition changes of Mn and Cr as a function of the shielding gas (Ar) flow rate for the sample material 1.4435 / 315821 at constant heat input (491 J/mm). Figure 5 (left) demonstrates the accumulated higher Mn content at the HAZ for any given flow rate with approx. 0.2 wt% increase. The significant increase occurs at the highest applied gas flow rate (20 l/min) with approx. 1 wt% difference to values measured at the base material and the weld bead (FZ). This result also indicated that the vaporized Mn is swept away from the area above the FZ faster and carried onto the HAZ, i.e. also accumulated on the HAZ faster therefore resulting in higher composition value.

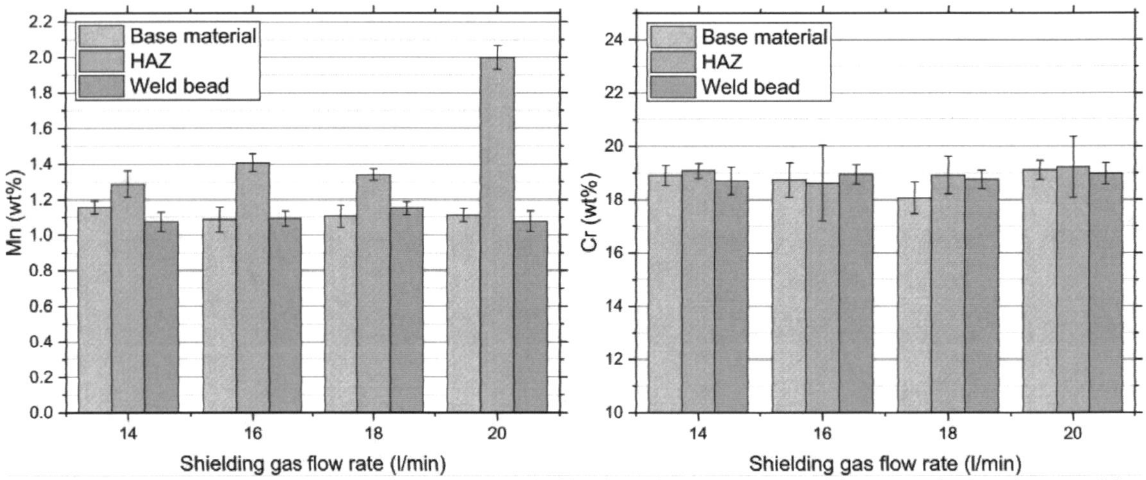

Figure 5. Various shielding gas flow rates and alloying element compositions on the bead-on-weld samples for Mn (left) and Cr (right). (EN 1.4435 / 315821 – Heat input = 491 J/mm constant).

Furthermore, previous in situ LIBS investigations had demonstrated the vaporization rate of the Mn is related to the welding input energy, hence the overall thickness of the accumulated Mn layer was enhanced at higher heat input values[10]. However, this result indicates that the transport model of the individual alloying elements in the welding arc plasma is also strongly related to the shielding gas flow rate, although the rates of vaporization is considered as independent of the gas flow rate in the arc plasma[3]. Considering the rate of vaporization is independent of the gas flow rate, it is possible to interpret this higher Mn content at 20 l/min setting to the faster transport kinetics in the welding arc plasma; not due to the higher vaporization rate.

Cr has demonstrated relatively stable results throughout the all selected flow rate settings for this sample material. The measured chemical composition of the Cr in various areas depend neither on the shielding gas flow rate nor the heat input. Moreover, comparing the latent enthalpy of vaporization values for Mn and Cr would make this assumption more comprehendible. Mn would require 225 kJ/mol to vaporize whereby Cr would need 347 J/mol[15]. Hence, this supports the possible Mn vaporization and consequent condensation to happen thermodynamically easier compared to Cr. Furthermore, Huber et al. reports that the temperature of the weld pool has a high impact on the metal vapor generation and the excitation degree of the individual alloying elements[16]. In this example, the welding heat input possibly lies below the critical level, therefore a distinguishable Cr composition difference on the weld metal was not observed.

4 Conclusions

Laser-induced breakdown spectroscopy method was used to investigate post TIG bead-on-plate welds and the possible metal vaporization and condensation phenomena due to the welding process. Key findings of the study are:

- Mn vaporize from the weld pool during welding and tend to be swept away from the welding arc along the weld bead. Instantaneously, it condenses on the HAZ and resulting in higher chemical compositions across both sides of the weld bead.
- Increasing the heat input results in higher Cr composition on the weld bead (FZ).
- Increasing the heat input results in higher Mn composition on the HAZ.
- Increasing shielding gas flow rate enhanced the Mn content accumulated at the HAZ.

We investigated post-weld individual trends of alloying elements chemical compositions for the selected stainless steel sample materials. The detected phenomena are in good agreement with the literature and the theory. Further research on online LIBS feedback for welding process regulation is in development.

Acknowledgements

Authors would like to thank Lasertechnik Berlin GmbH (LTB) for the analytical equipment. U.A. Taparli would like to thank Mr. Lars Jacobsen (LTB) for his contribution to the data processing.

References

[1] Park H, Trautmann M, Tanaka K, Tanaka M, Murphy A B. A computational model of gas tungsten arc welding of stainless steel: the importance of considering the different metal vapours simultaneously. 2018 *J. Appl. Phys.* **51**(39):395202.

[2] Blockbolten A, Eagar T W. Metal Vaporization from Weld Pools. 1984 *Metall Trans B*.15(**3**):461-9.

[3] Collur M M, Paul A, Debroy T. Mechanism of Alloying Element Vaporization during Laser-Welding. 1987 *Metall Trans B*.;**18**(4):733-40.

[4] Khan P A A, Debroy T. Alloying Element Vaporization and Weld Pool Temperature during Laser-Welding of AISI 202 Stainless-Steel. 1984 *Metall Trans B*.**15**(4):641-4.

[5] Yamamoto K, Tanaka M, Tashiro S, Nakata K, Yamazaki K, Yamamoto E, et al. Numerical Simulation for TIG Welding of Stainless Steel with Metal Vapor. 2008 *The International Conference on Computational & Experimental Engineering and Sciences*. **7**(1):1--6.

[6] Tanaka M, Tsujimura Y, Yamazaki K. Dynamic Behaviour of Metal Vapour in ARC Plasma During TIG Welding. 2012 *Weld World*.**56**(1):30-6.

[7] Tashiro S, Zeniya T, Yamamoto K, Tanaka M, Nakata K, Murphy AB, et al. Numerical analysis of fume formation mechanism in TIG welding. 2015 *Weld Int.* **29**(3):165-72.

[8] Tanaka K, Shigeta M, Tanaka M, Murphy AB. Investigation of the bilayer region of metal vapor in a helium tungsten inert gas arc plasma on stainless steel by imaging spectroscopy. 2019 *J. Appl. Phys.* **52**(35):354003.

[9] Taparli U A, Jacobsen L, Griesche A, Michalik K, Mory D, Kannengiesser T. In situ laser-induced breakdown spectroscopy measurements of chemical compositions in stainless steels during tungsten inert gas welding. 2018 *Spectrochim Acta B.* **139**:50-6.
[10] Taparli U A, Kannengiesser T, Cieslik K, Mory D, Griesche A. In situ chemical composition analysis of a tungsten-inert-gas austenitic stainless steel weld measured by laser-induced breakdown spectroscopy. 2020 *Spectrochim Acta B.* **167**:105826.
[11] Noll R. 2012 *Laser-induced breakdown spectroscopy fundamentals and applications*. (Berlin: Springer).
[12] Mueller M, Gornushkin I B, Florek S, Mory D, Panne U. Approach to Detection in Laser-Induced Breakdown Spectroscopy. 2007 *Anal Chem.* **79**(12):4419-26.
[13] Hahn D W, Omenetto N. Laser-Induced Breakdown Spectroscopy (LIBS), Part I: Review of Basic Diagnostics and Plasma—Particle Interactions: Still-Challenging Issues within the Analytical Plasma Community. 2010 *Appl Spectrosc.* **64**(12):335A-6A.
[14] Hahn D W, Omenetto N. Laser-Induced Breakdown Spectroscopy (LIBS), Part II: Review of Instrumental and Methodological Approaches to Material Analysis and Applications to Different Fields. 2012 *Appl Spectrosc.* **66**(4):347-419.
[15] Zhang Y M, Evans J R G, Yang S F. Corrected Values for Boiling Points and Enthalpies of Vaporization of Elements in Handbooks. 2011 *J Chem Eng Data.* **56**(2):328-37.
[16] Huber S, Glasschroeder J, Zaeh M F. Analysis of the Metal Vapour during Laser Beam Welding. Lasers in Manufacturing 2011 *Proceedings of the Sixth International WLT Conference on Lasers in Manufacturing* **12**, 712-9.

IOP Conf. Series: Materials Science and Engineering **882** (2020) 012024 doi:10.1088/1757-899X/882/1/012024

First principles density functional theory calculations on the elastic properties of Mo-Si based solid solutions

R S Touzani and M Krüger

Otto-von-Guericke University Magdeburg, Faculty of Mechanical Engineering, Institute for Materials and Joining Technology, Universitätsplatz 2, 39106 Magdeburg, Germany

rachid.touzani@ovgu.de

Abstract. The alloy system Mo-Si-B gained a lot of attention, as it showed superior elastic properties than currently used Ni based superalloys at elevated temperatures. In the Mo-Si-B alloy system, Mo based solid solutions play an important role as a compound and one can observe a solid solution with small amounts of Si and tiny, almost negligible, amounts of B. To improve the ductility of this alloy at higher temperatures, titanium is added and as a result Mo-Si-Ti solid solutions are formed. To get an idea of the elastic properties of new Mo based solid solutions, first principles density functional theory calculations serve as a powerful method for the prediction and qualitative, often almost quantitative results for the elastic moduli are achieved. Here we show the elastic properties like bulk, shear and Young's modulus as well as the Vickers hardness, calculated with first principles density functional theory (DFT), of Mo-Si-B-Ti based solid solutions with approximated atom concentrations Mo-1 at.% Si, Mo-3 at.% Si, Mo-2 at.% Si-1 at.% Ti, Mo-3 at.% Si-1 at.% B and Mo-3 at.% Si-1 at.% Ti.

1 Introduction

To improve the thermodynamic efficiency of aircraft engines, a slight increase of the gas inlet and therefore operating temperature would result in a significant efficiency gain. So, recent advances aim to overcome state-of-the-art Ni-based superalloys which are limited in their maximum operating temperatures with new high temperature materials [1]. Promising candidates for these applications are Mo-Si-B alloys, because they offer a balanced proportion of continuously distributed Mo solid solution phase and intermetallic phases (mainly the A15 and the T2 phase), which is important to obtain good high temperature and acceptable room temperature properties [2]. However, as Mo-Si-B alloys contain both ductile solid solution and brittle intermetallic phases, the distribution, volume fractions and properties of constituents must be tailored for specific load-dependent applications. Research on the Mo solid solution phase is aimed in avoiding embrittlement by interstitial elements as well as by solid solution elements likewise. Unfortunately, significant embrittlement is given intrinsically by the alloying element Si, which is dissolved up to concentrations about 3 at.% in technical alloys [3-5]. Another drawback is that the base material Mo, although a refractory metal with a high melting point of around 2600 °C, tends to have a high density (10.22 g/cm^3) [6], which is not particularly beneficial for aeronautic applications. To reduce the density and improve the ductility at higher temperature Ti is added to the Mo-Si-B alloy [7], similar to the alloying of Mo-Si-B with vanadium [8]. The effect of small amounts of Ti added to the Mo-Si solid solution was investigated recently in a study by Krüger et al. for a multi-phase Mo-Si-B-Ti alloy with a continuous Mo solid solution matrix [9]. Another study by

Schliephake et al. left out the boron and investigated the effect of large amount of Ti in the Mo-Si alloy system, where the T1 phase is formed instead of the T2 phase [10].

Density functional theory calculations on the elastic properties of the compounds in Mo-Si-B alloys were mainly focused on the intermetallic phases like $MoSi_2$, Mo_5Si_3, Mo_3Si, Mo_5SiB_2 and Mo_2B [11]. A study by Hütter et al. presented the elastic properties of bcc Mo, Mo_3Si and Mo_5SiB_2 [12]. DFT calculations on Ti_5Si_3 show a decrease of the bulk, shear and Young's modulus compared with Mo_5Si_3 [11, 13] indicating an improvement in ductility at low temperatures for titanium silicides. This finding is supported by a DFT study about $Nb_{5-x}Ti_xSi_3$. With increasing amount of Ti, the bulk, shear and Young's modulus is decreasing, improving the ductility at low temperatures [14]. Elastic properties via the DFT calculations on the quaternary T2 phase in Mo-V-Si-B alloys are available for $Mo_{5-x}V_xSiB_2$ borides and were presented recently by Touzani et al. They showed that the elastic properties improve with increasing amount of V [15]. Also, DFT calculations were used to explain the solid solution hardening (SSH) and weakening (SSW) effects by transition metals in bcc Mo [16-18]. It was shown that the electron concentration was the key factor to determine if alloying with a certain element would cause SSH or SSW. Also, as displayed in experiments, it was proven that SSW would only occur at low solute concentrations and with little change in the electron concentration at low temperatures ($T < 300$ K), while SSH is possible for higher solute concentration and temperatures ($T \geq 300$ K) [19]. However there is a lack of the DFT calculated intrinsic elastic properties of Mo-Si based solid solutions.

Therefore, in this work DFT calculations of the isotropic and anisotropic bulk, shear and Young's moduli as well as the Vickers hardness of Mo based solid solutions at 0 K will be presented. For this purpose models of the Mo-Si-B-Ti solid solutions Mo-0.781 at.% Si, Mo-3.125 at.% Si, Mo-2.344 at.% Si-0.781 at.%, Mo-3.125 at.% Si-0.781 at.% B, Ti and Mo-3.125 at.% Si-0.781 at.% Ti will be investigated.

2 Experimental procedure

To mimic the random distribution of atoms in the Mo-Si based solid solutions Mo-0.781 at.% Si, Mo-3.125 at.% Si, Mo-2.344 at.% Si-0.781 at.%, Mo-3.125 at.% Si-0.781 at.% B, Ti and Mo-3.125 at.% Si-0.781 at.% Ti special quasirandom structures (SQS) [20,21] with 128 atoms (129 for Mo-3.125 at.% Si-0.781 at.% B) per 4x4x4 bcc supercell made via the Monte Carlo SQS (mcsqs) code as implemented in the Alloy Theoretic Automated Toolkit (ATAT) [22] were used (see Figure 1). The SQS unit cells were then used as starting structures for further calculations. The first-principles calculation were carried out with Quantum ESPRESSO [23, 24] for the structural relaxation of solid Mo and the Mo-Si based solid solutions using PAW pseudopotentials [25] from the PSLibrary version 1.0.0 [26]. The kinetic energy cut-off of the plane waves were set to 100 Ry, while the cut-off for the charge density and potential was set to 400 Ry. The structural relaxation stopped until a total energy convergence of 10^{-5} Ry and a force convergence of 10^{-4} Ry/a_0 was reached. The Marzari-Vanderbilt cold smearing [27] and a Gaussian spreading of 0.01 Ry were chosen to account for the Brillouin-zone integration in metals. The k-mesh was divided by 4x4x4 for the Mo-Si solid solutions using the Monkhorst-Pack algorithm [28]. Exchange and correlation in this density functional theory (DFT) based method were treated with the generalized gradient approximation (GGA) functional as parameterized by Perdew, Burke and Ernzerhof (PBE-GGA) [29]. The elastic properties were determined with thermo_pw [30], a Fortran program using Quantum ESPRESSO routines as the underlying engine. In order to get the Voight-Reuss-Hill [31-33] approximated bulk, shear and Young's modulus, the standard algorithm and frozen ions were used. To calculate the Vickers hardness, Tian et al's formula [34] was used. In this work the isotropic and anisotropic elastic moduli will be presented. It is worth noting though that Tian's formula works best for solid state materials with an experimental Vickers hardness above 5 GPa [34].

For the sake of simplicity and to improve the readability in the following paragraphs Mo-0.781 at.% Si, Mo-3.125 at.% Si, Mo-2.344 at.% Si-0.781 at.% Ti, Mo-3.125 at.% Si-0.781 at.% B, and Mo-3.125 at.% Si-0.781 at.% Ti will be called Mo-1Si, Mo-3Si, Mo-2Si-1Ti, "Mo-3Si-1B" and "Mo-3Si-1Ti", respectively. The latter two are only hypothetical (hence the " ") and were not experimentally discovered and described yet.

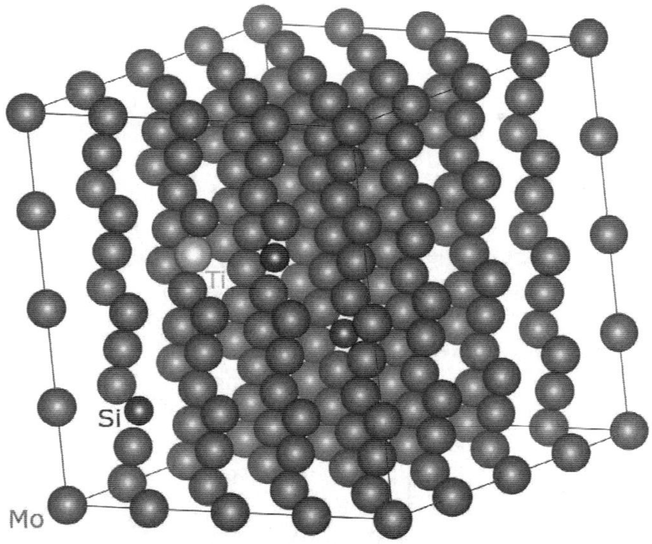

Figure 1. 128 SQS unit cell of the model of a bcc-Mo-2.3 at.% Si-0.8 at.% Ti (Mo-2Si-1Ti) solid solution.

3 Results and Discussion

3.1 Structural relaxation

First, the results of the structural relaxation will be discussed (See Table 1).

For the element molybdenum in the solid state the lattice parameter is 3.1610 Å. In the study by Sturm et al. [5] the experimental lattice parameter was reported to be 3.1478 Å. So the lattice parameter from our DFT calculation is in good agreement with the experiment. As the radius of the Si atom in crystals is smaller (Mo: 145 pm, Si: 110 pm), one expects a smaller lattice parameter with increasing Si concentration. This is indeed true as the lattice parameter of Mo-1Si and Mo-3Si are 3.1588 Å and 3.1537 Å, respectively. For Mo-2Si-1Ti the situation is a bit different. The radius of the Ti atom (140 pm) is about the same as of Mo, the lattice parameter should be between the ones of Mo-1Si and Mo-3Si as the concentration of Si is higher than in Mo-1Si, but lower than in Mo-3Si. In fact, the lattice parameter is 3.1554 Å, which is right between the lattice parameters mentioned above. The lattice parameter for the hypothetical "Mo-3Si-1B" (3.1592 Å) is larger than the lattice parameter of Mo-3Si (3.1537 Å) as expected due to the occupation of an interstitial site by the boron atom. However, the lattice parameter of the hypothetical "Mo-3Si-1Ti" is expected to be similar to the lattice parameter of Mo-3Si, as Mo is substituted by Ti. This is true, because the lattice parameter is 3.1536 Å while the lattice parameter of Mo-3Si is 3.1537 Å.

Table 1. Lattice parameter a, as well as the isotropic bulk modulus B, shear modulus G, Young's modulus E and Vickers hardness H^V of Mo, Mo-1Si, Mo-3Si, Mo-2Si-1Ti, "Mo-3Si-1B" and "Mo-3Si-1Ti".

Compound	a (Å)	B (GPa)	G (GPa)	E (GPa)	H^V (GPa)
Mo	3.1610	258.7	117.1	305.1	10.88
Mo-1Si	3.1588	260.1	116.8	304.8	10.77
Mo-3Si	3.1537	258.0	113.1	296.0	10.23
Mo-2Si-1Ti	3.1554	257.4	113.2	296.2	10.29
"Mo-3Si-1B"	3.1592	256.9	112.5	294.5	10.19
"Mo-3Si-1Ti"	3.1536	256.7	112.4	294.1	10.17

3.2 Elastic properties

In the next two paragraphs the DFT calculated isotropic and anisotropic elastic properties will be presented and discussed.

3.2.1 Isotropic elastic properties. In this paragraph the isotropic elastic properties e.g. bulk, shear, Young's moduli and Vickers hardness of Mo, Mo-1Si, Mo-3Si, Mo-2Si-1Ti, "Mo-3Si-1B" and "Mo-3Si-1Ti" (see also Table 1) will be discussed. The isotropic bulk moduli of Mo, Mo-1Si, Mo-3Si, Mo-2Si-1Si, "Mo-3Si-1B" and "Mo-3Si-1Ti" are very similar: 258.7 GPa, 260.1 GPa, 258.0 GPa, 257.4 GPa 256.9 GPa and 256.7 GPa, respectively. A slightly different situation is apparent for the shear and Young's moduli. Regarding these properties, it is possible to divide the six compounds into three groups of two: For Mo and Mo-1Si the isotropic shear moduli are 117.1 GPa and 116.8 GPa, respectively, and therefore very similar to each other. Lower shear moduli are found for Mo-3Si and Mo-2Si-1Ti as they are 113.1 GPa and 113.2 GPa, respectively, and therefore almost the same. A similar situation is present for the shear moduli of Mo-3Si-1B" and "Mo-3Si-1Ti". The shear moduli of these hypothetical compounds are 112.5 GPa and 112.4 GPa and very close to each other. The isotropic Young's moduli of Mo, Mo-1Si, Mo-3Si, Mo-2Si-1Ti, "Mo-3Si-1B" and "Mo-3Si-1Ti" are 305.1 GPa, 304.8 GPa, 296.0 GPa, 296.2 GPa, 294.5 GPa respectively. The isotropic Vickers hardness are 10.88 GPa, 10.77 GPa, 10.23 GPa, 10.29 GPa, 10.19 GPa and 10.17 GPa for Mo, Mo-1Si, Mo-3Si, Mo-2Si-1Ti, "Mo-3Si-1B" and "Mo-3Si-1Ti", respectively. The Vickers hardness of Mo-3Si and "Mo-3Si-1B" are very similar to each other (10.23 GPa, 10.19 GPa) and could be sign that interstitial atoms at low concentration do not have a large influence on the isotropic Vickers hardness of Mo-Si based solid solutions. Interestingly, "Mo-3Si-1B" (10.19 GPa) and "Mo-3Si-1Ti" (10.17 GPa) have similar Vickers hardness, despite a different mechanism of doping (interstitial site vs transition metal substitution).

The bulk, shear and Young's modulus of Mo according by a DFT study by Hütter et al. using LDA are 302 GPa, 135 GPa and 352 GPa, respectively [12]. The Vickers hardness using Tian et al's formula is 11.87 GPa. The calculated elastic properties are therefore higher than in our DFT study, which might be due to the well-known "overbinding" problem of LDA, overestimating the strength of chemical bonding.

3.2.2 Anisotropic elastic properties. In this paragraph the anisotropic shear, Young's moduli and Vickers hardness of Mo, Mo-1Si, Mo-3Si, Mo-2Si-1Ti, "Mo-3Si-1B" and "Mo-3Si-1Ti "(see also Figures 2-4) are discussed. Due to the cubic crystal structure, the bulk moduli of all the compounds are isotropic and were already described in the paragraph above. As the crystal structure of the Mo based solid solution compounds is very similar, the minimum and maximum of the other elastic properties shear moduli, Young's moduli and Vickers hardness are along the same crystallographic directions.

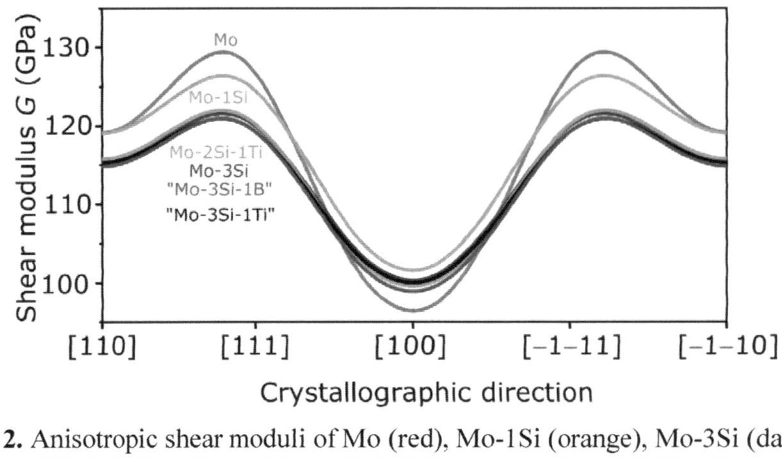

Figure 2. Anisotropic shear moduli of Mo (red), Mo-1Si (orange), Mo-3Si (dark blue), Mo-2Si-1Ti (light blue), "Mo-3Si-1B" (cyan) and "Mo-3Si-1Ti" (black).

The shear moduli of the Mo-based solid solutions (Figure 2) show their maximum along the vicinity of the [111] direction, while the minimum can be seen along the [100] direction. The shear moduli of Mo and Mo-1Si along the [110] direction are 119.2 GPa and 119.1 GPa, respectively and therefore very similar. The shear moduli of Mo-3Si, Mo-2Si-1Ti, "Mo-3Si-1B" and "Mo-3Si-1Ti" along the same direction are 115.5 GPa, 115.8 GPa, 115.3 GPa, 114.8 GPa, respectively, and very similar for these four compounds. The maxima of the shear moduli of Mo, Mo-1Si, Mo-3Si, Mo-2Si-1Ti, "Mo-3Si-1B" and "Mo-3Si-1Ti" are 129.4 GPa, 126.4 GPa, 121.6 GPa, 122.0 GPa, 120.9 GPa, and 121.0 GPa. The minima of the shear moduli are 96.4 GPa, 101.6 GPa, 99.3 GPa, 99.0 GPa, 97.9 GPa and 98.9 GPa for Mo, Mo-1Si, Mo-3Si, Mo-2Si-1Ti, "Mo-3Si-1B" and "Mo-3Si-1Ti", respectively. Unlike the isotropic shear moduli the maxima and minima of the anisotropic shear moduli of Mo and Mo-1Si differ from each other and more so for the shear moduli of Mo-3Si, Mo-2Si-1Ti, "Mo-3Si-1B" and "Mo-3Si-1Ti", which are very similar to each other. The anisotropic behavior of the shear moduli in general is very similar for Mo-3Si, Mo-2Si-1Ti, "Mo-3Si-1B" and "Mo-3Si-1Ti". Experimental values of the shear moduli are present for Mo and Mo-0.34 at.% Si [5]. The latter might be qualitatively compared to Mo-1Si of our DFT calculation. The shear moduli of Mo and Mo-0.34 at.% Si are 117.4 GPa and 119.9 GPa, respectively. These values are just between the maxima (129.4 GPa and 126.4 GPa) and minima (96.4 GPa and 101.6 GPa) of Mo and Mo-1Si, respectively. Unfortunately, the crystallographic directions during the experimental shear moduli measurements are unknown, making a valid comparison difficult. According to the study by Sturm et al. [5] the shear modulus seems to decrease with increasing Si concentration above 0.34 at.%.

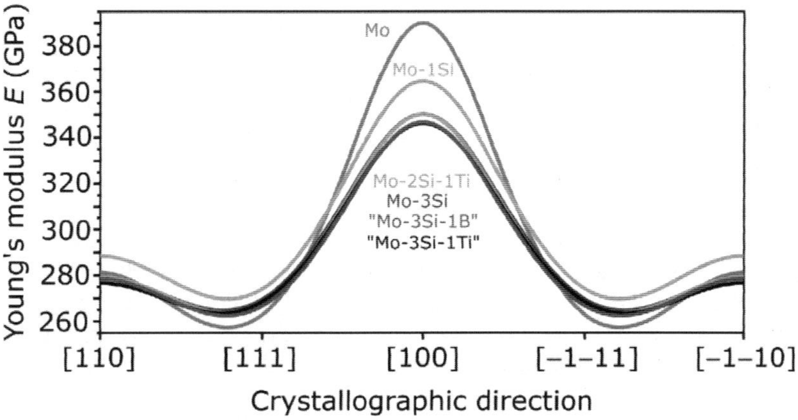

Figure 3. Anisotropic Young's moduli of Mo (red), Mo-1Si (orange), Mo-3Si (dark blue), Mo-2Si-1Ti (light blue), "Mo-3Si-1B" (cyan) and "Mo-3Si-1Ti" (black).

A different behavior for the anisotropic Young's moduli is observed (see Figure 3). The maxima are along the [100] direction, while the minima are in the vicinity of the [111] direction. Along the [110] direction the Young's moduli of Mo-1Si, is 288.4 GPa and therefore larger than the Young's moduli of other five compounds Mo (281.2 GPa), Mo-3Si (278.8 GPa), Mo-2Si-1Ti (279.5 GPa), "Mo-3Si-1B" (276.7 GPa) and "Mo-3Si-1Ti" (278.5 GPa), which are very similar. The maxima of the Young' modulus of Mo, Mo-1Si, Mo-3Si, Mo-2Si-1Ti, "Mo-3Si-1B" and "Mo-3Si-1Ti" are 390.0 GPa, 364.6 GPa, 348.3 GPa, 351.2 GPa, 347.5 GPa and 346.7 GPa, respectively. The minima of the Young's moduli are 257.2 GPa, 269.6 GPa, 264.7 GPa, 263.7 GPa, 263.5 GPa and 262.4 GPa for Mo, Mo-1Si, Mo-3Si, Mo-2Si-1Ti, "Mo-3Si-1B" and "Mo-3Si-1Ti", respectively. Similar to the shear moduli and unlike the isotropic Young's moduli, the minima and maxima of the Young's moduli of Mo and Mo-1Si differ from each other, while the anisotropic behavior as well as the minima and maxima of the Young's moduli of Mo-3Si, Mo-2Si-1Ti, "Mo-3Si-1B" and "Mo-3Si-1Ti" are very similar to each other. The experimental Young's moduli for Mo and Mo-0.34 at.% Si are 302.6 GPa and 307.5 GPa, respectively and lie between the maxima (390.0 GPa and 364.6 GPa) and minima (257.2 GPa and

269.6 GPa) of the DFT calculated anisotropic Young's moduli. Again, the crystallographic directions during the experimental Young's moduli measurements are unknown. Similar to the shear moduli the experimental Young's modulus seems to decrease with increasing Si concentration above 0.34 at.% [5].

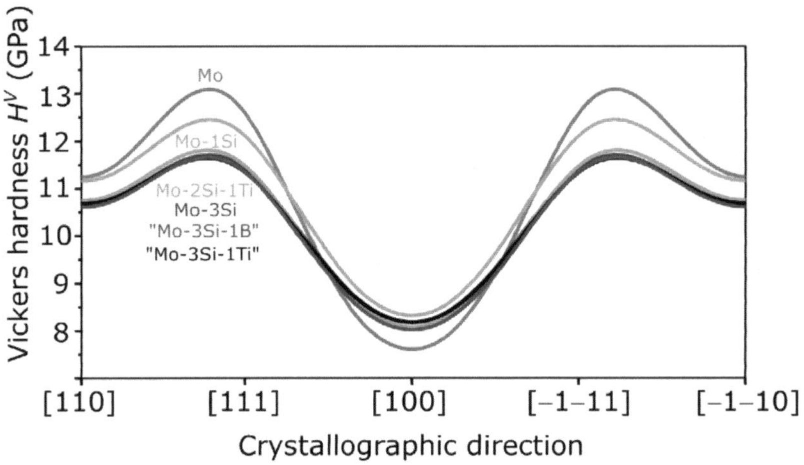

Figure 4. Anisotropic Vickers hardness of Mo (red), Mo-1Si (orange), Mo-3Si (dark blue), Mo-2Si-1Ti (light blue), "Mo-3Si-1B" (cyan) and "Mo-3Si-1Ti" (black).

As the Vickers hardness according to Tian et al. [34] is proportional to the shear modulus, a similar anisotropic behavior of the Vickers hardness and the shear modulus is expected. Indeed, the maxima of the Vickers hardness is along the vicinity of the [111] direction, while the minima are along the [100] direction (see Figure 4). The Vickers hardness along the [110] direction of Mo, and Mo-1Si are 11.25 GPa and 11.16 GPa, respectively, and very similar. For Mo-3Si, Mo-2Si-1Ti, "Mo-3Si-1B" and "Mo-3Si-1Ti" the Vickers hardness along [110] is lower and within the same range: 10.67 GPa, 10.75 GPa, 10.75 GPa, 10.68 GPa and 10.61 GPa, respectively. The maxima of the Vickers hardness of Mo, Mo-1Si, Mo-3Si, Mo-2Si-1Ti, "Mo-3Si-1B" and "Mo-3Si-1Ti" are 13.09 GPa, 12.45 GPa, 11.71 GPa, 11.81 GPa, 11.64 GPa and 11.66 GPa, respectively. The minima for Mo, Mo-1Si, Mo-3Si, Mo-2Si-1Ti, "Mo-3Si-1B" and "Mo-3Si-1Ti" are 7.60 GPa, 8.32 GPa, 8.18 GPa, 8.08 GPa, 8.18 GPa and 8.02 GPa, respectively. Similar to the other elastic moduli, the anisotropic behavior of Mo-3Si, Mo-2Si-1Ti, "Mo-3Si-1B" and "Mo-3Si-1Ti" for the Vickers hardness is very akin. However, one must take the results of the Vickers hardness calculations of Mo and Mo-1Si with a pinch of salt. The experimental Vickers hardness was reported to be 2.61 GPa [35] and 4.12 GPa [3], respectively, which is below 5 GPa and Tian et al's formula might be inaccurate in this case. Also it was shown in Ref. [19] that temperatures above 0 K will cause a decrease of the Vickers hardness in bcc Mo solid solutions, while DFT calculations consider only the ground state properties at 0 K. The experimental Vickers hardness at room temperature of Mo-2.8 at.% Si and Mo-1.8 at.% Si-0.6 at.%Ti are 8.51 GPa and 7.76 GPa, which are close to the results of the qualitative similar solid solutions Mo-3Si and Mo-2Si-1Ti used in the DFT calculations presented here (8.18 GPa and 8.08 GPa, respectively). The small disagreement might be due to the solid solution hardening (SSH) effect which might take place at room temperature or higher temperatures [19] and the DFT calculations do not include the SSH effect appropriately as mentioned above. Also, another reason for the deviation is the fact that the crystallographic directions during the experimental Vickers hardness measurements are unknown. The qualitative change of the Vickers hardness, however, is observable and displays in accordance with the experiment, the increase in hardness with increasing Si content and electron concentration. On the other hand, increasing the silicon concentration from Mo-2Si-1Ti to "Mo-3Si-1Ti" seems to lower not only the isotropic Vickers hardness but the anisotropic Vickers hardness, from 11.81 GPa/11.66 GPa to 8.08 GPa/8.02 GPa, respectively, as silicon is a substitute for molybdenum. Investigating the influence of interstitial atoms, "Mo-3Si-1B" shows a similar anisotropic Vickers hardness to Mo-3Si (Maxima: 11.71 GPa and

11.64 GPa, minima: 8.18 GPa and 8.18 GPa). This might be another sign, that interstitial atoms have only little influence on the Vickers hardness at low concentrations as already observed for the isotropic Vickers hardness. The same is true for the substitution of Si with Ti for Mo-3Si to Mo-2Si-1Ti as the anisotropic behavior of the Vickers hardness is very similar (Minima: 8.18 GPa/8.08 GPa, maxima: 11.71 GPa/11.81 GPa, respectively).

Interestingly, Mo shows the most anisotropic behavior in terms of the elastic moduli as its maxima and minima of the shear modulus, Young's modulus and Vickers hardness are the highest and the lowest, respectively, compared to Mo-1Si, Mo-3Si, Mo-2Si1Ti, "Mo-3Si-1B" and "Mo-3Si-1Ti" according to our DFT calculations.

4 Conclusions

DFT calculations to explain the elastic behavior of Mo solid solutions were performed using experimental and hypothetical alloy compositions including the elements Si, B and Ti. The calculated isotropic bulk, shear and Young's moduli as well as the Vickers hardness of Mo-Si-B-Ti solid solutions are within the groups of Mo/Mo-1Si, Mo-3Si/Mo-2Si-1Ti and "Mo-3Si-1B"/"Mo-3Si-1Ti" very similar and in decreasing order. However, the DFT calculations of the anisotropic elastic properties of these compounds show that the shear and Young's moduli as well as the Vickers hardness of Mo and Mo-1Si differ more than expected, while the anisotropic shear and Young's moduli as well as the Vickers hardness for Mo-3Si, Mo-2Si-1Si, "Mo-3Si-1B" and "Mo-3Si-1Ti" are very similar. In general, the Young's moduli of all the compounds investigated here are very anisotropic, while the shear moduli and Vickers hardness show less anisotropic behavior. However and quite surprisingly, Mo show the highest anisotropic behavior for all elastic moduli presented here.

Acknowledgements

The authors appreciate for funding by the Federal Ministry of Education and Research (BMBF) project FlexiDS (project number 05K16NMA) and by the MDZ Werkstoffprüfung e.V. Magdeburg. RT thanks Dr. Julia Becker (Otto-von-Guericke University Magdeburg) for fruitful discussions and the URZ OVGU Magdeburg for calculation power and time.

References

[1] Reed R C 2006 *The Superalloys: Fundamentals and Applications* (Cambridge: Cambridge University Press)
[2] Becker J and Krüger M 2015 Der Einfluss der Phasenverteilung auf die Bruchzähigkeit von hochtemperatur-beständigen Mo-Si-B- Legierungen *Practical Metallography* **52** 295-313
[3] Jain P and Kumar K S 2010 Dissolved Si in Mo and its effects on the properties of Mo-Si-B alloys *Scr. Mater.* **62** 1–4
[4] Krüger M, Jain P, Kumar K S and Heilmaier M, 2014 Correlation between microstructure and properties of fine grained Mo-Mo_3Si-Mo_5SiB_2 alloys *Intermetallics* **48** 10–18
[5] Sturm D, Heilmaier M J.H. Schneibel J H, Jéhanno P, Skrotzki B, and Saage H 2007 The influence of silicon on the strength and fracture toughness of molybdenum *Mater. Sci. Eng.* A. **463** 107–14.
[6] Binder H H 1999 *Lexikon der chemischen Elemente* (Stuttgart: S. Hirzel Verlag)
[7] Azim M A, Schliephake D, Hochmuth C, Gorr B, Christ H J, Glatzel U and Heilmaier M 2015 Creep Resistance and Oxidation Behavior of Novel Mo-Si-B-Ti Alloys *JOM* **67** 2621–28
[8] Becker J, Betke U, Hoffmeister M and Krüger M 2018 Density Reduction of Mo-Si-B Alloys by Vanadium Alloying *JOM* **10** 2574-81
[9] Krüger M, Becker J, Michael O and Touzani R S *submitted to Intermetallics*
[10] Schliephake D, Kauffmann A, Cong X, Gombola C, Azim M, Gorr B, Christ H-J and Heilmaier M 2019 Constitution , oxidation and creep of eutectic and eutectoid Mo-Si-Ti alloys *Intermetallics* **104** 133–42

[11] Aryal S, Gao M C, Ouyang L, Rulis P and Ching W Y 2013 Ab initio studies of Mo-based alloys: Mechanical, elastic, and vibrational properties *Intermetallics* **38** 116-25

[12] Hütter S, Hasemann G, Al-Karawi J, Krüger M and Halle T 2018 *Metall. Mater. Trans* A **49** 6075-83

[13] Poletaev D O, Lipnitskii A G, Kartamyshev A I, Aksyonov D A, Tkachev E S, Manokhin S S, Ivanov M B and Kolobov Yu R 2014 Ab initio-based prediction and TEM study of silicide precipitation in titanium, *Comput. Mat. Sci.* **95** 456-63

[14] Papadimitriou I, Utton C and Tsakiropoulos P 2017 The impact of Ti and temperature on the stability of Nb_5Si_3 phases: a first-principles study, Sci. Technol. *Adv. Mater.* **18** 467-79

[15] Touzani R S, Becker J and Krüger M 2020 Site preference of V and its influence on the elastic properties in the boride series VxMo5-xSiB2 as studied by first principles density functional theory, *J. Alloys Comp.* **819** 153041

[16] Medvedeva N I, Gornostyrev Yu N and Freeman A J 2005 Electronic Origin of Solid Solution Softening in bcc Molybdenum Alloys *Phys. Rev. Lett.* **94** 136402

[17] Medvedeva N I, Gornostyrev Yu N and Freeman A J 2005 Solid solution softening in bcc Mo alloys: Effect of transition-metal additions on dislocation structure and mobility *Phys. Rev.* B **72** 134107

18] Medvedeva N I, Gornostyrev Yu N and Freeman A J 2007 Solid solution softening and hardening in the group-V and group-VI bcc transition metals alloys: First principles calculations and atomistic modeling *Phys. Rev.* B **76** 212104

[19] Stephens J and Witzke W R 1972 Alloy hardening and softening in binary molybdenum alloys as related to electron concentration, *J. Less Common Met.* **29** 371-88

[20] Zunger A, Wie S-H, Ferreira L G and Bernard J E 1990 Special Quasirandom Structures *Phys. Rev. Lett* **65**, 353–6

[21] Wie S-H, Ferreira L G, Bernard J E and Zunger A 1990 Electronic properties of random alloys: Special quasirandom structures, *Phys. Rev.* B, **42**, 9622–49

[22] van de Walle A, Tiwary P, de Jong M, Olmsted D L, Asta M, Dick A, Shin D, Wang Y, Chen L-Q, and Liu Z-K 2013 Efficient stochastic generation of special quasirandom structures, *Calphad J.* **42**, 13–18

[23] Giannozzi P, Baroni S, Bonini N, Calandra M, Car R, Cavazzoni C, Ceresoli D, Chiarotti G L, Cococcioni M, Dabo I, Dal Corso A, de Gironcoli S, Fabris S, Fratesi G, Gebauer R, Gerstmann U, Gougoussis C, Kokalj A, Lazzeri M, Martin-Samos L, Marzari N, Mauri F, Mazzarello R, Paolini S, Pasquarello A, Paulatto L, Sbraccia C, Scandolo S, Sclauzero G, Seitsonen A P, Smogunov A, Umari P, and Wentzcovitch R M 2009 Quantum ESPRESSO: a modular and open-source software project for quantum simulations of materials *J. Phys. Condens. Matter* **21**, Art. 395502

[24] Giannozzi P, Andreussi O, Brumme T, Bunau O, Buongiorno Nardelli M, Calandra M, Car R, Cavazzoni C, Ceresoli D, Cococcioni M, Colonna N, Carnimeo I, Dal Corso A, de Gironcoli S, Delugas P, DiStasio R A, Ferretti A, Floris A, Fratesi G, Fugallo G, Gebauer R, Gerstmann U, Giustino F, Gorni T, Jia J, Kawamura M, Ko H-Y, Kokalj A, Küçükbenli E, Lazzeri M, Marsili M, Marzari N, Mauri F, Nguyen N L, Nguyen H-V, Otero-de-la-Roza A, Paulatto L, Poncé S, Rocca D, Sabatini R, Santra B, Schlipf M, Seitsonen A P, Smogunov A, Timrov I, Thonhauser T, Umari P, Vast N, Wu X, and Baroni S, Advanced capabilities for materials modelling with Quantum ESPRESSO, J. Phys. Condens. Matter, **2017**, 29, Art. 465901

[25] Blöchl P E 1994 Projector augmented-wave method *Phys. Rev.* B **50** 17953-79

[26] Dal Corso A 2014 Pseudopotentials periodic table: From H to Pu, *Comput. Mat. Sci.* **95** 337-50

[27] Marzari N, Vanderbilt D, De Vita A and Payne M C 1999 Thermal Contraction and Disordering of the Al(110) Surface, *Phys. Rev. Lett.* 82 3296-9

[28] Monkhorst H J and Pack J D 1976 Special points for Brillouin-zone integrations *Phys. Rev.* B, 13, 5188-92

[29] Perdew J P, Burke K and Ernzerhof M 1996 Generalized Gradient Approximation Made Simple, *Phys. Rev. Lett,* 77, 3865-8
[30] thermo_pw is an extension of the quantum ESPRESSO (QE) package which provides an alternative organization of the QE work-flow for the most common tasks. For more information see: https://dalcorso.github.io/thermo_pw/
[31] Voigt W 1928 *Lehrbuch der Kristallphysik* (Leipzig and Berlin: BG Teubner)
[32] Reuss A 1929 Berechnung der Fließgrenze von Mischkristallen auf Grund der Plastizitätsbedingung für Einkristalle *Z. Angew, Math. Mech.* **9** 49-58
[33] Hill R 1952 The Elastic Behaviour of a Crystalline Aggregate *Proc. Phys. Soc*, A **65**, 349-54
[34] Tian Y, Xu B and Zhao Z 2012 Microscopic theory of hardness and design of novel superhard crystals *Int. J. Refract. Met. H.* **33** 93-106
[35] Becker J, Betke U, Wessel E and Krüger M 2018 Alloying effects in Mo-5X (X=Zr, Ti,V) - microstructural modifications and mechanical properties, *Mater Today Comm.* **15**, 314-21

IOP Conf. Series: Materials Science and Engineering **882** (2020) 012025 doi:10.1088/1757-899X/882/1/012025

Microstructural Characterization of Quenched and Partitioned commercial Medium Carbon Steel

A Wilke[1], S Dieck[1], M Härtel[2], T Lampke[2] and T Halle[3]

[1] DeltaSigma Analytics GmbH, Mittagstraße 16p, 39124 Magdeburg
[2] Institute of Materials Science and Engineering, Chemnitz University of Technology, 09125 Chemnitz, Germany
[3] Otto-von-Guericke-University Magdeburg, Faculty of Mechanical Engineering, Institute of Materials and Joining Technology, Universitätsplatz 2, 39106 Magdeburg

alisa.wilke@deltasigma.de

Abstract. In order to achieve the desired properties, the microstructure of metals is often modified by heat treatments. A suitable combination of high strength and good ductility can be achieved by adjusting the amounts of martensite and retained austenite through a Quenching and Partitioning (Q&P) process. Controlling these material properties offers new potentials in the production of light weight steel parts. This paper presents a basic study of heat treatment conditions with subsequent microstructural characterization of quenched and partitioned commercial 0.54C-1.45Si-0.71Mn spring steel. Beginning from full austenitization, different quenching and partitioning parameters were applied. Microstructural characterization was performed using optical microscopy in combination with different etching methods as well as Scanning Electron Microscopy (SEM) and Electron Backscattered Diffraction (EBSD). It is shown that the intended dual phase microstructure with a specific phase fraction of austenite can be produced. Those results and consideration of hardness measurements show that the investigated commercial spring steel is a promising candidate for improving mechanical properties by a Q&P process.

1 Introduction

Advanced high strength steels offer new possibilities in lightweight construction. Especially for automotive parts increasing strength while keeping ductility at sufficient values can allow significant weight reduction. Over the last years quenching and partitioning (Q&P) process has become a popular heat treatment process for producing excellent materials properties, combining high strength and good ductility. The aimed properties are achieved by an individual material-dependent processing route, starting from full or partial austenitization followed by quenching to a defined temperature between martensite start (M_s) and martensite finish (M_f) temperature. The resulting microstructure contains supersaturated martensite as well as a fraction of retained austenite. During the following low temperature tempering (partitioning) the retained austenite is stabilized to room temperature by carbon diffusion from martensite [1-3]. Q&P heat treatment is either performed as one or two step process. Figure 1 shows the processing route for each variant [4]. Due to this heat treatment the materials properties develop from brittle (as quenched) to a high strength and ductile mechanical behavior (partitioned). During the low temperature tempering carbon escapes from the supersaturated martensite, which becomes formable with only a small decrease in strength. The carbon enriches the retained austenite. As a consequence, M_s of the retained austenite is lowered below room temperature. The austenite fraction enhances the formability of the material. Furthermore, the austenite transforms into martensite under mechanical loading (TRIP effect), which increases the materials strength [5, 6].

Steels used for Q&P heat treatment must contain a certain amount of carbon for hardenability. Additionally, Silicon is needed to prevent carbide formation and Manganese for expanding gamma phase area [7-10]. An appropriate Q&P heat treatment can lead to a unique combination of tensile strength and total elongation. For low carbon steel R_m = 1047 MPa and very high elongation of 15.5 % can be reached [11]. Medium carbon steels reach R_m = 2000 MPa and a total elongation beyond 10 % [12-15].

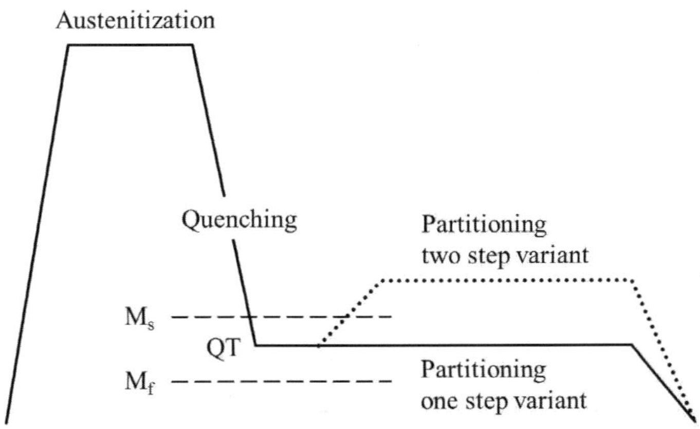

Figure 1. Schematic processing variants of Q&P heat treatments with quenching temperature QT defined as $M_s > QT > M_f$ [4].

0.54C-1.45Si-0.71Mn is a material that is already widely used in constructing automotive parts and may be a key for industrial application of the Q&P process. The chemical composition especially with a comparatively high amount of silicon makes this kind of steel a promising candidate for Q&P heat treatment. The present study examines appropriate heat treatment parameters for Q&P by the investigation of parameter dependent microstructure development.

2 Experimental procedure

The material investigated in this study is a 0.54C-1.45Si-0.71Mn spring steel and provided from our customer thyssenkrupp. Chemical composition is given in Table 1. Q&P heat treatment was performed as a dilatometric study. Dilatometric specimens were produced by electrical discharge machining (EDM) and had dimensions of 4 mm in diameter and 10 mm in length. The austenitization was performed under vacuum at 950 °C for 100 s. First, the transformation temperatures were measured by hardening the material to room temperature using He gas as quenching medium. Derived from the measured M_s and M_f temperatures for Q&P heat treatment the samples were quenched to either 175 °C or 200 °C to vary the fraction of retained austenite. For partitioning the samples were reheated to the partitioning temperature (PT) of 250 °C for QT = 175 °C and 300 °C and 400 °C, respectively, for QT = 200 °C. The partitioning time (Pt) was fixed to 600 s. An overview is given in Table 2.

The microstructure of each specimen was investigated by optical microscopy and scanning electron microscopy (SEM) using 3 % HNO_3 (Nital) etchant and Klemm I color etchant as contrasting method after grinding and polishing. Volume fractions of retained austenite were analyzed by electron backscatter diffraction (EBSD), whereas areas with poor signal quality (Confidence Index below 0.1) were excluded. For all EBSD measurements FEI Scios DualBeam scanning electron microscope was used. Vickers micro hardness testing was performed on a Buehler Wilson VH3300.

IOP Conf. Series: Materials Science and Engineering **882** (2020) 012025 doi:10.1088/1757-899X/882/1/012025

Table 1. Chemical composition (wt.%) of the investigated material.

C	Si	Mn	Cr	Ni	Cu	P	S	Fe
0.54	1.45	0.71	0.63	0.04	0.06	< 0.01	< 0.01	Bal.

Table 2. Overview of chosen parameters for Q&P heat treatment.

Specimen	QT [°C]	PT [°C]	Pt [s]
Hardened	RT	-	-
T2.1	175	250	600
T2.2	200	300	600
T2.3	200	400	600

3 Results

3.1 Transformation temperatures

As can be seen in Figure 2 the initial material is primary perlitic with embedded islands of ferrite. The dilatometric curve for the measurement of transformation temperatures is shown in Figure 3. Therefore, the hardening process was performed by quenching to room temperature. The investigated steel 0.54C-1.45Si-0.71Mn is characterized by M_s = 265 °C and M_f = 100 °C. Hence, for Q&P the quenching temperature was set to be 175 °C and 200 °C, respectively, to retain different amounts of austenite.

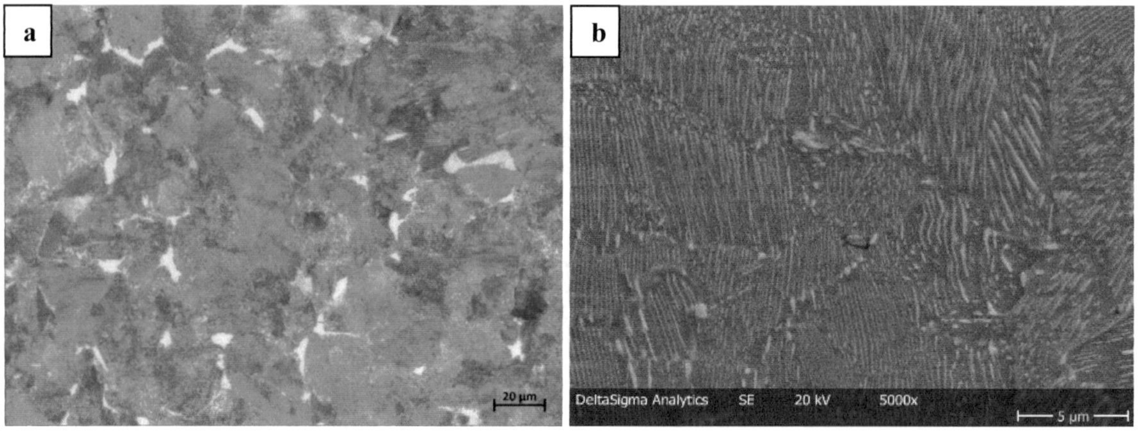

Figure 2. Micrographs of the initial material (**a** optical microscopy, **b** SEM) showing a primary perlitic matrix with ferritic islands.

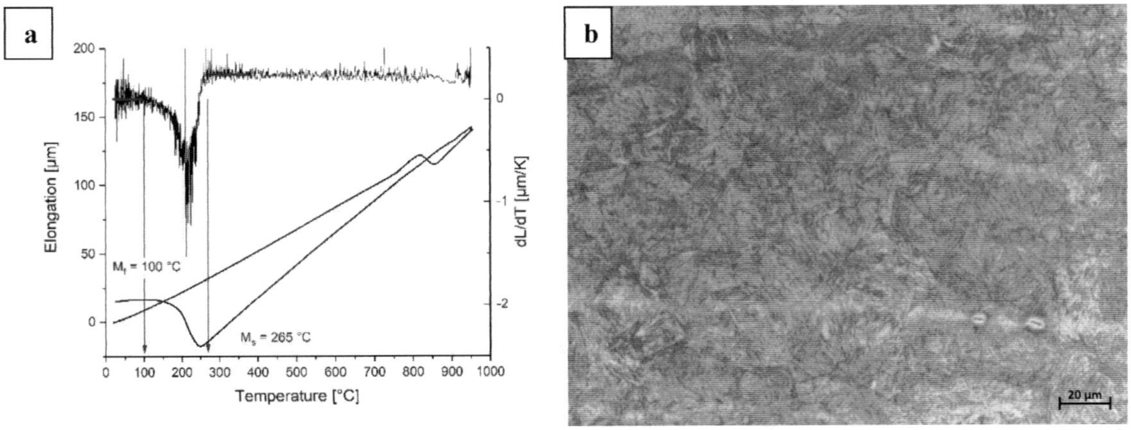

Figure 3. Dilatometric graph for quenching 0.54C-1.45Si-0.71Mn for austenitization at 950 °C for 100 s (**a**) and the resulting martensitic microstructure for reference (**b**).

3.2 Microstructural evolution

From the curves of dilatometer measurement (Figure 4) can be seen, that the sample with QT = 175 °C is characterized by a gradient in slope during final cooling, indicating a martensitic transformation. The specimen with QT = 200 °C shows a constant slope during final cooling.

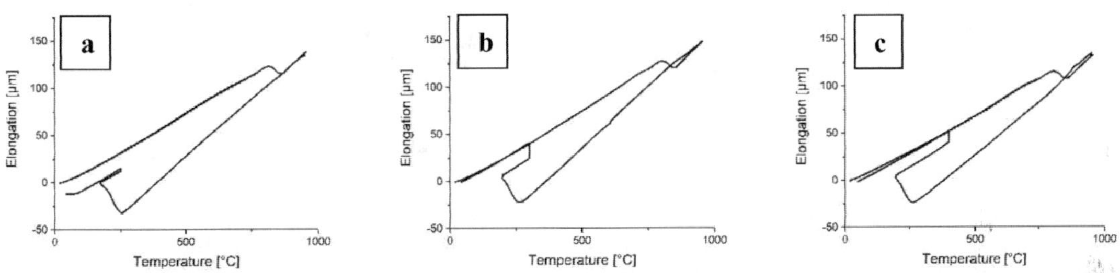

Figure 4. Temperature dependent elongation during Q&P (**a** QT = 175 °C, PT = 250 °C; **b** QT = 200 °C, PT = 300 °C; **c** QT = 200 °C, PT = 400 °C).

The microstructure after quenching (Figure 3 **b**) shows martensitic laths and, as expected, no evidence of austenite. Figure 5 shows optical micrographs of microstructures resulting from the selected Q&P parameters. The specimens quenched to 175 °C and partitioned at 250 °C do not show any sure indication of austenitic fractions. The sharp needle-like martensitic structure is comparable with the as-quenched stage.

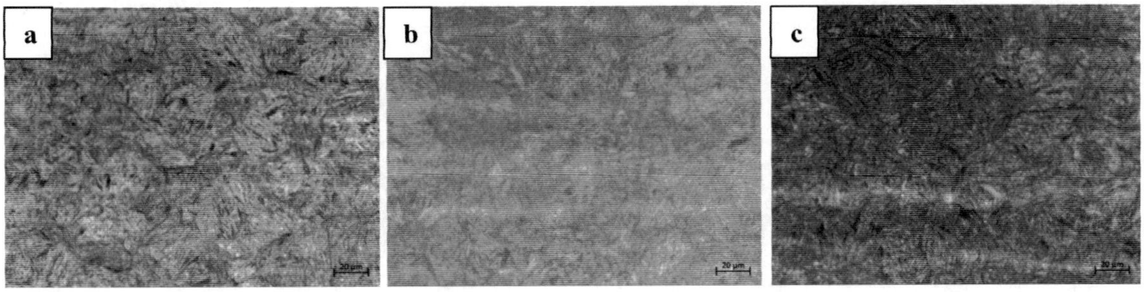

Figure 5. Optical micrographs of etched surfaces of 0,54C-1,45Si-0,71Mn after Q&P (**a** QT = 175 °C, PT = 250 °C; **b** QT = 200 °C, PT = 300 °C; **c** QT = 200 °C, PT = 400 °C).

For the specimen quenched to 200 °C the sharp needle-like martensitic structure is less pronounced compared to the 175 °C condition. The small bright regions between the martensitic laths can be observed in both specimens, see Figure 6, and can be attributed to the formation of austenite [16]. The fraction of austenite increases with higher partitioning temperature.

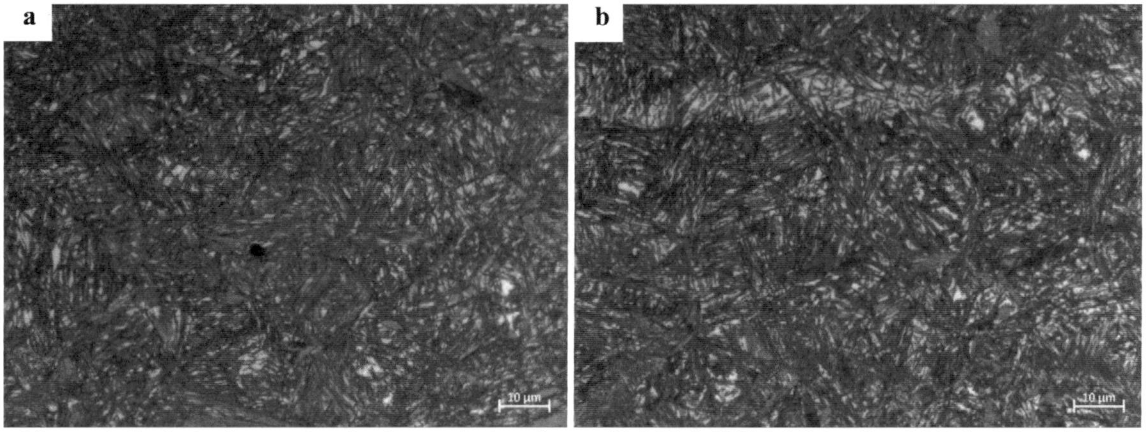

Figure 6. Optical micrographs of colour etched surfaces of 0,54C-1,45Si-0,71Mn after Q&P (**a** QT = 200 °C, PT = 300 °C; **b** QT = 200 °C, PT = 400 °C).

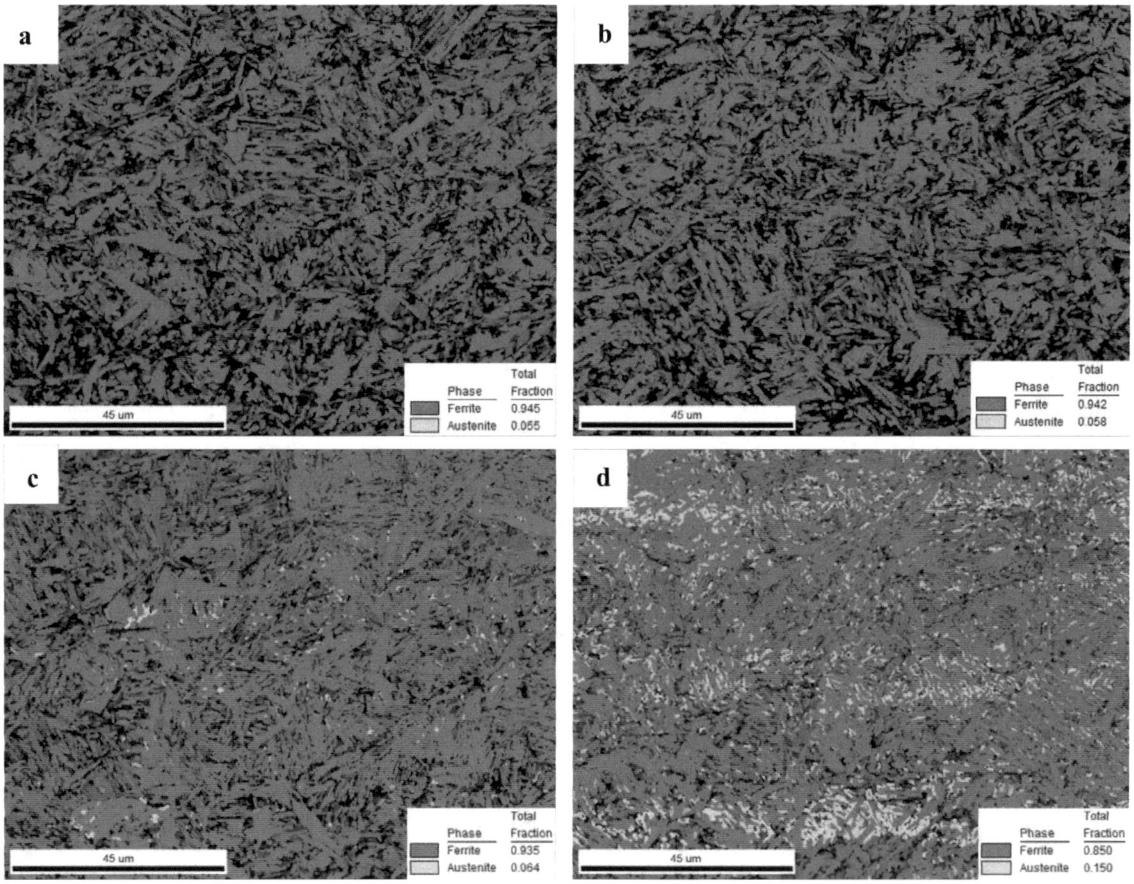

Figure 7. EBSD phase maps (CI > 0,1) of specimen **a** quenched to room temperature, **b** quenched to 175 °C and partitioned at 250 °C, **c** quenched to 200 °C and partitioned at 300 °C, **d** quenched to 200 °C and partitioned at 400 °C.

The EBSD measurements confirmed the assumptions made from optical microscopy investigations. Figure 7 shows corresponding EBSD phase maps. At lower quenching temperature (Figure 7 **b**) austenite fraction is comparable to the as-quenched stage (Figure 7 **a**). The phase fraction of austenite increases with higher partitioning temperature (Figure 7 **c** and **d**). Thereby, the austenite is located between the martensite laths. Furthermore, the fraction of areas with poor signal quality (CI < 0.1) is lowered for the Q&P samples quenched to 200 °C, leading to the assumption that the residual stresses are lowered compared to as-quenched and quenched to 175 °C.

3.3 Mechanical properties

The results of hardness measurements are summarized in Figure 8. As expected, there is a loss of hardness when the quenching is followed by a reheating. The hardness decreases with increasing partitioning temperature. The associated hardness distribution maps can be seen in Figure 9. Each hardness profile is comparatively homogeneous.

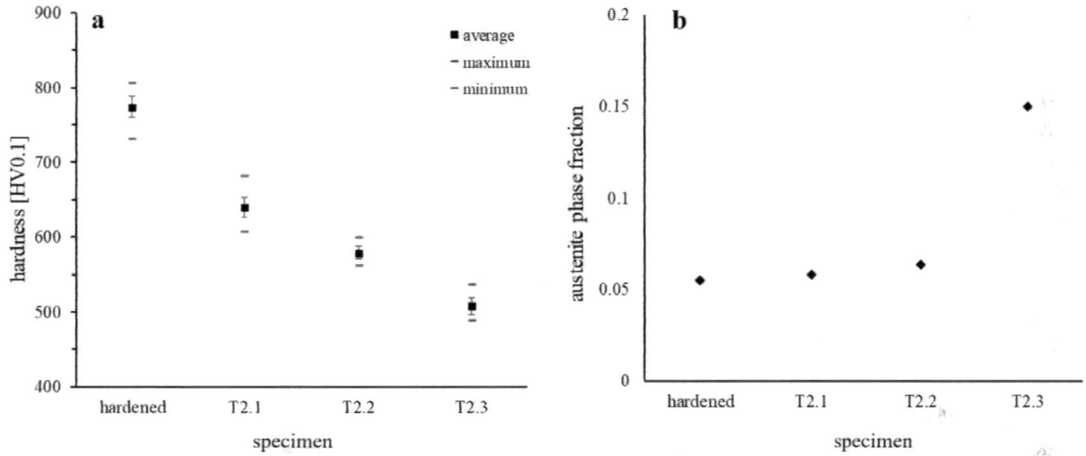

Figure 8. Vickers hardness (**a**) and fractions of austenite (**b**) in comparison.

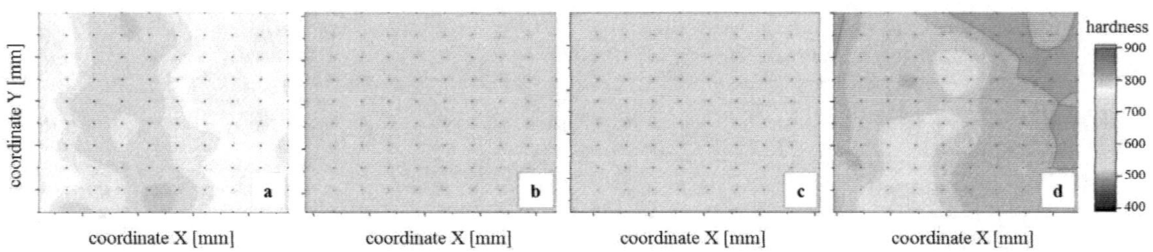

Figure 9. Vickers hardness distribution maps of specimens **a** quenched to room temperature, **b** quenched to 175 °C and partitioned at 250 °C, **c** quenched to 200 °C and partitioned at 300 °C, **d** quenched to 200 °C and partitioned at 400 °C.

4 Conclusions

Different Q&P processing routes were tested on commercial 0.54C-1.45Si-0.71Mn spring steel. Volume fractions of retained austenite and Vickers hardness were determined.

The microstructure of the specimen with QT = 175 °C is comparable with the as-quenched stage, meaning that the chosen parameter for Q&P are insufficient. It is possible, that the amount of retained austenite is already too low at QT = 175 °C (QT might be too close to M_f). Another possible explanation is that PT = 250 °C could be too low for sufficient stabilisation of retained austenite due to limited carbon diffusion. Regarding the temperature dependent elongation of the sample, the last option seems to be the probable explanation, since transformation of fcc to bcc seems to take place during final cooling. Similar behaviour has been observed by DIECK et al. [15]. The specimens with QT = 200 °C are characterized by a distinct austenite fraction. In the case of PT = 400 °C an amount of 15 % could be stabilized. This leads to the assumption, that this parameter setup is appropriate for sufficient stabilising retained austenite. The carbon diffusion further causes the decreasing hardness. The amount of austenite is a function of PT and can be adjusted. Whether this stabilized austenite enhances the mechanical properties of the material will be checked by mechanical testing in further investigations as well as the impact of hydrogen embrittlement [17].

It was shown that commercial 0.54C-1.45Si-0.71Mn is a very promising candidate for excellent enhancement of mechanical properties through Q&P heat treatments. The formation of microstructures with different amounts of retained austenite was shown successfully. By designing individual processing routes the microstructure can be adjusted precisely to the desired properties.

Acknowledgements
The authors gratefully acknowledge department 9.4 of BAM Berlin for performing dilatometric studies and the MDZWP e. V. for hardness measurement.

References
[1] Speer J, Matlock D K, De Cooman B C and Schroth J G 2003 *Acta Mater.* **51** 2611-22
[2] Clarke A J, Speer J G, Matlock D K, Rizzo F C, Edmonds D V and Santofimia M J 2009 *Acta Mater.* **61** 149-52
[3] Frint P, Kaiser T, Mehner T, Bruder E, Scholze M, Mašek B, Lampke T and Wagner M F-X 2019 *Sci Rep* **9** 17023
[4] Fonstein N 2015 *Advanced High Strenth Sheet Steels* (Switzerland: Springer) chapter 10 p 328
[5] Arlazarov A, Bouaziz O, Masse JP and Kegel F 2015 *Mater. Sci. Eng.* A **620** 293
[6] Arlazarov A, Ollat M, Masse JP and Bouzat M 2016 *Mater. Sci. Eng.* A **661** 79
[7] Wang L and Speer J G 2013 *Metallogr. Microstruct. Analy.* **2** 268
[8] Santofima M J, Zhao L and Sietsma J 2011 *Metall. Mater. Trans.* A **42** 3620
[9] Hsu T Y, Jin X 2011 *Advanced Steels* (Berlin: Springer) pp 67-73
[10] Jirková H, Kučerová L and Mašek B 2012 *Mater. Sci. Forum* **706–709** 2734–9.
[11] Liao X S, 2010 Dr degree thesis, Shanghai Jiao Tong University
[12] Wang X D, Zhong N, Rong Y H, Hsu T Y and Wang L 2009 *J. Mater. Res.* **24** 260
[13] Dieck S, Ecke M, Rosemann P and Halle T 2018 *Proc. Int. Conf. on Martensitic Transformations (Chicago)* pp 123-28
[14] Dieck S, Ecke M, Rosemann P, Fritsch S, Wagner M F-X and Halle T 2020 *Plasticity, Damage and Fracture in Advanced Materials* pp 35-42
[15] Dieck S, Rosemann P, Kromm A and Halle T 2017 *IOP Conf. Ser.: Mater. Sci. Eng.* **181** 12034
[16] Angeli J, Füreder E, Panholzer M and Kneissl A 2006 *Practical Metallography* **43** 489
[17] Mehner T, Scharf I, Frint P, Schubert F, Mašek B, Wagner M F-X and Lampke T 2019 *Mater. Sci. Eng.* A **744** 247-54

IOP Conf. Series: Materials Science and Engineering **882** (2020) 012026 doi:10.1088/1757-899X/882/1/012026

Pyroelectric X-ray Generator for XRF in Combination with Optical Microscopy

M Wilke[1], L Hanns[2], K Harnisch[1], W Knapp[3], M Ecke[1] and T Halle[1]

[1] Otto-von-Guericke-University Magdeburg, Faculty of Mechanical Engineering, Institute of Materials and Joining Technology, Universitätsplatz 2, 39106 Magdeburg, Germany
[2] DeltaSigma Analytics GmbH, Mittagstraße 16p, 39124 Magdeburg, Germany
[3] Knapptron GmbH, Privatweg 3, 39291 Möser, Germany

markus.wilke@ovgu.de

Abstract. X-ray fluorescence spectroscopy (XRF) is a method for elemental analyses of materials. In comparison to other analysing methods such as energy dispersive X-Ray spectroscopy (EDS) within a scanning electron microscope (SEM) or optical emission spectroscopy (OES) there are several advantages. This paper contains experimental investigations to improve the emission behaviour of a pyroelectric X-ray source regarding X-ray energy and output intensity. Pyroelectric lithiumtantalate ($LiTaO_3$) was used to emit and accelerate electrons and to generate primary X-rays subsequently. Geometrical conditions were altered and the design of the X-ray source was optimized. All steps of optimization have been performed with respect to an application in a spectroscopic device in combination with an optical microscope (OM). Based on this purpose, the miniaturization of all applied components is of major importance. Different pure metals, alloys, and a coated specimen were analysed to demonstrate potential applications and current limitations of the designed system. Materials with a wide range of necessary excitation energy for chemical analyses of 1.5 to 10 keV for the K_α and L_α X-ray fluorescent lines were used to demonstrate the technical relevance of this type of application. As a result, a prototype of a spectrometer that combines XRF and OM was designed and tested.

1 Introduction

The pyroelectric effect is a well-known phenomenon by now. It is based on a change in polarization of pyroelectric crystals such as lithiumniobate ($LiNbO_3$) and lithiumtantalate ($LiTaO_3$) caused by a change of crystal temperature[1]. Under atmospheric conditions this change in polarization does not cause an electrical field due a compensation of the polarization by surface charges [1, 2]. The compensation is suppressed by applying vacuum conditions and a strong electrical field of up to $1.35 \cdot 10^7$ V/cm for $LiNbO_3$ can be measured[3]. The generated potential difference ΔU in dependence on the temperature change ΔT can be estimated by the following equation:

$$\Delta U = \frac{p \cdot z}{\varepsilon_r \cdot \varepsilon_0} \cdot \Delta T \quad (1)$$

z is the dimension of the crystal in direction of polarization, p is the pyroelectric coefficient of the used pyroelectric material, ε_r is the relative permittivity of the crystal and ε_0 is the permittivity of the vacuum ($8.854 \cdot 10^{-12}$ As/Vm)[4].

The strength of the electrical field caused by the crystal under suitable conditions can be used to ionize residual gas molecules and emit electrons. These electrons are accelerated by the surface charges of the crystal subsequently. The direction of the acceleration depends on the direction of ΔT, heating or cooling respectively[5]. The energy and intensity of the accelerated electrons is sufficient high to generate X-rays, which was first described in 1992 [6]. In the following numerous possible applications were discussed and first commercial pyroelectric X-ray generators were produced (*Cool-X*, Amptek, Bedford MA, USA) and tested for potential X-ray fluorescence analyses (XRF) [7, 8].

A significant increase of intensity and a controllable maximum output energy of a pyroelectric X-ray source could be reached by selectively optimizing working conditions [9, 10]. The existence of high intensities is fundamental for further applications of pyroelectric X-ray sources in µ-XRF spectrometers[10]. Aim of the studies described here is the transfer of the knowledge of possible optimization steps to miniaturize necessary components by providing basic requirements such as high intensities and adjustable energy in order to combine a pyroelectric X-ray source with an optical microscope for elemental analyses of materials.

2 Experimental

The experiments are divided in four major developing steps as shown in Figure 1.

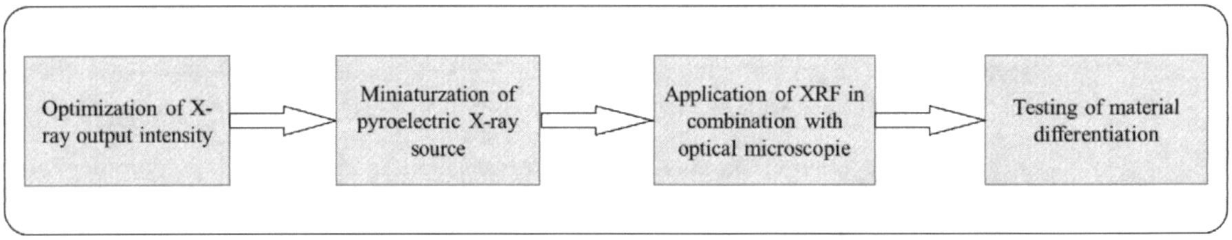

Figure 1. The stepwise development of a combined XRF and optical microscopy system.

The pyroelectric X-ray source was optimized in terms of high output intensity. Results of previous experiments with regard to crystal type, geometrical parameters and optimized pressure were used [9, 10]. This is essential for the integration into an optical microscope because of the reduction of primary X-ray intensity due to the application of smaller crystals in a miniaturized device. After optimization and miniaturization, the X-ray source was integrated into an optical microscope to demonstrate a possible application for XRF during microscopic investigations. Finally, the developed system was tested with different samples. XRF and microstructural investigations were performed.

2.1 Optimization of X-ray output intensity

In the latest studies different possibilities of an optimization of output intensity were investigated. Different arrangements between crystal and target were compared regarding X-ray generation in remission and transmission respectively. The influence of the target material, target angle referring to the direction of the electrons, crystal size and crystal material were described as well [9, 10]. As a result of these improvements, an X-ray output intensity for primary X-rays of up to $1.0 \cdot 10^7$ counts in a period of 600 s for the Cu K_α peak has been achieved using a $5 \cdot 5 \cdot 6$ mm³ $LiTaO_3$ crystal. A remission setup for crystal, target and window with a target angle of 12° and a working pressure of $2 \cdot 10^{-4}$ mbar was identified as optimum. The maximum endpoint energies measured were up to 60-70 keV [9].

XRF with different materials was performed to evaluate potentials and limitations for an application as spectrometric tool in materials science. A vacuum test rig was used. An overlay of different spectra as results of XRF investigations is shown in Figure 2.

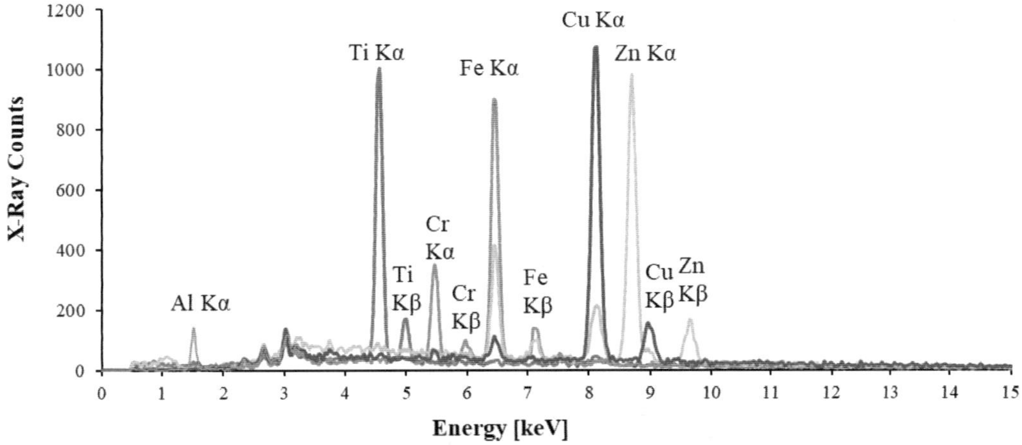

Figure 2. Results of XRF analyses with optimized pyroelectric X-ray source. Five sheets with different chemical composition were analyzed using the optimized X-ray source to determine the detection range. Distinct peaks of aluminum (Al), titanium (Ti), chromium (Cr), iron (Fe), copper (Cu) and zinc (Zn) are apparent.

An energy dispersive X-ray detector (Amptek X-123 silicon drift x-ray detector) was used to collect the X-ray spectra. The crystal was heated from room temperature up to 140 °C using a 5 Ω resistance heater. The spectra were collected for 4 to 9 heating-cooling-cycles of 600 s per cycle. The number of cycles was increased to reach evaluable intensities in the XRF spectra for elements with low atomic numbers such as aluminium (Al). Sheet metals of different composition with dimensions of $50 \cdot 100 \cdot 1.5$ mm³ were used as specimens. Different characteristic X-ray fluorescent transitions from 1.48 keV for aluminium K_α to 9.56 keV for zinc K_β were excited by the optimized X-ray source.

2.2 Transferring optimized conditions into a miniaturized X-ray source

For the combination of a pyroelectric X-ray source with an optical microscope the miniaturization of the arrangement described above is necessary. It had to be considered to reach sufficient output energy to excite fluorescence of all technical relevant materials and highest possible output intensity to lower the duration of the XRF analyses. The following setup was chosen:

- copper target in combination with a lithiumtantalate crystal ($9 \cdot 9 \cdot 5$ mm³)
- target angle of 12 ° and a distance of 11 mm between target and crystal surface
- steel tube with a working pressure of $2 \cdot 10^{-3}$ mbar and a 25 µm Kapton® X-ray window
- 5 Ω resistance heater
- Pt-1000 temperature sensor
- 3D-printed polymer mount for internal components
- external vacuum pump
- Amptek® X-123 silicon drift X-ray detector

To implement all conditions into the miniaturized X-ray source the polymer insert shown in Figure 3 was produced by additive manufacturing.

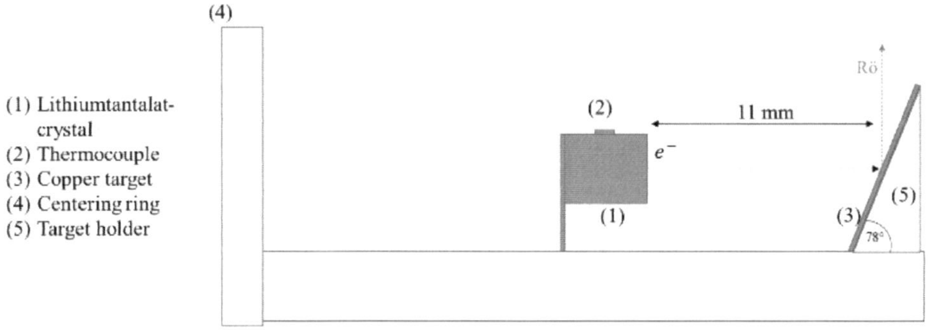

Figure 3. Schematic construction of the additive manufactured insert which implements the optimized parameters in the miniaturized pyroelectric X-ray source.

The shown device is inserted into a cylindrical stainless steel case. This case is connected to a mechanical rotary vacuum pump to evacuate the system. After evacuation, the systems pressure is about $2 \cdot 10^{-3}$ mbar. The case has a hole with a diameter of 3 mm which is covered by a polyimide film and used as an exit window for the primary X-rays. First XRF analysis of a brass sample was made using the developed X-ray source in combination with a silicon drift detector (SDD). The used arrangement is schematically shown in Figure 4.

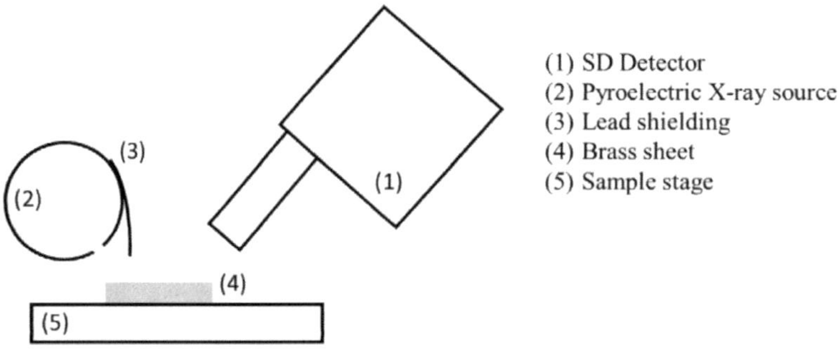

Figure 4. Schematic figure of the arrangement used for testing the lead shielding in order to prevent the detection of X-ray fluorescence emitted by the X-ray sources steel case which could affect XRF results.

The results show characteristic peaks for copper (Cu) and zinc (Zn). Additionally a peak for iron (Fe) was measured which should not appear for the analysis of this type of brass sample. Obviously, the primary X-rays cause an emission of characteristic X-rays at the edge of the 3 mm hole of the steel case. To avoid the detection of this radiation a lead shielding was added. To prove the function of the shielding the brass sample was analysed again. The spectra were collected for 600 s with shielding and without using a lead shielding respectively. An overlay of both spectra is shown in Figure 5.

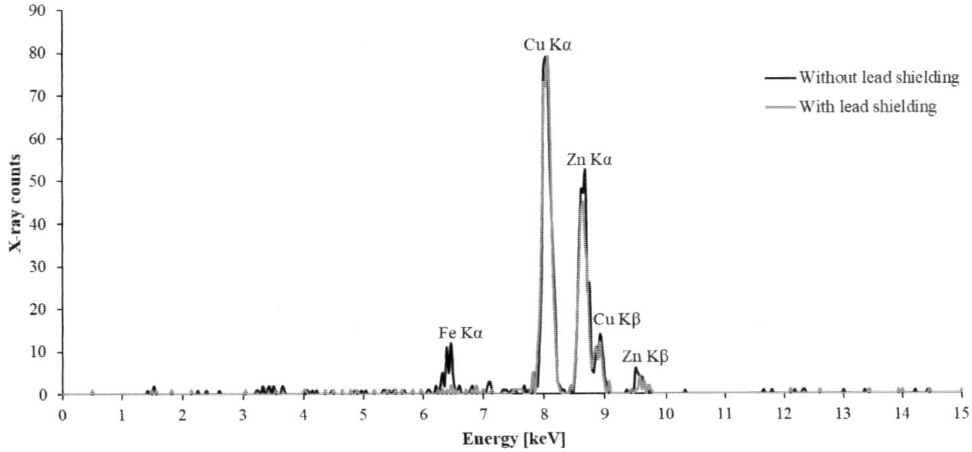

Figure 5: Spectra of XRF analyses of brass sheet. It verifies that the lead shielding is sufficient to prevent the detection of X-ray fluorescence emitted from the X-ray sources steel case during XRF analyses.

The copper K_α– and K_β– lines as well as the zinc K_α– and K_β– lines show similar intensities in both cases. The iron K_α– line disappears when using the arrangement with a lead shielding. The lead shields the X-ray fluorescence of the steel case without decreasing the primary intensity and guarantees correct results.

2.3 Integration of the pyroelectric X-ray source into an optical microscope

A suitable geometrical arrangement was figured out for the optical microscope (Zeiss Stemi 305 with AxioCam ERc5s), the X-ray source and the SDD. It is necessary to ensure the sample to be in the focal point for both methods, optical microscopy and XRF respectively. This offers the possibility of microstructural investigations and XRF at the same time. For this reason an electrically movable stage was implemented. It allows to focus the sample surface for optical analyses as well as for chemical analyses without changing settings of the system itself. The arrangement is shown in Figure 6 A. Furthermore, the complete prototype system is shown in Figure 6 B.

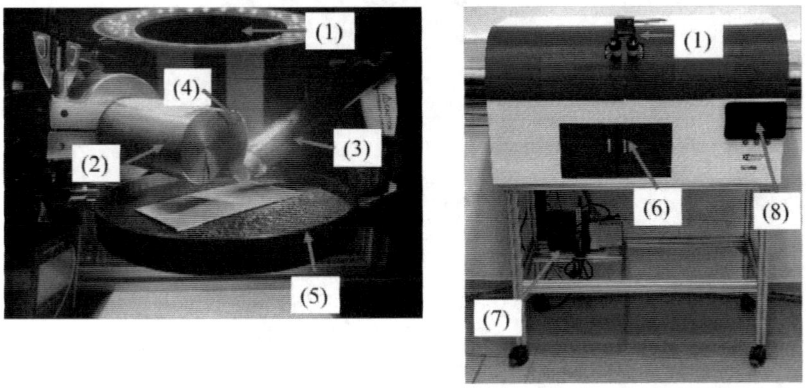

(1) Optical microscope
(2) Pyroelectric X-ray Source
(3) X-ray detector
(4) Lead shielding
(5) Stage
(6) Sample chamber
(7) Pump
(8) Control touchscreen

Figure 6. A: Arrangement of pyroelectric X-ray source, SD detector and optical microscope. **B:** Developed prototype system for XRF and optical microscopy with its components.

IOP Conf. Series: Materials Science and Engineering **882** (2020) 012026 doi:10.1088/1757-899X/882/1/012026

2.4 Testing of combined analyses system

Different samples were analysed using the system described above. Three sheets made of titanium, stainless steel and brass were used. The spectra generated were collected within 900 s and are shown in Figure 7.

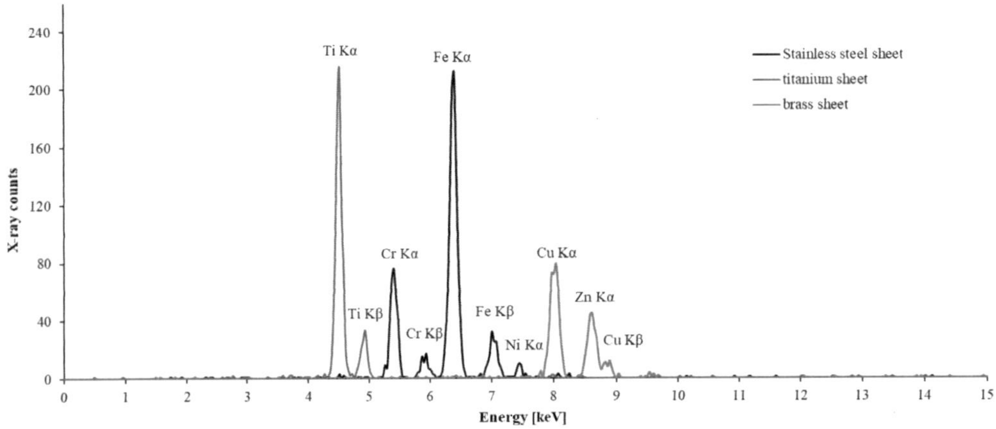

Figure 7. Spectra from XRF analyses of titanium, brass and stainless steel sheets with miniaturized pyroelectric X-ray source to proof that several elements can be detected with the developed arrangement.

Referring to the characteristic lines at 4.5 keV and 4.93 keV in the spectrum it is possible to identify the sheets as titanium and stainless steel with associated lines at 5.4 keV for chromium, 6.39 keV for iron and 7.47 keV for nickel. The copper and zinc lines at 8.04 keV and 8.63 keV in the spectrum belong to the brass sheet. As a consecutive step two parts of medical implants of unknown chemical composition were analysed. A microscopic characterization of the samples surface with a simultaneous chemical analysis was performed. Figure 8 A and B, are giving an overview. Figure 8 C and D are showing micrographs of each implant part applying a magnification of 25x.

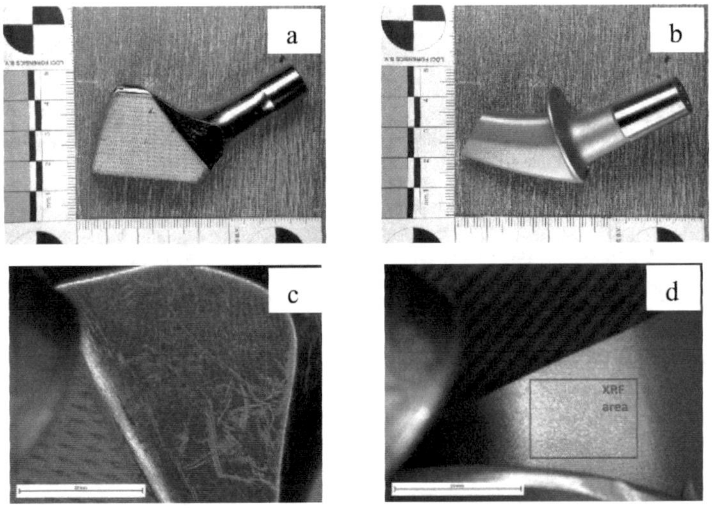

Figure 8. a: Overview of implant 1. b: Overview of implant 2. c: Micrograph of implant 1 at a magnification of 25x. d: Micrograph of implant 2 at a magnification of 25x. XRF analyses were applied on the red framed areas.

After the optical analysis an XRF analysis with the developed pyroelectric X-ray source was applied on both implant parts (red framed XRF area). The spectra were collected for about 900 s and are shown in Figure 9.

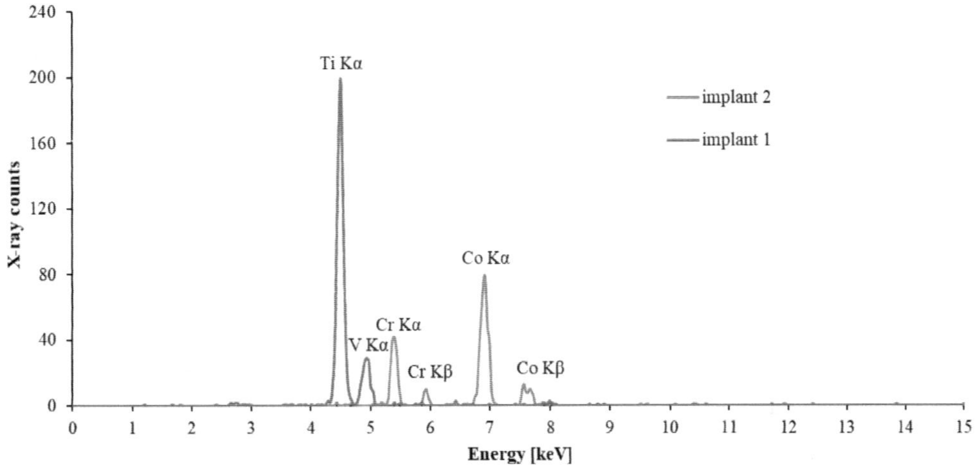

Figure 9: Spectra of XRF analyses with miniaturized pyroelectric X-ray source of implant 1 (red) and implant 2 (blue) to distinguish both samples with respect to their chemical composition.

The results of XRF show that implant 1 is made of a titanium vanadium alloy (lines at 4.5 keV for titanium and 4.94 keV for vanadium) and has a different chemical composition compared to implant 2, which is made of a cobalt chromium alloy (lines at 5.4 keV and 5.95 keV for chromium and 6.92 keV and 7.65 keV for cobalt). The lower part of implant 1 shows a different surface structure. An image was taken with the optical microscope and a chemical analysis was applied to this area. It shows a calcium (Ca) rich coating of this part of implant 1. The micrograph and spectrum are shown in Figure 10.

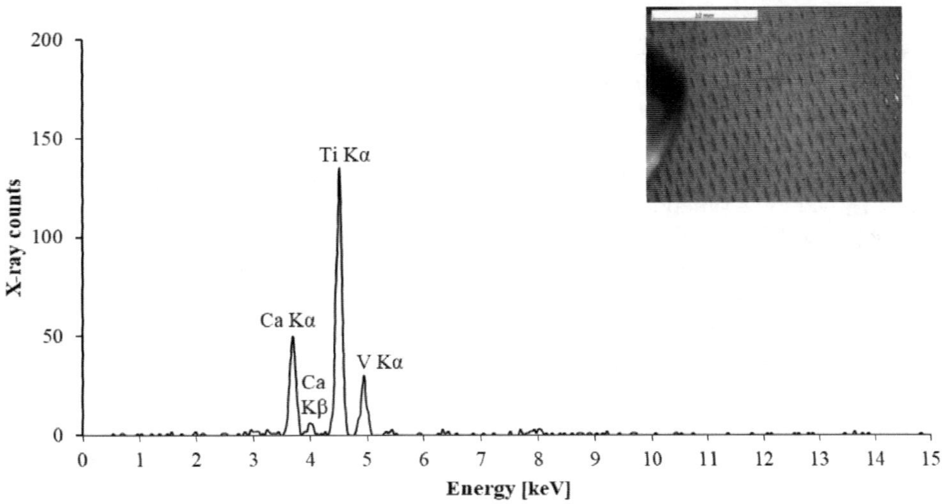

Figure 10: Spectrum from XRF analysis of hydroxyapatite coating on a TiAlV implant (top right corner) with miniaturized pyroelectric X-ray source. Distinct peaks of the substrate (Ti, V) and of the coating (Ca) can be separated.

Calcium K_α (3.7 keV) and K_β (4.01 keV)-Lines were detected. Titanium and vanadium are also detected caused by the interaction with the substrate. By identifying calcium, the coating is assumed to be hydroxyapatite being a common coating for prosthesis besides Ti(Nb)N, different ceramic alternatives and polymer coatings. Hydroxyapatite also contains phosphorus. This element could not be detected using this arrangement.

3 Summary and conclusion

A functional prototype combining an optimized pyroelectric X-ray source and an optical microscope was developed and tested. Different materials were characterized by XRF and optical microscopy. It was shown that a wide range of technically used metals and alloys can be analyzed due to their elemental composition. The developed system is a suitable tool for extending the opportunities of characterization of an optical microscope. During the analyses, limitations of detectable elements were determined. The detectable element with the lowest atomic number Z in this arrangement was Calcium. By a further increase of the primary X-ray intensity, a measuring time reduction for faster investigations might be achieved and the detection of characteristic X-rays of lower energy might be enabled as well. For local XRF (μXRF) the focusing of primary X-rays is conceivable. This might be accomplished by individually designed glass fiber capillary lenses[9].

Acknowledgments

This research was sponsored by the European Regional Development Fund (ERDF) and the Investment Bank of Saxony-Anhalt. The authors are grateful for the support of the Methodisch-Diagnostisches Zentrum Werkstoffprüfung e.V. (MDZWP e.V.).

References

[1] Lang S B 2005 Pyroelectricity: From Ancient Curiosity to Modern Imaging Tool *Phys. Today* **58** 31.
[2] Geuther J A and Danon Y 2005 Electron and positive ion acceleration with pyroelectric crystals *J. Appl. Phys.* **97** (10).
[3] Rosenblum B, Bräunlich P and Carrico J P 1974 Thermally stimulated emission from pyroelectric LiNbO3 *Appl. Phys. Lett.* **25** 17.
[4] Hanamoto K, Kawabe A, Sakoda A, Kataoka T, Okada M and Yamaoka K 2012 Pressure dependence of x rays produced by a LiTaO3 crystal at low pressures *Nucl. Instrum. Methods A* **669** 66.
[5] Brownridge J D and Shafroth S M 2005 Using static charge on pyroelectric crystals to produce self–focusing electron and ion beams and transport through tubes *J. of Electrostatics* **63** 249-259.
[6] Brownridge J D 1992 Pyroelectric X-Ray Generator *Nature* **358** 287-288.
[7] Ida H and Kawai J 2004 Identification of Glass and Ceramics by X-ray Fuorescece Analysis with a Pyroelectric X-ray Generator *Anal. Sciences* **20** 1211-1215.
[8] Ida H and Kawai J 2004 X-ray fluorescence analysis a mg l^{-1} level with an x-ray source powerd by a dry battery *J. Anal. At. Spectrom.* **19** 1524-1528.
[9] Wilke M, Harnisch M, Knapp W, Ecke M and Halle T 2019 Focusing of x-rays emitted by a pyroelectric x-ray generator for micro x-ray fluorescence *J. Vac. Sci Technol. B* **37**, 011203.
[10] Wilke M, Harnisch M, Knapp W, Ecke M and Halle T 2018 Optimization of pyroelectric electron sources for the generation of x-rays for x-ray fluorescence applications *J. Vac. Sci Technol. B* **36**, 02C101.

AUTHOR INDEX

Institute of Physics
Dirac House, Temple Back
Bristol BS1 6BE UK

ISSN: 1757-8981
ISBN 978-1-7138-2809-9

CURRAN ASSOCIATES INC.
proceedings.com